AF443748

PHOTONICS
FOR SAFETY AND SECURITY

# PHOTONICS

# FOR SAFETY AND SECURITY

Editors

**Antonello Cutolo**
University of Sannio, Italy

**Anna Grazia Mignani**
CNR – Institute of Applied Physics 'Nello Carrara', Italy

**Antonella Tajani**
CNR, Italy

 **World Scientific**

NEW JERSEY · LONDON · SINGAPORE · BEIJING · SHANGHAI · HONG KONG · TAIPEI · CHENNAI

*Published by*

World Scientific Publishing Co. Pte. Ltd.

5 Toh Tuck Link, Singapore 596224

*USA office:* 27 Warren Street, Suite 401-402, Hackensack, NJ 07601

*UK office:* 57 Shelton Street, Covent Garden, London WC2H 9HE

**British Library Cataloguing-in-Publication Data**
A catalogue record for this book is available from the British Library.

**PHOTONICS FOR SAFETY AND SECURITY**



ISBN 978-981-4412-96-4

Printed in Singapore by World Scientific Printers.

*One of the authors wants to dedicate this book to his women:
Maria Alessandra, Maria Teresa and Emilia, and to his parents (AC)*

*To Penny, who will always live in my heart (AT)*

# Preface

The tremendous amount of technology available today would allow us to guarantee the perfect safety and security of any site with reliability very close to 100%. This dream might be realized under two extreme conditions:

- no budget limitation has to be taken into account,
- no legal constrains would limit the technological choices.

In everyday life the above mentioned constrains would strongly restrict the choices effectively available. An example over all is the terahertz technology. It might offer the possibility of the perfect scan of any person in order to easily discover the presence of explosives, drugs and also non metallic weapons.

Anyhow, one of the major challenges of this technology is just its power and its capability of performing the accurate scan of any body. In fact, each person would appear completely naked thus showing any personal defect or anomaly. This is a typical case in which the possibility offered by technology is strongly limited by the natural need of privacy.

Another case very similar to this is offered by the possibility of registering the main identifying elements of each person passing through an assigned gate. Many of us, entering the United States or many Middle East countries, have been scanned in the eye, pictured and have had their finger prints registered. This approach is totally forbidden in many European countries.

In addition, we observe that, as each physical quantity can be measured in many different ways, the final choice is not only a matter of price or performance but very often is limited by the boundary conditions of the measurement. As an example, we can refer to very high temperature measurements, which can be made only by infrared detectors. Of course, no sensor or no measurement method can be the jolly of any practical situation.

With the previous considerations in mind a modern sensor engineer would be able to design a safety and/or security system with the classical goal of maximizing its performance for a given total available budget. This means that this kind of engineer must know the basic characteristics of any sensor in order to integrate it according to the specific requirements.

## Why photonics?

Photonics is a remarkable technological field able to support a multi-billion Euros per year market.

It is essentially based on the photons – matter interaction. It plays a fundamental role in safety and security because of its capability to supply sensors which can be sensitive only to a specific parameter being totally insensitive to any other agent. A specific example is offered just by optical fiber sensors which, unless specific tools are intentionally used, are characterized by a total independence of any other environmental parameter. Free space propagation, on the other side, is the fundamental tool for a variety of contactless and not invasive diagnostic techniques. Indeed, this last property is shared with all measurement techniques based on the use of waves, either acoustic or electromagnetic.

Today, based on the use of extremely miniaturized devices, photonics is able to supply compact and user friendly systems easy to be embedded almost in any apparatus. In addition, the possibility to separate the diagnostic sensor from the electrical apparatus makes photonic devices ideal candidates in many dangerous or extreme environments likes explosive substances tanks, low or high temperature environment, high electromagnetic field systems. It is enough to think about measurements to be performed inside cryogenic environments where the optical fiber is often the only suitable candidate, or high temperature furnaces for semiconductor processing where light is the unique possibility for real time online measurements. And also explosive or flammable substance pipes, tanks where any parameter measurements must provide the total absence of any electrical signal.

Besides the previous cases, another field of interest is that of medical and food application where the perfect chemical inertial properties of the glass play a fundamental role for using optical fibers for a large variety of different measurements.

In this volume, the authors aim to give a survey of the wide range of applications of photonics in the fascinating field of safety and security. The book continues the series initiated with the Handbook *An Introduction to Optoelectronic Sensors*, containing the state-of-the-art of optoelectronic technologies for sensing.

*Photonics for Safety and Security* is a collection of up-to-date contributions by renowned experts in the field. After a chapter introducing the principal concepts which underpin the realization of photonic systems, a large variety of photonics applications in different areas of safety and security is illustrated.

The overview of the different applications ranges from the use of Fiber Bragg Grating sensors for diagnostic and control in civil structures, remote sensing techniques for earth and territory monitoring, cultural heritage, food industry, detection of chemicals, drugs and explosives, environment, underwater monitoring, aerospace devices, medical and clinical diagnostics, quantum cryptography, night vision and many others such as the ever so popular forensic sciences.

Antonello Cutolo          Anna Grazia Mignani          Antonella Tajani

# Contents

# WHAT IS PHOTONICS?


Brian Culshaw

*Department of Electronic & Electrical Engineering, Strathclyde University*
*Royal College Building, Glasgow G1 1XW*
*E-mail: b.culshaw@eee.strath.ac.uk*



This chapter outlines in brief the principal concepts which underpin the realization of photonic systems. The aim in a few pages is to present a qualitative introduction supported by a bibliography which will hopefully enable the reader to seek more quantitative information and additional conceptual depth.


## 1. Introduction

Photonics, like many generic technical terms, has yet to be blessed with an unambiguous definition. Sometime during the last few decades the combination of the word "photon" and the word "electronics" emerged to describe science, art and technology of appreciating and utilizing that part of the electromagnetic spectrum which manifests itself predominately as photons rather than predominately as electric currents or magnetic fields; loosely, indeed very loosely defined as "working with light". We could attempt to be more precise. If an electromagnetic wave is to appear in some form of detection circuit as a particle rather than an induced current, then intuitively the energy of that particle must comfortably exceed the thermal energy of the particles within the detector. In this context the detector can be a photodiode, a photographic emulsion, a CCD array or even your eye. All will respond to photons.

Continuing this argument and observing that the photon energy for electromagnetic radiation at 1 µm in the near infra-red is about 1.2 eV and the equivalent thermal energy at 300 K is about 0.025 eV then – very roughly – anything that deals with electromagnetic radiation of wavelength less than around 5 μm can be considered to be "photonics".

This, of course, leads into a somewhat paradoxical world.

Electromagnetic radiation is still electromagnetic radiation so everything about waves still applies. The wave concept[1] is extremely useful, even indispensable, when looking at propagation through media with dimensions much larger than the wavelength. Indeed it still applies at dimensions comparable to a wavelength where all the well known wave guiding and diffraction ideas can be reliably brought to bear.

It is only then at the detection process – when this electromagnetic radiation interacts with a material – that the photon comes into its own.

This, of course, covers a huge range of well known phenomena, not only photodetectors themselves but everything to do with spectroscopy and even concepts like radiation pressure, photothermal phenomena and non-linear scattering. The "photonics" label embraces other nuance as well. If we make things small enough - much smaller than the wavelength - then instead of wave phenomena or even molecular properties determining what happens to the electromagnetic wave, we can begin to exploit the concepts of electronic circuits leading into nanophotonics[2] leading to even more esoteric ideas, like negative index materials, using a slight modifications to all the conceptual underpinning which circuit designers have been using for decades without ever appreciating this alternative perspective.

So the aim of this short chapter is to very briefly introduce the concepts (figure 1) behind the generation, detection and manipulation of electromagnetic signals in the photonic region of the spectrum.


## 2.   Photons – Waves or Particles?

Weighty volumes[3] have been written on this with no real conclusion apart from that it is all a matter of convenience. Detection and generation in the photonic region of the spectrum is a photon based affair and is treated as such, though as nanophotonics gains ground perhaps circuit based generation concepts will emerge. Indeed, with suitably tight structural tolerances, the concepts of the microwave travelling wave[4] tube could easily be scaled up in frequency though whether this might be valuable is an entirely different question. Similarly the concepts of the interaction between electromagnetic radiation in this part of the spectrum and materials are also more simply viewed in the photon domain[5].

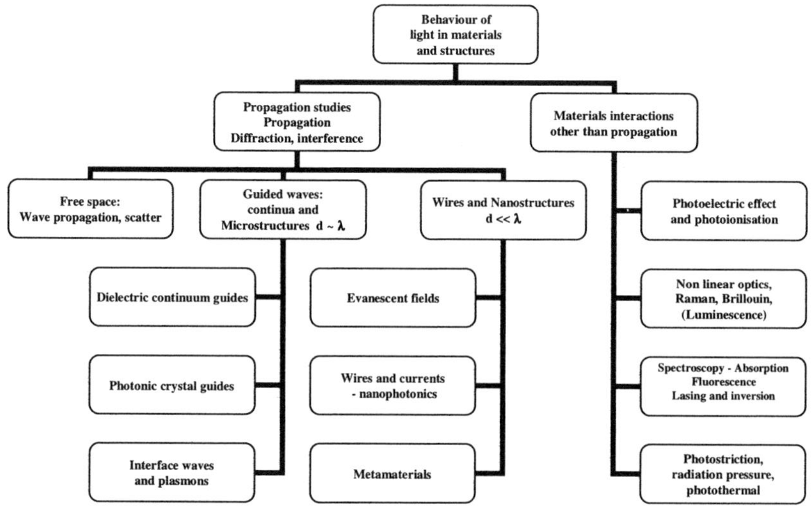


Fig. 1. What is Photonics? This illustrates the principal concepts.

However, when it comes to propagation and the interaction with structures rather than materials, then provided that the structural features are much larger than or comparable to a wavelength then the ideas of waves principally diffraction, interference and polarization are all that is required.

This works fine when the total energy in the system we are considering corresponds to lots and lots and lots of photons. It gets a little shakier when very few photons are involved, or does it? Long ago, the idea that, for example, an interference pattern is really an expression of a photon arrival probability distribution[6] emerged and became philosophically accepted. Consequently the idea of "lots and lots" of photons corresponds to having enough (hundreds, preferably thousands) to plot out the distribution. With few photons around, the ideas of "entanglement"[7] lead into apparently more and more mysterious phenomena. However, for many, perhaps all, of the situations where this mystery prevails, thinking of the generation and detection process as photons and the propagation in-between as waves, goes a long way to intuitively grasping what it is really all about.


## 3.   Manipulating Photons: Structures and Materials

The basic phenomena are essentially determined by the dimensions of the structure compared to the wavelength of the propagating (or even entrapped) radiation.

For very large structures much greater than a wavelength in dimension free space propagation prevails and it is here that the concept of the refractive index or dielectric constant is arguably the most important contribution. What is particularly important is that the photonic region of the spectrum by definition includes frequencies at which the molecules of the material itself are resonant – they absorb. The classical simplistic treatment of the dielectric constant through the Clausius Mosotti[8] equation, which also emerges more or less unscathed from a quantum mechanical treatment, gives behavior typified by the sketch in figure 2. The rapid variations in index at particular frequencies are an inevitable consequence of the resonances and these clearly introduce dispersion and can even when tuned and tweaked to be reinvented as "slow" light[9].

So fundamentally the refractive index invariably has considerable structure within it. This inevitably reflects into the material's propagation properties and indeed the features within it which can be exploited. Moreover, a truly homogeneous medium is at best elusive and perhaps impossible to realize so there will be variations in density and consequently index throughout the material often over very small dimensions. These variations also make their mark and manifest themselves as scatter[10], an effect which gives us our blue skies and also our white clouds, but which can also be invaluable as a measurement tool, for example, to detect size the distribution of particles in air or of tiny creatures in water[11]. When the features become comparable to the wavelength then we encounter two very important domains of photonics both of which implicitly exploit diffraction and interference.

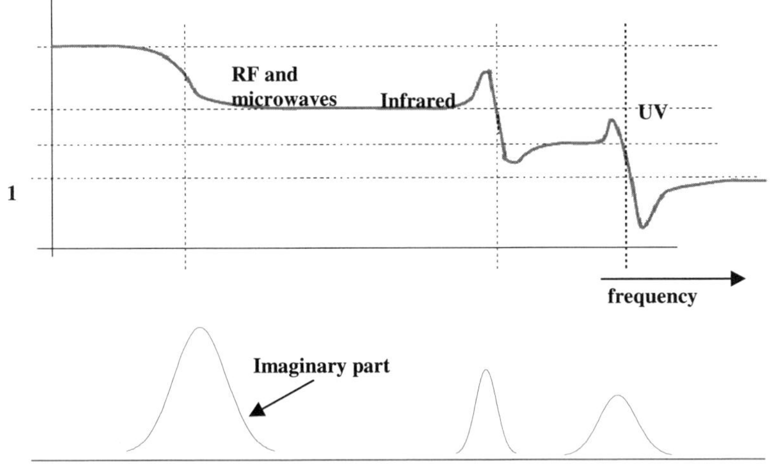



Fig. 2. A very schematic representation of the behavior of dielectric constant vs frequency with polar, rotational, molecular stretch, electronic, and in the X ray region, nuclear contributions. The number of terms in the Clausius Mossotti equation which contribute falls with frequency until eventually we observe dielectric constant <1 at very high frequencies (and sometimes, just close to resonances, as observed in some forms of "slow" light).

The first of these concerns realizing artefacts to confine and manipulate photonic signals, the best known of which is the solid core optical fiber responsible in no small part for, among other things, realizing the broadband internet. The idea is simple (figure 3) - a high index core surrounded by a lower index cladding and total internal reflection at the interface. The diffraction and interference contributions to the story emerge as we consider the propagating modes and the necessity for a stable field distribution across the guide.

More recently the basic waveguide has evolved into the photonic crystal version where essentially holes are introduced into the structure to enable a wider range of index differences (figure 4) or, in a slight variation on the theme, to produce a totally hollow waveguide which guides through reflection from the surface somewhat akin to the well known microwave guide of many years ago.

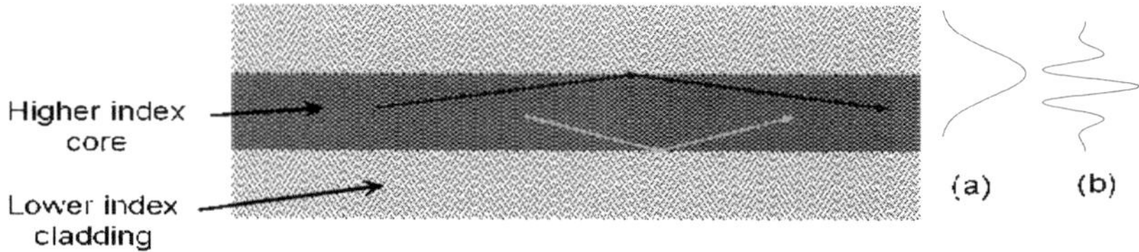


Fig. 3. Index guided optical waveguides showing ray paths and indicative transversal field distributions for (a) the lowest order mode and (b) a slightly higher order mode corresponding to interference between 'rays' at a steeper angle of incidence.

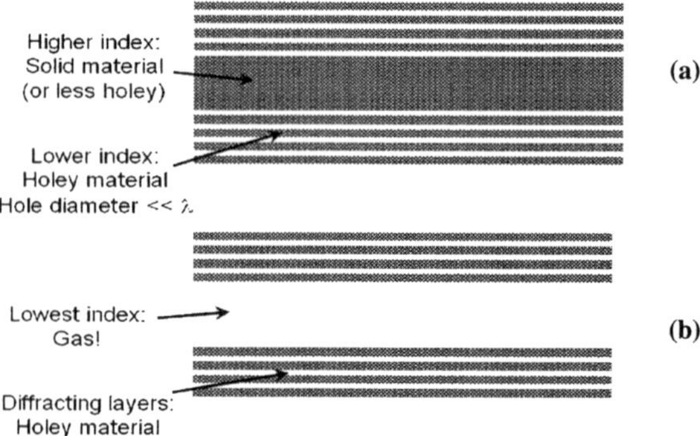


Fig. 4. Longitudinal sections through (a) a photonic crystal waveguide designed for index and /or birefringence control and (b) a 'Bragg guiding' hollow optical fiber. Guiding wavelength ranges determined by Bragg diffraction from the cladding.

There is, though, a slight difference. Metal guides at photonic frequencies are quite amazingly lossy as becomes evident from a simple extrapolation of the skin effect, so our reflector needs to be a low loss dielectric one, in other words a Bragg mirror. The reflection process itself is then wavelength selective so guides of this type have quite well defined and characteristic transmission spectra.

The other important domain where diffraction and interference from structures comparable to a wavelength is, often subconsciously, commonly utilized lies in imaging. It is self evident in a moment's reflection that very fine structure in an object will diffract light at a greater angle than coarse structure (figure 5). Consequently any imaging system can only produce an image with structure characterized by its capacity to collect the diffracted light[17] with important consequences for microscopes and telescopes alike, the most evident of which is the need for large aperture optics.

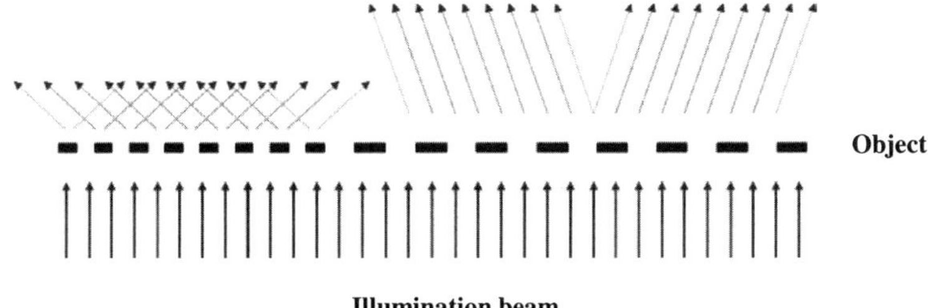


Fig. 5. Diffraction in imaging: the effect of collection lens aperture on resolution – illumination on the finer detail (left) is deflected more and therefore needs a large aperture lens to ensure accurate images are formed.

Both imaging and waveguiding are critical enablers for photonic systems in safety and security. Waveguiding supports the whole technology of fiber optic sensing much applied to structural health monitoring and infrastructure assessment. Imaging supports everything in remote monitoring, everything in microscopy and a whole host of other applications. When the structural dimensions become much less than a wavelength we enter the domain of nanophotonics. At a very simplistic level, this enables familiar system concepts from the RF and microwave domains to be implemented optically. This does, of course, assume that the technology to fabricate reasonably precise structures with dimensions significantly less than an optsical wavelength exists and is available and it also assumes that the properties of materials at photonic frequencies are compatible with the simplistic circuit concepts.

Even if both of these apply there are some fundamental constraints which must be recognized most important of which is arguably the properties of conducting wires.

For example, if we do a skin effect analysis (figure 6) of an imaginary gold wire one wavelength long and $1/10^{th}$ of a wavelength in diameter then we find its resistance increases rapidly as the wavelength decreases. Whilst this is a somewhat contrived example it indicates that losses are very important, even prohibitive, in photonic circuits. There are, of course, real operating examples of which the most well known is surface plasmons (figure 7) where the metal film is much thinner than the skin depth. These have been studied and used for the past half century or thereabouts as both polarizers and as the basis of refractive index (and therefore chemical or biochemical) sensing systems[13,14].

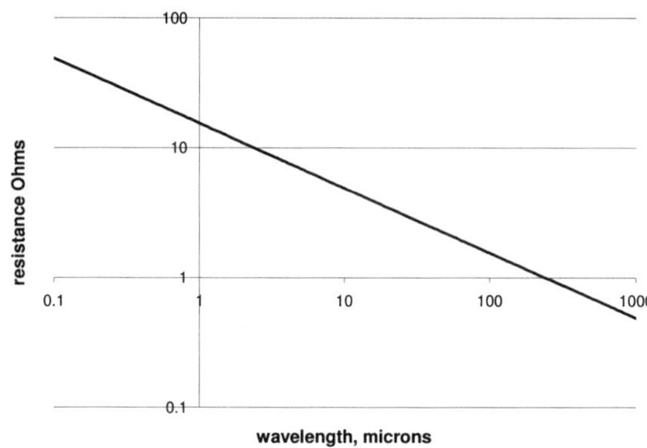


Fig. 6. An illustrative indication of the behaviour of 'conventional' conductors at photonic frequencies. This shows the resistance of an imaginary wire l/10 in diameter and 1 in length indicating that very large ohmic losses are inevitable in all but the smallest of nanophotonic systems.

These "circuits" can also facilitate significant modifications to the refractive index. If these small circuits are assembled in an array they act, in effect, as synthetic molecules.

In a normal material the molecules are set in motion by an incoming electromagnetic field and reradiate a little while later, effectively in phase quadrature.

The bigger the oscillations the more of the incoming field is stored and reradiated and the larger the refractive index. If we can contrive a circuit, particularly some form or other of LC resonator, with dimensions significantly less than a wavelength, then this phase relationship becomes, in principle, under our control and this, in a very simplistic way is essentially the principle behind the whole science of metamaterials[15] which, among other things, enable, at least conceptually, materials with negative index (figure 8) albeit for only a limited range of frequencies. These materials, among other things refract in entirely the 'wrong' direction and are also capable, in principle, of realizing point to point (figure 9) super resolution imaging.

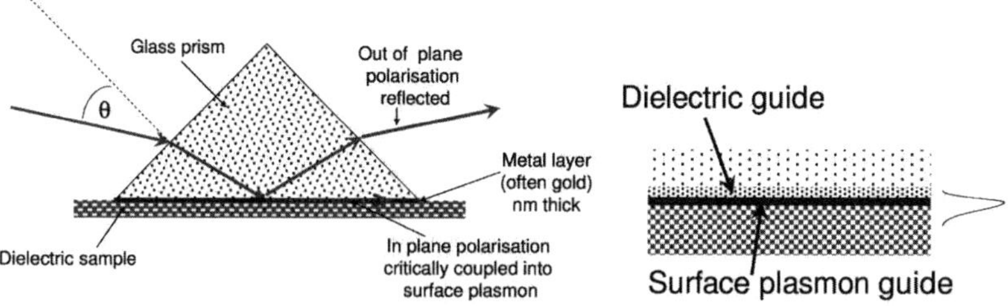


Fig.7. Guided surface plasmons in very thin (sub skin depth) metal films. The Kretschmann configuration (left) for chemical (i.e refractive index) sensing by detecting the critical angle as illustrated and (right) a waveguide equivalent which couples from the upper all dielectric guide into the surface plasmon guide comprising a thin metal field and a dielectric substrate when the propagation constant in the two coincide.

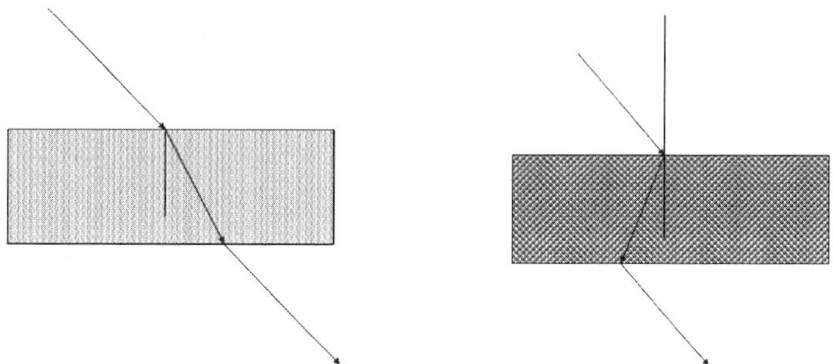

Fig. 8. Contrasting refraction in a conventional optical material with refractive index $n>1$ with (left) refraction in a negative index ($n<1$) synthetic metamaterial (right).

Super resolution imaging stimulates philosophical chagrin among those brought up with Abbé resolution criteria but has never once concerned an electrical circuit engineer measuring electric fields as a function of position on a printed circuit board. The former is a wave field image; the latter is the plot of voltage vs. position. In principle, our nanophotonic circuit approach facilitates something somewhat similar to

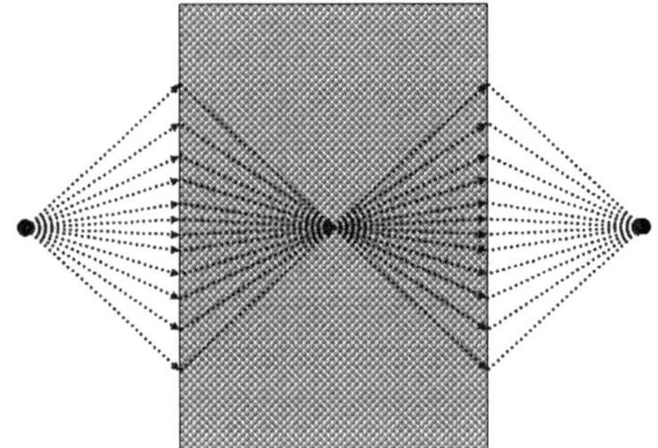

Fig. 9. The basics of perfect unit magnification imaging using a metamaterial lens with $n = -1$.

the latter. For those involved in the nanophotonics domain there is little or no mention of either waves or photons. We are here in the domain of voltages, currents and circuits though as those who have worked at VHF and microwave are all too well aware of impact of radiation phenomena on circuit design and performance, effects far more prevalent in the nanophotonics domain.

## 4. Photonics and Materials

In the photonic region of the spectrum the electromagnetic radiation manifests itself in materials as photons with very important consequences, many of which have a direct application in photonics for sensing and security.

Photodetection and photogeneration are, in effect, reciprocals of each other. For photodetection a photon is absorbed by a material and creates a free carrier (an electron, or in a semiconductor an electron-hole pair) which is swept by an electric field applied to the detector and thereafter manifests itself as an electric current. Photogeneration is its inverse in that an applied stimulus, using an electric field or sometimes simply applying heat, excites the molecules in the structure and these excited states thereafter decay with the release of a photon. There are, of course, myriad phenomena which can impede or even arrest this process. Photogeneration typically results from a small percentage of the excited states as they drop down to a stable level - more likely the process will generate heat rather than light.

Photoionization is a similar effect to photodetection in that free electrons are generated but here the molecule is completely stripped of one or more of its electrons by absorbing the photon. The electron becomes totally free of its host atom. The distinction here is that sometimes, most notably in biological systems and in many polymeric systems this process can, and often does result in a (bio) chemical modification to the original molecule. Photoionization is particularly evident for exposure to photons of energy around 5 eV and above, in other words the UV extending into x-rays. This has extensive implications ranging from applications in UV curable epoxies to the well known decay of many polymers in strong ultraviolet light, to the increased susceptibility to skin tumors for sunbathers.

Photodetection and photoionization can also be viewed as one of the many manifestations of the spectroscopic properties of materials.

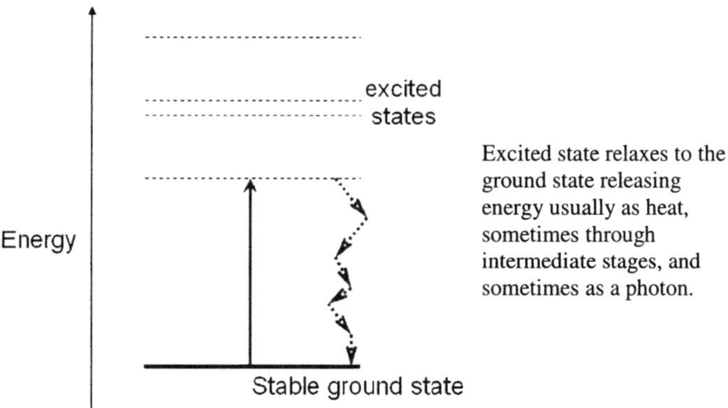


Fig. 10. The principles of absorption spectroscopy exemplified in this much simplified energy level model. The absorption process removes the absorbed photons (colours) from the illuminating radiation, whilst reflecting the remainder, hence giving objects their characteristic colour.

Spectroscopic analysis is an immensely powerful tool through which the chemical constituents of the material may be inferred. The basic idea is simple[16]. A molecule has a well defined set of energy levels (figure 10) and a photon incident on this molecule with an energy equal to or very close to this energy level will be absorbed. Some of those absorbed photon levels will re-emerge in the material as heat, some will be remitted as photons at the same or different energies usually radiating in random directions.

Visually, this absorption process corresponds to extracting some of the wavelengths from the white light which normally illuminates the world around us and consequently giving materials their characteristic color.

There are also some important ancillary effects. If the molecules are relatively far apart then the light which is absorbed is at very specific wavelengths and this, in fact, typifies the behavior of gases at reasonably low pressures - up to around 1 atmosphere depending on the gas species.

Liquids and solids in contrast have their molecules very close to each other and so the exclusion principle comes into play and the energy levels separate (figure 11) with the consequent broadening of the

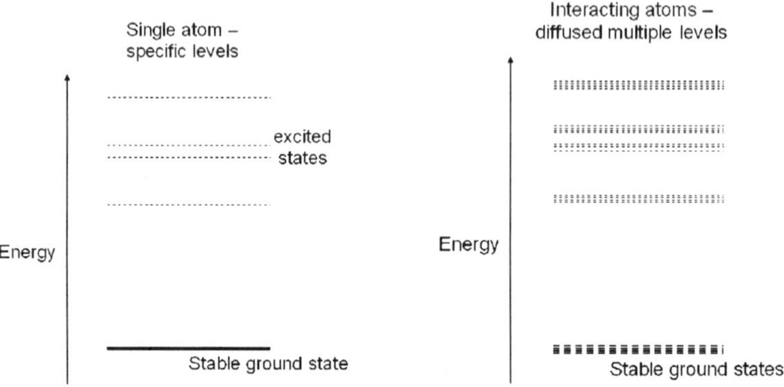


Fig. 11. A simplified energy level model of a single atom – approximating to a gas at low pressure (left) and a group of atoms (right) typified by energy levels in liquids and solids. The energy level diagram's structure is characteristic of a specific element and is unique to that element. Molecules also have similar energy level diagrams – but invariably with significantly more detail therein.

spectrum. This broadening makes spectroscopy in solids and liquids frequently more difficult than the equivalent process in gases. The energy level differences of different materials tend to overlap significantly for solids and liquids. However as will be indicated later in this volume, there are many ingenious data processing approaches to separate out the influence of difference species with overlapping absorption spectra.

Direct absorption of this type is very important tool. There are also many indirect spectroscopy approaches which utilise the fact that the absorbed photon can be emitted at a different wavelength (figure 12).

Of these tools probably Raman spectroscopy and fluorescence spectroscopy are the most important. Raman spectroscopy produces a characteristic signature which is in effect a map of the energy levels within a solid distributed around the ground state. This map is a characteristic of the material itself and also of the conditions which it is experiencing. The signature here is the difference in energy between the incident photons and the re-emitted photons and this difference is broadly independent of the incoming photon.

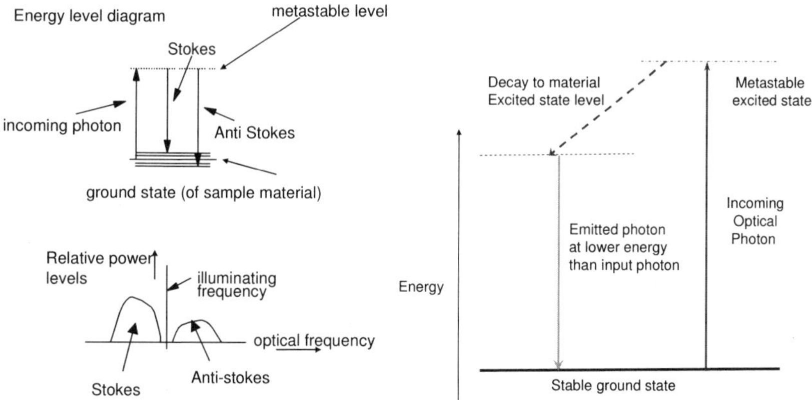


Fig. 12. The elements of Raman (left) and fluorescence (right) spectroscopy. In Raman spectroscopy some of the incoming light is frequency shifter by an amount corresponding to the very characteristic energy distributions around the material ground state. In fluorescence spectroscopy, the re-emitted light emerges at a wavelength determined by the molecular energy levels in the material itself.

An important variation on this theme uses thin metallic overlays to enhance the surface electric fields and therefore enhance the Raman signal, which usually needs relatively high optical flux densities, so called SERS - Surface Enhanced Raman Spectroscopy. Fluorescent spectroscopy targets the presence of a particular material. It is particularly evident in biochemical and biomedical analysis where a fluorescent label is added to a molecule of interest in a particular reaction sequence. Thereafter the presence of this molecule is detected by fluorescence from the label, a technique of such import that the originators of the now well known green fluorescent protein[17] were recognised as Nobel Laureates (Chemistry 2008).

## 5.   Some other Manifestations of Photonics

Photonics is a very versatile technology, arguably the most versatile of technological enablers. There are many more subtleties associated with the concepts we have mentioned, and a few concepts which we have yet to mention.  One of these is the somewhat surprising realization that optics can exert useful forces. The idea of radiation pressure[18] has been around for a considerable period of time. If a photon has a specified energy then, associated with the energy there is also a momentum.  If the photon is deflected from its path, i.e. reflected, then the object causing the deflection feels a force as a direct result of the change in momentum of the reflected photon.  Usually these forces are too tiny to be important, e.g. a 1W laser beam being 100% reflected towards into its incident direction produces a force of only a few nano Newtons.  However, when the objects get small enough then the force is useful and the result (figure 13) is optical tweezers[19] which have evolved from being a laboratory curiosity into an important manipulative tool particularly for biosciences. A single cell can be effectively transported using nothing more than light. The basic concepts we have already described also have multiple nuances.  As one example, suppose we start with a solid which is capable of fluorescing[20], in other words electrons can be excited to a higher

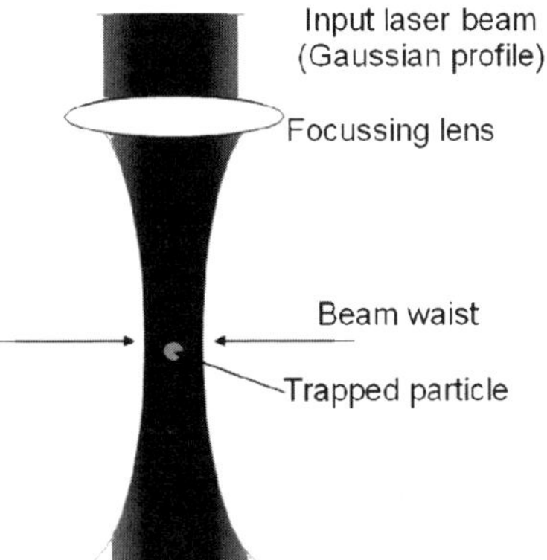



Fig. 13. Optical tweezers – the essential elements.  The higher power density towards the beam axis introduces net radiation pressure forces which keep the trapped particle on axis and slightly beyond the focus.

energy level (in distribution of levels) and can emit photons by relaxing from the higher energy level to a band of lower energy levels - then if we provide appropriate optical feedback in principle we have a laser (figure 14).

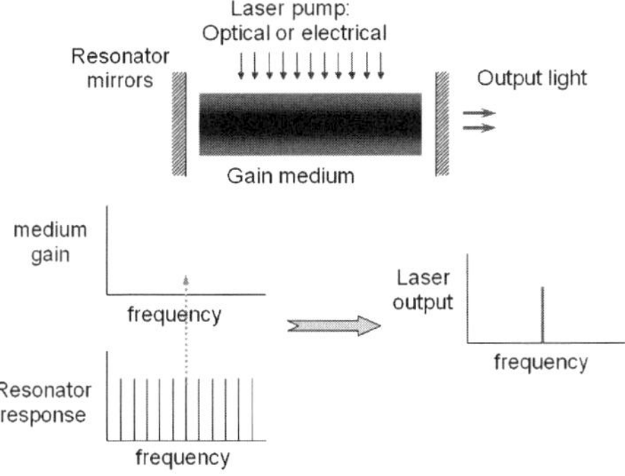



Fig. 14.   Lasing action –the cavity resonances effectively selects the exact energy level (frequency) of the photons emitted by the laser.

In principle, any of the higher levels could fall to any other of the lower levels and so we could get a broad spectrum (which is what happens in light emitting diodes for example). However, two factors conspire to select a wavelength. One is internal and determined by the statistics of the gain process which differs between different energy pairs and the second is the influence of an external structure – the resonant cavity without which there would be no feedback and no lasing. It is the cavity (figure 14) which determines the exact characteristics of the output radiation. Suppose though we go one stage further. The energy levels actually within the material itself can be analyzed in terms of quantum mechanical wave functions (as opposed to optical wave functions) confined in a cavity - though now referred to as a potential well, which is determined by the local environment in which the molecule finds itself. As we have already mentioned, this can range from the very precise levels in a single molecule well isolated from its companions to the spread of levels in a large volume of dense material.

However, if the structure itself becomes small enough to influence the quantum mechanical wave functions, then this, in turn will also influence the energy levels within the system. This gets us to the so-called "quantum dot" which has many curious features, the most obvious of which is its color depends upon its size. A common and well known example of this modification of the conventional world from the microbe to the nano scale, is the gold nanosphere which has been used as pigments for stained glass for centuries and for which the 100 nm and less dimensions determine the color. Typically a colloidal suspension of gold nanospheres will be red rather than gold and the smaller the nanosphere[21] the more blue it gets. Gold nanospheres too have found applications in biological and chemical sensing and measurement by observing changes in coatings affecting changes in perceived color using field enhancement mechanisms somewhat related to those exploited in SERS.

The photonic properties of materials and structures clearly offer an immense diversity of tools and concepts which can be manipulated into intriguing applications. At the most basic conceptual level everything is determined by interactions between waves, materials and structures. The basic materials interact with the light through energy levels and photons - but paradoxically the very importance concept of the refractive index can be very effectively appreciated through classical electromagnetism. The mechanical dimensions of photonic structures, defined through ingenious machining processes and designed to manipulate electromagnetic waves at photonic frequencies typically control light through diffraction and interference but can even be small enough to affect the quantum wave mechanical characteristics of the baseline material which in turn will influence its optical properties. However, when detected these electromagnetic waves will only manifest themselves in quantised energy units - one by one - as photons!

## 6. Some Final Observations

These few pages have attempted to highlight the principal concepts underpinning what has become known as photonics. There is much more to study and hopefully the bibliography will help to support these very basic ideas. In the context of safety and security[22], much of what is required centers upon utilizing photonics in measurement whether of physical parameters (does the structure of interest get longer or shorter, often monitored through one form or other of change in optical delay) or chemical parameters usually measured through some form or other of spectroscopic technique which reduces conceptually to monitoring the changes in the refractive index of a sample. Photonic systems also contribute immensely to the communication network required to ensure that those who need to know are kept aware of whatever safety or security issues are of interest.

The evolution of photonics has been immensely stimulated through the laser which conceptually was an entirely different approach to realizing an oscillating circuit. The gain mechanism in the laser or the maser is a suitably stimulated material rather than an electronic amplifier though both, of course, need resonant feedback. This was a huge conceptual step again resulting in the initiators' being recognized as Nobel Laureates (Physics 1964). Interestingly though much, arguably all, of the remainder of photonics is conceptually well established, harking back to the microwave era for circuits, antennas and waveguides (including dielectric systems) and even further back into spectroscopy and photochemistry. In addition to the laser, the importance of which cannot be underestimated, there has though been a much quieter but possibly equally important evolution in the ways of making things. Very little in our current photonic toolbox would be possible without the availability of very high precision fabrication and assembly techniques which were inconceivable half a century ago. These have facilitated not only more and more transistors on a single semiconductor chip but the realization of the precision structures which feature in many of our photonic devices. This increasing precision lends[23] optimism to those who view the nanophotonic circuit and the metamaterial as something which will be practically realizable as opposed to a laboratory curiosity within a generation or two.

Photonics has immense potential. It is a tool for communications, for sensing and measurement, for diagnostics and therapies, for moving and machining and modifying[24].

Many of these capabilities offer promise to meet the needs of an emerging society over the coming decades (figure 15) by combining wave phenomena with material properties with particles and circuits with ever increasing ingenuity. Perhaps its claim to be the electronics of the 21st century is entirely justifiable.

1. Make solar energy economical
2. Provide energy from fusion
3. Provide access to clean water
4. Reverse-engineer the brain
5. Advance personalized learning
6. Develop carbon sequestration methods
7. Engineer the tools of scientific discovery
8. Restore and improve urban infrastructure
9. Advance health informatics
10. Prevent nuclear terror
11. Engineer better medicines
12. Enhance virtual reality
13. Manage the nitrogen cycle
14. Secure cyberspace

Fig. 15. The Grand Challenges in Engineering as elucidated by the US National Academy of Engineering.

# References


1. J. R. Pierce, *"Almost all about waves"*, Original published by MIT press, 1976, reissued by Dover Books (2006).
2. S. V. Gaponenko, *Introduction to nanophotonics,* Cambridge University Press (2010).
3. C. Roychoudhuri, A.F. Kracklauer and K. Creath, *The nature of light: What is a photon?* CRC Press (2008).
4. R. Kompfner (1964), *The Invention of the traveling-wave tube*, San Francisco Press, and also J. R. Pierce (1950), *Traveling-Wave Tubes*, D. van Nostrand Co.
5. A. Rogers, *"Essentials of photonics",* 2$^{nd}$ Ed., CRC Press (2009).
6. B. E. A. Saleh and M.C. Teich, *"Fundamentals of photonics"*, 2$^{nd}$ Ed. Wiley-Interscience, New York (2007).
7. A. Bokulich and G. Jaeger, *Philosophy of quantum information and entanglement*, Cambridge University Press, (2010).
8. C. Kittel, *"Introduction to solid state physics"*, 8$^{th}$ Ed. John Wiley and Sons (2005).
9. Focus edition on slow light in Nature Photonics, Vol. **2** No. 8, pp. 447-509, *Nature Photonics* (2008).
10. J. C. Stover, *Optical Scattering: Measurement and Analysis,* SPIE Press (1995).
11. R. J. Adrian, *Selected Papers on Laser Doppler Velocimetry*, SPIE Milestones Series (June 2006).
12. J. W. Goodman, *Introduction to Fourier Optics*, 3$^{rd}$ Ed. Roberts and Company (2007).
13. S. A. Maier, *Plasmonics: Fundamentals and Applications*, Springer Ed. (2007).
14. J. Homola, *Chem. Rev.*, **108**, pp. 462 – 493, (2008).
15. J. B. Pendry and D. R. Smith, *Scientific American*, **295**, 60 (July, 2006): D. R. Smith, *Superlens breaks optical barrier, Physics World*, **18**, 23 (2005).
16. J. M. Hollas, *Modern Spectroscopy,* Wiley Ed. (2004).
17. N. C. Shaner, G. H. Patterson and M. W. Davidson, *Advances in fluorescent protein technology, J. Cell Sci.*, **120**, 4247-60 (2007) and also A. Tromans in Bright future for GFP, *Nature Reviews Molecular Cell Biology*, **5**, 865 (November 2004).
18. R. Loudon and C. Baxter, *Proc. Royal Society A*, **468**, pp., 1825-1838 (2012).
19. K. C. Neuman and S. M. Block, *Sci. Instr.*, **75**, 9, pp. 2787-2809 (2004).
20. A. E. Siegman, *Lasers*, University Science Books (1990).
21. H. Huang, P. K. Jain, I. H. El-Sayed and M. A. El-Sayed, *Nanomedicine*, **2**, pp. 618-693, (2007).
22. *Second Strategic Research Agenda in Photonics*, European Technology Platform Photonics21 (2010) and also *Harnessing Light,* currently being updated (2012 with the 1$^{st}$ Ed. published by US National Academies Press (1998).

23. P. Van Zant, *Microchip Fabrication: A Practical Guide to Semiconductor Processing 5$^{th}$ Edition*, McGraw Hill Ed. (2004).
24. P Schaaf Ed., *Laser Processing of Materials, Applications and Developments*, Springer (2010).

# STRUCTURAL HEALTH MONITORING IN BUILDINGS, BRIDGES AND CIVIL ENGINEERING


Alfonso Martone[1], Mauro Zarrelli[1], Michele Giordano[1*]
and José Miguel López-Higuera[2]

*Institute for Composite and Biomedical Materials – IMCB*
*National Research Councyl of Italy - CNR*
*P.le Tecchio, 80 - 80125Naples (Italy)*
**E-mail: gmichele@unina.it*

*[2]Photonic Engineering Group of the University of Cantabria,*
*Avda Los Castros s/n, 39005, Santander, Spain*



Monitoring the in-service structural behavior of engineering structures has become a primary requirement on assessing their safety and integrity. Owing to their outstanding features, such as high sensitivity, multiplexing, capability, lightweight, easy application, noise immunity and harsh environmental resistance, Fiber Bragg Grating (FBG) have emerged as attractive sensors for monitoring, diagnostic and control in civil structures.

In this work, we present recent research and development activities in structural health monitoring using FBG sensors. A brief survey on FBG based sensors, suitable for in-line monitoring, is carried out. Several implementations are described in order to illustrate the use of quasi distributed system as reliable, non destructive in situ monitoring tool. Finally, limitation and market barriers related to the use of FBG sensors have been argued.


## 1. Challenges for Structural Health Monitoring

Civil infrastructures like bridges, buildings or dams, are conceived to accomplish design requirements over decades. These and similar structures, during their long-time service life, are subjected to the concurrent action of environmental corrosion, fatigue loads and material aging, which lead inevitably to a progressive damage accumulation and, in some cases, to their collapse.

Lots of these structures needs smart monitoring system to assure structural integrity and durability. Furthermore an adequate monitoring system is needed, also, when the structure is affected by external works or during modifications or to verify the state of health subsequently natural disasters[1].

An ideal SHM system should accomplish some basic requirements:

- since damage is essentially a local phenomenon sensors may be integrated in the structure without distortions in its behavior over wide length and time scales;
- generally structural integrity decays slowly, small flaws propagate during long periods, sometimes, in adverse environments, thus it is essential that sensors provides accurate and reliable measurements;
- as electronic devices are subjected to failures, the damage recognition algorithm should be robust, namely it must be able to self adapt to dynamic modifications of the sensor network (when a sensor fails), therefore a proper redundancy should be account in designing the sensor network.

The final challenge is the complete integration between the monitoring system (sensors and damage recognition algorithm) and the structure management, providing economic benefit over current maintenance approach and a step forward for the life safety of civil infrastructures by means of damage tolerant structures.

Fiber Bragg Gratings (namely FBG) sensors are gathering the interest of the SHM community due to their amazing properties, FBGs are immune to electromagnetic interferences, lightweight and they are characterized by tiny physical dimensions which make them ideal candidates to be embedded within the observed structure. Furthermore, FBGs do not need a closed web to be interrogated; in fact, only reflected signal is important for demodulation and thus these sensors could be multiplexed by shifting the characteristic wavelength[2]. FBGs have been successfully used to monitoring bridges and civil structures; compared with standard Non Destructive Techniques (NDI) they could monitor the structure remotely without the constraint of keep physically the instrumentation located on the structure.

Main application of fiber optic sensor involves bridges and concrete structures, where sensors are bonded to the structure or embedded, where possible, in order to monitor the deformation and dynamic response[3]. In concrete structures embedding FBG within the cement allows to evaluate the shrinkage due to the water evaporation, responsible for small cracks which could propagate within the structure weakening it.

## 1.1. *Implementation of SHM*

Except for certain types of public or private housing, civil infrastructures are unique, hence frequently, there are not baseline derived from expensive qualification procedures as in the case of aerospace structures, moreover the overall structures has to be geared and monitored towards a long-term period.

In short terms, Structural Health Monitoring (SHM) concerns to the capability of evaluate trough a physical or parametric model the state of the structure using time depended data.

Farrar and Worden[4] described the SHM as a four step process which includes:
- Operational Evaluation,
- Data acquisition, normalization and cleansing,
- Feature selection and information condensation,
- Statistical model development for feature discrimination.

This latter approach, named *"The Statistical Pattern Recognition Paradigm"*, collects from an operative point of view all the phases behind the monitoring the status of a structure, in fact each of the previously enumerated steps answers to questions about the overall health helping to define step by step the proper monitoring system. The first stage concerns to the opportunity of using an SHM system and the definition of parameters to be recorded, this step focuses the attention on the study of the structure to establish physical parameters to be monitored.

The second step pertains to the hardware definition and data management. While, the third stage focuses to the damage identification on the structure developing a suitable procedure to extract features from the acquired data, i.e. the shift in vibration frequencies or amplitude. The last stage reports the attempt to formulate a statistical approach to the structure behavior with the aim to identify "false positives" in the diagnostic system developed. Since, an extensively description of each phase is out of the topic of this chapter, the interested reader could find more information in the Farrar and Worden works[4-6].

Complementarily, as SHM can be understood as the system that includes the integration of sensing intelligence and, possibly, also actuation devices which allow the loading and the damaging conditions of the structure to be recorded, analyzed, localized and predicted, in a way that non-destructive testing becomes an integral part of the structure[7].

According to the functionality and degree of complexity, the SHM systems can be classified in five levels following what can be named as the staircase of the SHM systems being at the higher level the one corresponding to the higher complexity and functionality. The SHM systems of Level I, are capable only to detect the presence of damage without locating it on the structure; the Level II, in addition to the damage detection, is able to located it; Level III systems, add to the capabilities of the level II, also the possibility to perform diagnosis or to estimate the severity of the damage; Level IV SHM systems are able to realize the prognosis or to estimate the remaining service life. Finally, level V system are constituted by very complex hardware, customized algorithms and the related software to enable, by itself, the diagnosis and/or the prognosis and, in some cases, even to adopt healing functions[7].

## 2.  Optical Sensor for Structural Health Monitoring

Many sources could induce deformation within an engineering material, as external or thermal loads (structural strain) as well as residual stresses or rheological effects. In all cases, strain components generally occur simultaneously and the total strain within the material is the sum of these components:

$$\varepsilon_{tot} = \varepsilon_{Elastic} + \varepsilon_{Plastic} + \varepsilon_{Thermal} + \varepsilon_{Creep} + \varepsilon_{Shrink} + \varepsilon_{other} \quad (1)$$

Due to their properties, sensors based on fiber optic are the ideal candidates to measure local deformation of materials.

In the following, firstly, the working principle of FBG as strain sensors will be briefly described then some common sensors for SHM, like accelerometers and inclinometers, set by the usage of FBG will be reviewed and analyzed.

## 2.1. *Strain Measurement using FBG Sensors*

Bragg gratings are optical fibres with periodic alteration of refraction index extended approximately ten millimeters, when a tunable light source is led into the fiber containing the grating only the wavelength corresponding to the grating will be reflected while all the others will pass indisturbate[8,9]. The use of FBG based sensors gives the great benefit of unnecessary closed measuring loop, since the measure is obtained by the reflected light of the grating.

The reflected signal is a narrow spectral band centered at Bragg wavelength $\lambda_B$ according to the following first order condition:

$$\lambda_B = 2\, n_{eff}\, \Lambda \quad (2)$$

where $n_{eff}$ is the effective refractive index of the optical fiber and $\Lambda$ the grating period. If the fiber is stretched and/or exposed to temperature variations, a wavelength shift will be induced, which is proportional to the solicitation.

The wavelength shift is function of the strain physically applied to the glass fiber and the temperature changes, according to the following expression reporting the change in central spectrum wavelength;

$$\frac{\Delta\lambda_B}{\lambda_B} = \left(1-\rho_e\right)\varepsilon + \left(\alpha+\xi\right)\Delta T \tag{3}$$

where the first term correspond to the alteration of wavelength due to longitudinal strain while the second term is the shift due to temperature changes. In the latter formula, $\rho_e$ is the effective photo-elastic constant and $\xi$, is the thermo-optic coefficient of the fiber while $\varepsilon$ is the longitudinal strain applied to the fiber and $\alpha$ the thermal expansion coefficient.

Typical wavelength shift for grating having pitch at 1550 nm is 0.0012 nm/$\mu\varepsilon$ (only longitudinal strain applied to the fiber) and 0.013 nm/°C[10]. By determining the peak shift in the reflected signal, the strain and/or the temperature can be evaluated.

Detailed explanation of the shift phenomenology is out of the scope of this work.

### 2.1.1. Temperature Compensation

FBGs show high temperature dependence and cannot provide self-compensation of apparent strain, therefore to apply FBGs as reliable sensors for strain measurements, it is mandatory to find a solution for separating the two contributions of the overall Bragg wavelength shift (Eq. 3).

A number of solutions have been proposed to meet the latter requirement, the widely accepted method uses two FBGs, one as reference grating and another one located at the measurement point. The simplest case is to locate the reference grating where the same thermal load is acting but, at same time, the FBG is free of stresses. This technique, based on temperature compensation, allows the evaluation of the strain at the measurement point as difference between peak shift in measurement point and compensation point:

$$\varepsilon_m = \frac{1}{\left(1-\rho_e\right)}\left[\left(\frac{\Delta\lambda_B}{\lambda_B}\right)_m - \left(\frac{\Delta\lambda_B}{\lambda_B}\right)_c\right] \tag{4}$$

where subscript m refers to the grating located at measuring point and subscript c to the *compensation* grating[11].

Although, this procedure is commonly accepted and implemented for many applications, not always is possible to find a location of the examined structure characterized by a free stresses condition, therefore, in this cases, the reference grating is used as pure temperature sensor (fiber is not stressed, fixed at only one end). The peak shift on the reference grating is then related only of temperature changes:

$$\left(\frac{\Delta\lambda_B}{\lambda_B}\right)_T = \left(1-\rho_e\right)\alpha_{GLASS}\,\Delta T + \left(\alpha+\xi\right)\Delta T \qquad (5)$$

Combining the signals of both gratings, the strain at the measuring point can be written as follows:

$$\varepsilon_m = \frac{1}{\left(1-\rho_e\right)}\left[\left(\frac{\Delta\lambda_B}{\lambda_B}\right)_m - \left(\frac{\Delta\lambda_B}{\lambda_B}\right)_c\left(\frac{\left(1-\rho_e\right)\alpha_{spec}+\left(\alpha+\xi\right)}{\left(1-\rho_e\right)\alpha_{GLASS}+\left(\alpha+\xi\right)}\right)\right] \qquad (6)$$

Even though, this latter technique is very simple and direct, it may led to a drop in measurement resolution and it needs the interrogation of two fibres, at least.

### 2.1.2. Multiplexing and Networking of FBG Sensors

Taking advantages of the FBG optical response, a variety of multiplexing techniques based on different modulation formats have been developed.

They generally fall into one of the following categories: Wavelength Division Multiplexing (WDM), Time Division Multiplexing (TDM), Frequency Division Multiplexing (FDM), Coherence Multiplexing (CM) and Polarization Division Multiplexing (PDM)[12]. In addition, as shown in Figure 1, it can be also used simultaneously the operation of two different modulation formats within the same network (hybrid approaches)[13].

Sensor networks can be conceived using only passive fibers (without utilizing optical gain) or introducing optical amplification in some key parts of the networks and, hence, passive or active networks can be developed[13,14].

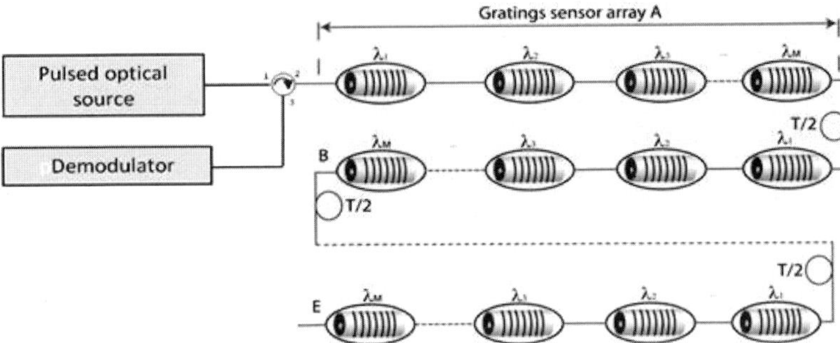


Fig. 1. A hybrid passive time/wavelength division multiplexed FBG sensor serial bus network basic scheme. The network is integrated by a number of identical sections (A, B, C,...) each one containing M FBG sensor properly WDM multiplexed ($\lambda_1$, $\lambda_2$, ..., $\lambda_{M-1}$, $\lambda_{M-1}$). The successive sections are connected by fiber delay lines and the response of each section is then delayed two times. The reflective responses are launched to the input of the WDM demodulator and hence each M output is time demodulated on the time division demodulator.

In order to build-up optical fiber sensor networks using multiplexing concepts, four basic tasks must be observed: a) launching an optical signal into the network having a suitable power, spectral distribution, polarization and modulation; b) detection of the signal, coded or modulated by the sensor element, and sent back to the optoelectronic unit by transmission or reflection; c) uniquely identify the information corresponding to each sensor of the network, by means of proper addressing, polling and decoding; d) evaluation of the generated optical signals by modulating each sensor with a calibrated electrical signal[14].

In addition, to design sensor networks, it must be consider such aspects as: i) modulation and coding format of the optical signal, ii) network topology, iii) inclusion or not of optical amplification technologies, iv) decoding method for the received signal, v) types of sensors multiplexed in the same network and finally, vi) economic conditions, which would eventually determine the most appropriate network[15].

It must be taken into account that the light traveling through optical components of the typical networks is tapped-off from the spine and leaded to the sensors by directional couplers inducing a signal attenuation and limiting the number of couplers which can be used.

Then, optical amplification (lumped or distributed) is commonly used to compensate these losses and even to increase the number of sensing points. The losses can be compensated by slightly doping the spine with Erbium and using a pump source as it is illustrated in the active network reported in the figure 2[12].

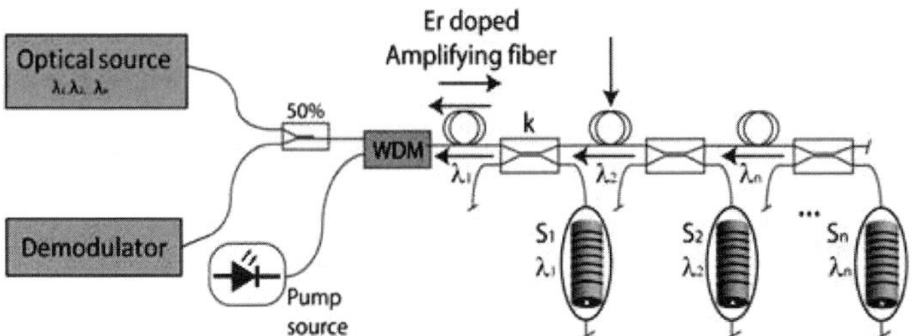


Fig. 2. Illustration of a bus topology with distributed amplification for WDM sensor multiplexing. $S_1$, $S_2$,..$S_n$ are the sensor transducer 1, 2,….. n.

There are two main motivations for including optical amplification in the sensor networks: a) to increase the number of sensors multiplexed in a single network while ensuring good signal quality, and b) to enable the possibility of remote sensing.

Sensor multiplexing networks have to deal with the problem of power losses due to the distribution architecture. Using optical amplification the splitting losses can be compensated and the number of sensors per fiber on the network could be, significantly, increased[16].

To summarize this section, we can conclude that a variety of techniques to multiplex FBG sensors can be employed. For each specific application, a wide set of parameters must be considered to determine the suitability of a sensor network.

The wavelength-dependent transfer function of FBGs makes them appropriate for wavelength division multiplexing. Time division multiplexing is simple in principle, following well-established electrical domain techniques. A single pulsed source for all sensors on network and hence a good signal to noise ratio is currently used. It needs fast electronics, but can measure, rapidly, variant parameters. Frequency division multiplexing provides low noise and wide bandwidth operation at the expense of a comparatively costly and complex network.

By using hybrid or combined multiplexing schemes (especially the hybrid time/wavelength division multiplexing approach) a higher significant multiplexing gain can be obtained, and thus several hundreds of optical point sensors can be supported. In addition, optical amplification can play a critical role in FBG multiplexed systems if remote sensing operation is desired or the network has to serve to a big number of sensors.

Finally, sensor networks can be designed using only passive fibers or introducing optical amplification in some key parts of the networks and hence passive or/and active networks can be developed and constructed.

## 2.2. *FBG-based Sensors*

In the last three decades, FBG has proven to be a reliable technology for temperature and strain sensing, actually they own the great advantage that information are contained in the spectral peak position, which leads to a signal immune by power fluctuation and prone to multiplexing.

Nevertheless, in the last decade have been proposed sensors based on FBGs wherein the property which would be measured is correlated by transforming strain of a properly designed device in the desired variable.

In the following paragraph a brief description of sensor based on FBGs, suitable for SHM, will be reported.

### 2.2.1. *Accelerometers*

In order to use an FBG sensor as accelerometer, the acceleration has to be correlated to the deformation of a dedicated device; finally the wavelength shift is calibrated to the level of acceleration. The structure of the accelerometer consists in an inertial mass connected to a supporting structure where the grating is bonded, the simplest schematic of such a sensor is an optical fiber acting as a spring attached to a mass, as shown in Fig.3. When exposed to external acceleration, the inertial mass move in vertical direction imposing a contraction/expansion of the optical fiber which induces a shift in the grating wavelength.

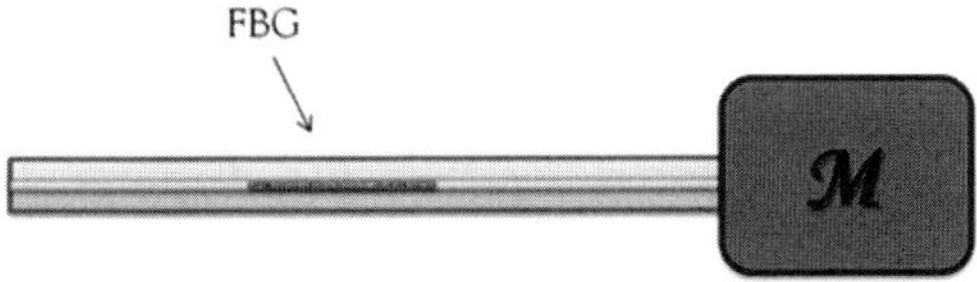


Fig.3.  FBG-based accelerometer: basic principle.

In this case the mechanical scheme is a cantilever beam which first natural angular frequency is:

$$\omega_0 = \sqrt{\frac{K_{eq}}{M}} \qquad (7)$$

where $K_{eq}$ is the equivalent elastic constant of the system and, M is the inertial mass.

The acceleration measured by the device is related to the deformation induced in the grating by the inertial mass movement:

$$\varepsilon \propto \left(\frac{M}{E\,A}\right) a \qquad (8)$$

the peak shift is linearly proportional to the applied strain, in that case the interrogator employed controls the measurement sensitivity of that device. The final characteristic of the sensor, such as linear response, range in frequency, maximum acceleration and sensitivity depend on properly designed devices in term of constituent physical properties and relative positioning of the elements. Since civil infrastructures present main fundamental eigenfrequencies in the range 0-20 Hz the device should have natural frequencies above this range but adequately to assure the needed sensitivity.

First examples of optical accelerometers were made by embedding the grating between two plates, one as supporting plate and another as inertial mass, then the system is clamped on one side to resemble like a cantilever beam[18-20], these devices could reach natural frequencies much higher than 100 Hz and sensitivity of 10 $\mu\varepsilon$/g[19].

This kind of accelerometers could not balance the deformation induced by thermal changes, thus monitoring the temperature, in some applications, is a mandatory requirement although the system cost will inevitably increases. In order to make the device insensitive to thermal changes some authors[20,21] successfully attempted to measure the reflected optical power, instead of the correlation with the peak shift in a cantilever beam structure; in that case, as the shift in the wavelength is not accounted, thermal changes would not affected the measurements.

These systems are able to detect accelerations up to 8g[20].

Other authors solved the problem by using two or more gratings[22-25] in order to keep accounting the temperature compensation. Antunes *et al.* based their accelerometer on an L-shaped cantilever beam where two Bragg gratings are placed side by side. As the inertial mass is excited, one FBG is contracted by the movement while the other is expanded therefore the FBGs are affected by the same temperature and thus the system become immune to thermal effects. A different technical solution is to use one fiber with two Bragg gratings. Morikawa *et al.* proposed a device where the inertial mass is hold on a fiber between two gratings in this way the fiber acts as a spring, moreover, in this case, FBGs are also at the same temperature state but subjected to opposite strains.

### 2.2.2. Inclinometers

Tilt sensors and inclinometers generate an artificial horizon and measure angular tilt with respect to this horizon. They have been used widely to monitor ground movements in tunnels, foundations, platform leveling, indeed anywhere tilt requires measuring.

Conventional inclinometer consists in rigid cases installed vertically to the ground to measure the local tilt, commonly used sensors are accelerometer, liquid capacitive, gas bubble in liquid, and pendulum.

Fig. 4 reports a schematic diagram of a device based on FBG to measure uniaxial rotations. The sensor consists of a pendulum clamped to a rigid frame, two FBG glued on shaft linking the mass to the frame.

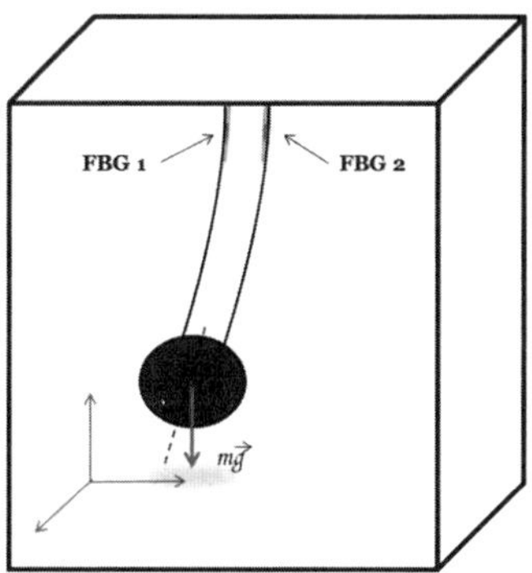


Fig. 4. Tilt sensor based on FBG - schematic diagram.

When the sensor rotates with an angle $\vartheta$ the force applied to is $mg$ $sin\vartheta$ where $m$ is the mass of the pendulum; according to the elastic beam theory the deflection in a point along the length of the shaft could be determined by knowing the strain field.

$$\varepsilon = \frac{mgL}{Ebh^2}\sin\vartheta \tag{9}$$

where L is the length of the cantilever, E the Young modulus of the shaft, b and h its dimensions. By using two close-located FBGs, the peak shift due to temperature could be assumed identical, but shifts due to deformation are in opposite direction with the same absolute value. Based on the differential measurement method the difference between the wavelength shift is directly related to the tilt angle[26]:

$$\Delta\lambda = \frac{\Delta\lambda_1 - \Delta\lambda_2}{\lambda_0} = \frac{12(1-\rho_e)mgL}{Ebh^2}\sin\vartheta \tag{10}$$

Devices based on this simple principle have been successfully realized[26-28] with the accuracy of 0.1° and resolution better than 0.007°.

The main limitation encountered in this kind of sensors (accelerometers and inclinometers) is the inability to have a single device allowing multidimensional measurements even if commercially are available properly designed stands to hold more than one sensor to keep they along desired planes.

## 2.3. *Distributed Strain Sensing*

Continuous spatially distributed sensing is of great interest for a very wide range of real applications. In addition, as it can be observed in the section of OFS market of this chapter, a higher growth revenues tax per annum is expected for the distributed optical fiber sensors. Technically, the intrinsic properties of optical fibers, enables their use as both as the optical channel and as the distributed optical transducer. With this powerful technology the measurement can be performed anywhere along the fiber length with a given resolution.

To build-up distributed fiber sensors several physical sensing principles can be used in conjunction with Optical Reflectometric techniques such as linear (Rayleigh) and non-linear (Brillouin, or Raman) scattering and Optical Kerr effects.

In order to interrogate the sensor, continuous and pulsed optical waves either in the time or in the frequency domain, with some variations, are currently being used. So, Optical Reflectometric Techniques like Optical Continuous Wave Reflectometry (OCWR)[29], Optical Time Domain Reflectometry (OTDR)[30], Optical Frequency Domain Reflectometry (OFDR)[31], Low Coherence Reflectometry (LCR)[32], variations of the above-mentioned techniques such as, Polarization Time Domain Reflectometry (POTDR)[33], Raman Optical Time Domain Reflectometry (ROTDR)[34], Brillouin Optical Time Domain Reflectometry (BOTDR)[35], Brillouin Optical Time Domain Analysis, (BOTDA)[36], and Brillouin Optical Correlation-Domain Reflectometry, (BOCDR)[37], among others, have been demonstrated and used.

To enhance the general performances of the distributed sensing technologies some "variations" have been published.

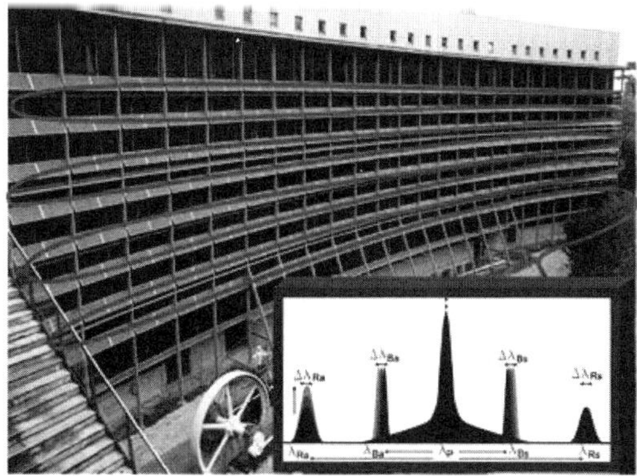

Fig. 5. The distributed fiber transducer (red color) and the Optoelectronic Unit of the Distributed fiber Sensor on the photovoltaic  panels on the principal wall of a corporative building. Courtesy of the Photonic Engineering Group of the University of Cantabria.

One of those on BOTDA schemes relies on, a combination of the controlled optical intensity before the main pulse[38] and a subtraction of the background Brillouin backscattering[39], the so called the *pre-pumping idea*. The light source -basically formed by a pre-pumped- pair pulse and by subtracting the couple Brillouin backscattering it is possible to use larger base pulses to obtain stronger backscattered signals or to reach larger distances in the fiber. Other ways to enhance the preexistence of acoustic wave are such the $\pi$-phase pulse techniques[40].

Nevertheless, in these techniques, the acoustic wave partially decays during the pulse duration, and a second attenuated response appears beyond the acoustic lifetime that may hides the measurement of submeter perturbations on the fiber. One way to suppress or attenuate the impact of this background Brillouin response is based on the subtraction of two Brillouin gain spectra obtained from pulses shifting in time or pulses with different widths[38]; in this technique, the rising and the falling times of the pulses define the spatial resolution. The DPP-BOTDA technique[41] is defined as a simplification of the Brillouin Echo Technique, in which the phase shift pulse is replaced by a dark pulse, and not by a bright pulse.

In conventional BOTDA techniques, if the spectral width of the pulse exceeds the Brillouin line width, the gain spectrum broadens and the measurement deteriorates. Nonetheless, in BOCDA if the spectral width of the pump and probe increases the resolution of correlation technique also increases. Spatial resolution up to 1.6 mm, as well as improvements on the sampling rate of 1 kHz and measurement range of 1 km[42] were achieved.

Brillouin Optical Frequency Domain Analysis (BOFDA) is based on the measurement of a complex transfer function that relates the amplitudes of counter-propagating pump and probe waves along a fiber. The continuous probe wave is modulated in intensity with a sinusoidal signal over a range of frequencies, whilst in the pump wave an intensity modulation is induced. This induced signal has an alternating component (AC component/part) due to the interaction with the counter-propagating probe wave. By measuring the changes in the AC pump wave component the Brillouin frequency shift profile is determined via the complex baseband transfer function. Once the base band transfer function is determined, the impulse response is calculated by applying the inverse Fourier transform (IFFT) to the function. Thus the temperature or strain can be determined from the Brillouin profile along the fiber[43].

Recently, the concept of Brillouin Dynamic Grating (BDG) has been newly implemented in Polarization Maintaining Fibers (PMFs) and single mode fibers[44], since the operation principle strongly depends on the local birefringence of the medium. Acoustic waves generated during the process of stimulated Brillouin scattering, in one polarization, are used to reflect an orthogonally polarized wave (*probe wave*) at a different optical frequency from the pump. BDG has been used as strain and temperature sensors[45] and birefringence sensor[46] and they can be applied also to enhance the spatial resolution of an ordinary Brillouin Optical Time-Domain Analysis (BOTDA) system by replacing the probe with the reflection from the BDG.

To reach very high spatial resolution, a Rayleigh backscatter and an Interferometric hybrid technique have been demonstrated[47]. The Optical Backscatter Reflectometer (OBR) uses Swept Wavelength Interferometry (SWI) to measure the Rayleigh backscatter as a function of optical fiber length. The strain causes temporal and spectral shifts in the local Rayleigh backscatter pattern. By measuring these shifts a distributed temperature or strain measurement profile can be obtained. The SWI approach enables robust and practical distributed temperature and strain measurements in standard fiber with centimeter-scale spatial resolution up to 70 meters of fiber with strain and temperature resolution as fine as 1 microstrain.

Finally, by using a digital signal generator which enables fast switching among 100 scanning frequencies, a new technique for the fast implementation of Brillouin Optical Time Domain Analysis (BOTDA) has been very recently demonstrated[48]. By this technique, which converts the classical BOTDA method to the dynamic sensing domain, a truly distributed and dynamic measurement of a 100m long fiber with a sampling rate of ~10kHz, limited only by the fiber length and the frequency granularity, was reported. A standard deviation of the measured strain of approximately 5 με was noticed.

## 3.  Applications of FBG in Structural Sensing

### 3.1. *Civil Construction*

Masonry structures exhibit usually complex geometries as result of possible interventions (specially historical buildings), damage in such structures relates to crack, foundation settlements, material degradation; when a crack occurs they are often localized splitting the structure in macro-areas.

The current practice for monitoring the health condition are based mainly on periodic visual inspections but, in last decade, software and hardware development have led to the possibility of continuous installation and monitoring of hundreds of sensors and therefore, now, the attention is focused on managing the enormous amount of data available.

An important phase of the monitoring work in the case of already existing structures is to identify its structural health behavior to properly set appropriate warning levels. In a first phase, the structural response of the structure is analyzed by collecting data from sensors to create a reference structural condition (Structural behavior at time "zero" recognition). Further, the structure will be instrumented by a limited number of sensors related to the system behavior positioned in key members in order to set early warning conditions[49,50].

FBG based monitoring system often will be preferred instead of canonical since it guarantees, in addition to the previously listed advantages, the minimal visual impact requirement. Antunes et al.[51,52], to fulfill the minimum visual impact issue bonded, the sensors directly to blocks, sensors were interrogated by an unit allowing a final displacement resolution varied from 1.1 μm to 2.3 μm.

They bonded 19 sensors to a decorative arch of an historical church in Aveiro (Portugal), sensors mounted on the center of the arch experienced a distinct behavior respect to sensor on the extremities. The center was identified as the most sensitive region of the arch since displacements occurred as an application of load over the arch.

In addition to seismic events, one of the major reasons for damage of stone masonry structures is the weathering through a series of complex chemical and biological processes. Frequently, cycles of temperature changes could lead to possible mechanical degradation due not only to temperature variation but also by the presence of moisture.

Masonry structures needs the continuous monitoring of both temperature and moisture, FBG technology allows the measure of both these latter variables by fixing flaws related to the conventional technology (electrical resistance, gravimetric etc...). The use of optical sensors permit a direct measure of the investigated property bypassing the trouble of chemical changes in porous structures and/or the presence of salt dissolved[50,54]. The sensing principle relies on the strain induced on the Bragg grating by a coating material's swelling sensitive to moisture changes[56].

Among the different members of a structure, piles represent indeed the primary elements as they carry the weight of the foundation and transmit the loads to the rest of the structure. The design and the structural monitoring of these members are crucial for all civil structures.

The implementation of a reliable health monitoring system, on these parts, need to account several aspects such as expected loads (traction/compression, bending moment), material properties, ground characteristics and environmental conditions. System based on FBGs, encouraged the use of this technology for the monitoring of these structures[55]. The pile construction of a 13-storey building has been monitored by welding in the piles cage a rebar equipped by sixteen FBG sensors[57]. In order to protect sensor form lateral stresses due to the concrete and to made them impermeable to moisture, sensors were protected by a composite packaging in order to grant the sufficient flexibility for transport and installation. During the concrete pouring, the sensor measured initially a compressive load correlated to concrete level within the bare-hole, then a tensile strain was experienced due to the curing phase. Piles were monitored during the construction of building's floors highlighting that the progressive cure change the structural response while other floor are added on the system.


## 3.2. *Bridges*

Long span bridges are critical to a region's economic vitality thus research and application of health monitoring and safety assessments are fundamental in improving bridge engineering.

A monitoring system for bridge application has to be capable to acquire, transmit and store large data measurement. However, data could not be directly correlated to the structural conditions and then used to generate early warning signals. Environmental actions like wind, temperature excursions, earthquake etc… and daily usage may affects the bridge behavior, thus the dynamic damage technique results difficult to applied due to the complexity of bridge structures and the insensitivity to local damages.

Fatigue life consists of two periods: crack initiation and crack propagation. The crack initiation covers the formation and growth of micro-sized crack, which could not be detected by direct inspection (eye or NDI), while crack propagation refers to the formation and growth of macroscopic cracks. In order to determine the need of structural intervention, on a new bridge, the first phase is carried out primary during the study of its dynamic behavior[45].

The development of the structural health monitoring provides an efficient way to get an accurate evaluation of stress history of key members of the structure, the key substructures of a bridge to be monitored are the hanger cables, the box girder and the beam ends.

Fig. 6 reports a possible route for managing a monitoring system of a bridge, in reason of data managing process. Firstly, raw data of weather action and structural response of the selected members, during a periodic observation window, are acquired; then, the action of environmental conditions, on the structure, are analyzed while the dynamic behavior is studied as function of time history data and system modal response. All the previous analyses should contribute to the formulation of a baseline model for the structural behavior of the system, in order to identify at which environmental condition a warning should be triggered and to indicate when maintenance is needed[59].

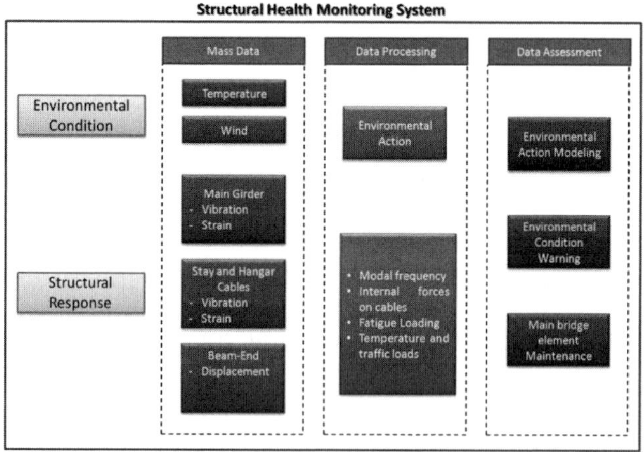


Fig. 6. Structural Health Monitoring Route for a Bridge.

An SHM system on a bridge requires the permanent installation, in various sections, of some accelerometers, strain gauges, displacement transducers, level sensors, anemometers and temperature sensors along with both a data acquisition and processing system.

A basic implementation of the structural monitoring system is to use a damage warning index based on the baseline model formulated.

$$e = \left| f_m - f_s \right| \tag{11}$$

where $f_m$ is the measured structural warning condition parameter, and $f_s$ is the simulated value, corresponding to the actual solicitation.

The use of optical strain sensor has been successfully proven on the Tsing-Ma bridge (Hong Kong, RPC), where the acquired data of the optical monitoring system were compared with results of a sophisticated long-term monitoring system (WASHMS, Wind and Structural Health Monitoring System)[60]. In order to facilitate the installation and, also, to protect the sensors from dust and moisture, FBGs were mounted on thin nitinol strips (~7.5 μm) and encapsulated between two Teflon sheets, finally the installation were covered by and ABS enclosure. Sensors were installed on three strategic locations: hanger cables, rocker bearing and in a selected section of the supporting deck. The optical system used has a resolution of 1 pm and accuracy of 10 pm with a sampling rate adjustable up to 20 kHz. To monitor selected members of the structure only three strands of fiber, containing 21 FBGs, were employed.

Schultz and his coworkers used twenty-six FBG strain sensors to monitor the Horsetail Falls Bridge (USA) mounted directly on concrete structure or on composite reinforcing wraps, which were added further from bridge built (1914) to strengthen the structure,. The optical system used in this application is able to measure up to 1 kHz of dynamical strain with a resolution of 0.1 microstrain[61].

### 3.3. Geodynamic

The measurement of ground movement is an essential part of many engineering operational, i.e. to monitor landslides or tunnel stability. FBGs sensors could be very useful in this field to monitor rock deformation, ground movements (optical geophones), or seismograph[62].

Many laboratory applications encouraged the investigation on devices based on optical fiber, since they have got the primary advantage of being immune to electromagnetic interferences, resistant to water damages and corrosion in harsh environment[63].

A seismic monitoring system based on the inverse pendulum concept was proven as a valuable device for monitoring ground landslide and early earthquake warning[64]. On this device four FBG arrays were mounted on a Plexiglass® cylinder with a seismic mass on top, and interrogate with a sampling rate of 100 kHz for each channel, this structure has got a natural frequency of about 10Hz and a flat response range up to 0.10 Hz proven it as a valuable optical technology for seismic monitoring.

Another application of FBG based devices for seismic purpose was fabricated and tested by Correia and coworkers[28]. In this device, two arrays of FBG were bonded on a tubular structure to serve as an inclinometer.


## 4.  OFS Market

Excluding biomedical, medical and the image sensors, the market of Optical Fiber Sensors (OFS) is expected to grow strongly over the second decade of this century. A Compounded Annual Growth Rate (CAGR) of 9.8% and revenues achieving the important figure of $1.95 billion by 2020 are expected[65].

Over the past recent years, the distributed fiber optic sensor (included the quasi-distributed) market has experienced very strong demand from both commercial and government sectors. OFS sensors are increasingly installed along pipelines to monitor for leaks, pressure and positional changes, and they are deployed along power lines and structures to detect breakages and shifts. Distributed OFS sensors can be an integral part of buildings to provide information on structural health. Similarly, they can be used to monitor the integrity of dams, bridges, roads, and tunnels. These sensors also offer various perimeter security solutions, where they can be either permanently installed or deployed as needed in the field.

The forecast for distributed OFS over the present decade (2009-2020) is strong and achieves a CAGR of 11.8% with revenues approaching $1.4 billion.  The growth rate of the market does slow down towards the end of the decade and reduces to a 2015-2020 CAGR of 6.2%.


## Acknowledgments

This work was co-supported by the Spanish Ministry of Education and Science (today Economics and Competitiveness) under the TEC2010-20224-C02-02 project.

# References


1. J. M. W. Brownjohn, *Phil. Trans. R. Soc. A.*, 365, 589-622 (2007).
2. M. Majumder, T. K. Gangopadhyay, A. K. Chakraborty, K. Dasgupta, and D. K. Bhattacharya, *Sensors and Actuators A: Physical*, **147**, 150-164 (2008).
3. H.-N. Li, D.-S. Li and G.-B. Song, *Engineering Structures,* **26**, 1647-1657 (2004).
4. C. R. Farrar and K. Worden, *Phil. Trans. R. Soc. A.*, 303-15 (2007).
5. H. Sohn, C. R. Farrar, F. Hemez and J. Czarnecki, *Los Alamos Nat. La,* 1-7 (2001).
6. K. Worden, C.R. Farrar, G. Manson and G. Park, *Phil. Trans. R. Soc. A.*, 463, 1639-1664 (2007).
7. J. M. Lopez-Higuera, L. Rodriguez Cobo, A. Quintela Incera and A. Cobo, *Journal of lightwave technology,* **29**, 587–608 (2011).
8. G. Meltz, W. Morey and W. Glenn, *Opt. Lett.,* **14**, 823-825 (1989).
9. A. D. Kersey, M. A. Davis, H. J. Patrick, M. LeBlanc, K.P. Koo, C. G. Askins, M. A. Putnam, E. J. Friebele, *J. of Lightwave Technology*, **15**, 1442-1463 (1997).
10. Y. J. Rao in *Optical Fibre Sensor Technology*, Vol.**2**, Chapman & Hall, London, 355-389 (1998).
11. F. M. Haran et al. *Meas. Sci. Technol.,* **9** 1163 doi:10.1088/0957-0233/9/8/004 (1998).
12. A. D. Kersey in *Fiber Optic Smart Structures* E. Udd (Ed). pp. 409-444 John Wiley & Sons (1995).
13. M. López-Amo, in *Advanced Photonic Topics*, pp169-186. J.M. Lopez Higuera (Ed), pp. 169-186, University of Cantabria (1997).
14. A. Dandridge, C. Kirkendall in *Handbook of Optical Fibre Sensing Technology,* J.M. López-Higuera (Ed.) New York: John Wiley & Sons (2002).
15. S. Abad, M. López-Amo, I. R. Matias in *Handbook of Optical Fibre Sensing Technology,* J.M. López-Higuera (Ed.) New York: John Wiley & Sons (2002).
16. M. López-Amo and J. M. López-Higuera, in *Fiber Bragg Grating Sensors: Recent Advancements, Industrial Applications and Market Exploitation*, Bentham Science Publishers Ltd, pp 99-115 (2011).
17. T. Storgaard-Larsen, S. Bouwstra and O. Leistiko, *Sensor and Actuators A: Physical*, **52**, 25-32 ISSN 0924-4247, doi: 10.1016/0924-4247/96/80121 (1996).
18. T. A. Berkoff and A. D. Kersey, *IEEE Photonics Technology Letters,* **8**, 1677-1679 (1996).
19. M. D. Todd, G. A. Johnson, B. A. Althouse and S.T. Vohra, *IEEE Photonics Technology Letters,* **10**, 1605-1607 (1998).
20. Y. Zhu, P. Shum, C. Lu, B. M. Lacquet, S. Member, P.L. Swart, S. Member and S. J. Spammer, *IEEE Photonics Technology Letters,* **15**, 1437-1439 (2006).
21. W. Zhou, X. Dong, C. Shen, C.-liu Zhao, C. C. Chan and P. Shum, *Microvawe and Optical Technilogy Letters,* **52**, 2282-2285 (2010).
22. P. Antunes, H. Varum and P. André, *Measurement*, **44**, 55-59 (2011).

23. P. Antunes, R. Travanca, H. Rodrigues, J. Melo, J. Jara, H. Varum and P. André, *Sensors,* **12**, 6629-6644 (2012).

24. S. R. K. Morikawa, A. S. Ribeiro, R. D. Regazzi, L. C. G Valente, A. M. B Braga, *Optical Fiber Sensors Conference Technical Digest*, Vol.1, 95- 98 (2002).

25. M. M. Nogueria, S. R. Morikawa, C. C. Kato, L. C. G. Valente and A.M. Braga in *Structural Health Monitoring*, ed. F.K. Chang, DEStech, 205-212 (2005).

26. B.-jin Peng, Y. Zhao, Y. Zhao and J. Yang, *IEEE Sensors Journal,* **6**, 63-66 (2006).

27. B.-O. Guan, H.-Y. Tam and S.-Y. Liu, *IEEE Photonics Technology Letters,* **16**, 224-226 (2004).

28. J. Li, R. Correia, E. Chehura, S. Staines, S.W. James and R.P. Tatam, *Proceedings of SPIE, 7653* (2010).

29. F.P. Kapron et al., *J.of Lightwave Tech.*, Vol. **7**, N° 8, pp. 1234-41 (1989).

30. R.H.West et al., *J.of Lightwave Tech.*, Vol. **12**, N° 4, pp. 614-20 (1994).

31. S. A. Kingsley and Davies, *Electr. Lett.*, Vol. **21**, pp. 434-435 (1985).

32. W. V. Sorin, *9$^{th}$ OFS Proc.*, pp. 243-46 (1993).

33. A. J. Rogers, *Appl. Opt.*, Vol. **20**, pp 1060-1074 (1981).

34. J. P. Dakin, *Electron. Lett.* Vol. **21**, pp.569-570 (1985).

35. X. Bao, et al., *9$^{th}$ OFS Postdeadline* (1993).

36. Kurashima et al., *Optics Letters*, Vol. **15**, 1038 (1990).

37. Y. Mizuno, W. Zou, Z. He and K. Hotate, in *Proceedings of SPIE* (2008).

38. K. Kishida and C.-H. Li, *Structural Health Monitoring and Intelligent Infrastructure*, **471**-477 (2006).

39. C. A. Galindez and J. M. Lopez-Higuera, *IEEE Sensors J.,* **11**, 2344-2348 (2011).

40. L. Thevenaz and S. F. Mafang, in *19$^{th}$ International Conference on Optical Fibre Sensors,* 70043N-70044, (2008).

41. W. Li, X. Bao, Y. Li and L. Chen, *Opt. Express,* **16**, 21616-21625 (2008).

42. M. Belal and T. P. Newson, *Opt. Lett.,* **36**, 4728-4730 (2011).

43. R. Bernini, A. Minardo and L. Zeni, *Sensors and Actuators, A: Physical,* **134**, 389-395 (2007).

44. K. Y. Song, *Opt. Lett.,* **36**, 4686-4688 (2011).

45. W. Zou, Z. He and K. Hotate, *Opt. Express,* **17**, 1248-1255 (2009).

46. Y. Dong, L. Chen and X. Bao, *Opt. Lett.,* **35**, 193-195, (2010).

47. S.T. Kreger, D. K. Gifford, M. E. Froggatt, B. J. Soller and M.S. Wolfe, *OSA proceedings of OFS,* Cancun, México (2006).

48. Y. Peled, A. Motil and M. Tur, *Optics Express*, 8584, Vol. **20**, No. 8, 9 (2012).

49. L.F. Ramos, L. Marques, P.B. Lourenço, G. De Roeck, A. Campos-Costa and J. Roque, *Mechanical Systems and Signal Processing,* **24**, 1291-1305 (2010).

50. G. Carpineri and A. Lacidogna, *Materials and Structures,* **39**, 161-167 (2006).

51. P. Antunes, H. Lima, N. Alberto, L. Birlo, P. Pinto, A. Costa, H. Rodrigues, J.L. Pinto, R. Nogueira, H. Varum and P. S. Andrè, *New Development in Sensing Technology for SHM*, S.C. Mukholopadhyay Ed., Springer (2011).

52. P. Antunes, R. Travanca, H. Rodrigues, J. Melo, J. Jara, H. Varum and P. André, *Sensors*, **12**, 6629-6644 (2012).

53. C.Y. Lee, G. B. Lee, *Sensor Letters* **3**, 1-14 (2005).

54. T. Sun, K.T.V. Grattan, S. Srinivasan, P.A.M. Basheer, B.J. Smith, and H.A. Viles, *IEEE Sensors Journal* 12, 1011-1017 (2012).

55. H.-N. Li, D.-S. Li  and G.-B. Song, *Eng. Structures,* **26**, 1647-1657 (2004).

56. T. L. Yeo, T. Sun, K. T.V. Grattan, D. Parry, R. Lade and B.D. Powell, *Sensors and Actuators B: Chemical* **110**, 148-156 (2005).

57. G. Kister, D. Winter, Y.M. Gebremichael, J. Leighton, R. A. Badcock, P. D. Tester, S. Krishnamurthy, W.J.O. Boyle, K.T.V. Grattan and G.F. Fernando, *Eng. Structures,* **29**, 2048-2055 (2007).

58. Z.X. Li, T.H.T. Chan, and J.M. Ko, *Int. J. of Fatigue,* **23**, 45-53 (2001).

59. A. Li, Y. Ding, H. Wang and T. Guo, *Science China Technological Sciences*, **55**, 1-13 (2012).

60. T. H. T. Chan, L. Yu, H. Y. Tam, Y. Q. Ni, S.Y. Liu, W.H. Chung and L.K. Cheng, *Eng.Structures*, **28**, 648-659 (2006).

61. W.L. Schulz, J.P. Conte, E. Udd and J. M. Seim, *Proc. of SPIE* (2000).

62. P. Ferraro and G. De Natale, *Optics and Lasers in Engineering,* **37**, 115-130 (2002).

63. F. Pingue, S. M. Petrazzuoli, F. Obrizzo, U. Tammaro, P. De Martino and G. Zuccaro, *Measurement,* **44**, 1628-1644 (2011).

64. A. Laudati, F. Mennella, M. Giordano, G. D'Altrui, C. Calisti Tassini and A. Cusano, *IEEE Photonics Technology Letters,* **19**, 1991-1993 (2007).

65. *Global Optoelectronics Industry Market Report and Forecast: Fiber Sensors, Optoelectronics Industry Development Association*, 1220 Connecticut Avenue, NW - Washington, DC 20036.

# REMOTE SENSING MONITORING

Daniele Riccio

*University of Napoli Federico II*
*via Claudio 21, 80125 Napoli, Italy*
*E-mail: Daniele.Riccio@unina.it*

Remote sensing of the Earth is entering a new era. A new generation of satellite sensors is becoming operational and information on the entire surface of our Planet is going to be acquired more frequently, with better resolutions, on larger portion of the electromagnetic spectrum, by employing a much wider set of operational modes, thus ensuring increased flexibility to the technique. More importantly, the remote sensed data distribution policy is also significantly changing: a significant number of products from sensors belonging to major space agencies is going to be freely distributed, thus definitively enlarging the number of potential end users. In addition, availability of reliable software codes (including open source) to read the remote sensing data and perform personal data analysis is also significantly increasing. Accordingly, the style of this chapter is in favor of enlarging the applicative community that could be interested in remote sensing techniques, and now deserves being informed on the upcoming news in this scientific area. In this chapter space borne sensor configurations are considered, being related to Earth global coverage revisited every few days and belonging to historical series now involving some decades. Attention is posed to the activity in Europe, similar considerations holding for extra-European space agencies.

## 1.  Introduction

Remote Sensing is one of most attractive technology for practical environmental monitoring and control. The entire solar system is being investigated with remote sensing missions dedicated to almost any planet, and focus also on most intriguing moons, thus revealing spectacular data and information. The activities in the field are

Fig. 1. Optical Image of the Earth's southern hemisphere taken from a photographic camera onboard the National Aeronautics and Space Administration - NASA - Apollo 17 spacecraft. It represents the first complete image of the South Pole ice cup, and it is a very remarkable precursor of spaceborne remote sensing sensors by dramatically showing the ability to synoptic monitor with unprecedented characteristics land, water, and atmosphere of our Planet (NASA Image).

continuously growing: as a matter of fact, national and international space agencies are continuously planning, designing and implementing new mission to be operational in the next years and decades. Obviously, most of the remote sensing missions are devoted to monitor the Earth, the Planet we live in. The entire Earth environment is being monitored, thus including land, oceans, atmosphere and (some) subsurface features.

Both passive and active systems are employed: the former produce images and data by collecting the electromagnetic radiation emitted or reflected by the Earth, thus using the Sun as a primary source of illumination; the latter radiate electromagnetic fields toward the Earth and collect the scattered radiation by means of appropriate receiver(s) (that can be eventually co-located with transmitter), thus permitting a better control of the transmitted source. Data and images are acquired with reference to any part of the electromagnetic spectrum allocated to remote sensing sensor and conveniently supported by the appropriate technology issues.

Sensors are sometimes being arranged in constellations of satellites to increase the Earth portion monitored at any specific time, or, equivalently, reduce the time to acquire a data set on a specific area of interest. The new generation of sensors is being designed to operate in several operational modes, thus increasing the sensor flexibility. Remote sensing techniques are proposed to convert the acquired raw data into reliable and useful products (data and images) of interest for the applications[1]. Both operational modes and innovative techniques are continuously upgraded and proposed to improve performances of the remote sensing tools. Finally, the logical connection among remote sensing designers (essentially, engineers) and products end users (e.g., geologist, oceanographer, etc...) is increasing to better relate the technological outcomes (i.e., remote sensing products) to users requirements that are generally focused on quantitative monitoring of entities holding a clear physical meaning.

This attractive scenario is continuously evolving[1] also because technological constraints on satellites, and consequently sensors, limits missions duration time to not more than one decade: this continuous complete change of the available sensors is compulsory for technological constraints, yet advisable for up to date sensors availability.

Two key consequences can be then deduced. First, it turns out that a complete description of even major remote missions and sensors would require books, not a mere book chapter; and, more importantly even the description of present remote sensing satellites would be out of date in few years which is not a very large amount of time for works published in books.

This chapter is then devoted to present major capabilities, offering from the remote sensing community to the technical and practitioner arena, including also Earth resources administrators.

Fig. 2. Conceptual view of one of the two satellites belonging to the European Space Agency – ESA - Sentinel-1 constellation (ESA Image). This constellation is expected to be fully deployed within year 2015. Each satellite is equipped with a C-band Synthetic Aperture Radar whose antenna points toward the Earth. The solar panels are always pointed toward the sun to acquire the energy for the satellite to properly work.

Accordingly, the focus is posed on actual major remote sensing missions that are expected to be somehow replicated (with appropriate updating) also in the future (beyond, at least, one decade), and to the upcoming generation of satellites. To further limit the presentation, the attention is posed on missions proposed by the European Space Agency (ESA)[2-6] supported by national space agencies[7-9]: similar activities are expected to be arranged in different missions by the United States National Aeronautics and Space Administration (NASA), and in part being present in plans from other Space Agencies.

                    *D. Riccio*

Fig. 3. COSMO/SkyMed synthetic aperture radar image of the Vesuvius - Mount Somma volcanic complex and its close surroundings. Side looking view with a 31 degrees looking-angle. (Italian Space Agency – ASI - Image). Major old lava flows are visible over the Vesuvius cone. Suburban areas can be identified at the image borders.

For a wider dissertation, the interested reader can refer to the huge available literature which is continuously updated for novelties, which is very appropriately supported by continuously updated remote sensing data that are continuously proposed, and updated, by major Space Agencies on dedicated web sites.

Fig. 4. COSMO/SkyMed synthetic aperture radar image of the central area of Napoli (ASI Image). Airport (top-right), city centre (middle) and harbor (bottom) can be easily identified. It is worth noting the different textures relevant to the historical centre (middle-left) and the modern business area (middle-right). Some ships are clearly identified (bottom centre) in the harbor and sea area.

Surfing those web sites is not that immediate, yet exciting: remote sensed images are spectacular and easily accessible, the information stratified for different levels of competences and supported by any sort of modern multimedia technologies.

## 2. Global Monitoring for Environment and Security

Global Monitoring for Environment and Security[2] (GMES) is the program led by the European Commission and performed under the aegis of the European Space Agency (ESA), for coordinating action to deploy remote sensing instruments and integrate the obtained data (available for the entire Planet) with other information derived from *in situ* (airborne, seaborne and ground station) sensors. The program main idea is to combine use of remote sensing data with (different types of), already available or to be developed, models, so to quantitatively support environmental studies for reliable Earth monitoring and global security issues. In particular, the program is in favor of policymakers and public authorities providing them with global, timely and easily accessible data relevant to land, ocean and atmosphere. These data are also conceived to support preparation and monitoring of the legislation on environmental issues. By creating the infrastructure for a global monitoring of the Earth, the program also provides the fundamentals for a significant growth of private companies in the field; these firms are expected, in general, to generate new value added services and, more specifically, appropriate solutions to specific, eventually local or limited in time, private user needs. Major topics involved in the Earth monitoring are detailed in the following subsections with reference to different Earth domains (land, water, and atmosphere).

## 2.1. *Land*

- Determination of land *cover* and land *use*; this information supports production of databases amenable to be exported in Geographical Information Systems (GIS) to simplify any applicative issue.
- Monitoring of *food productions*, with information on ability to produce, locations of production, estimation of expected yield and expected quality.
- *Forest monitoring*, concerning actual forest preservation in terms of extension and quality, as well as support to prevention of deforestation.
- Estimation of *soil sealing*, focusing on identification of area subtracted to their original natural use, like forestry or man-made agricultural activities, and actually devoted to constructions and urban planning.

- *Spatial planning*, allowing also integration with non-synoptic information for creation of GIS as appropriate tools to be employed for smart data storing and decision makers support.

## 2.2. *Water*

- Determination of water *content*, concerning also lakes and rivers.
- Estimation of water *quality*, relevant to the ability to support the human beings and life in general, including also issues concerning the water pollution.
- Water *availability*, concerning the possibility to reliably employ water for the many specific applications that require its use.
- Monitoring of the *ocean*, also in support of a better knowledge of the Earth environment on a global scale, and involving currents and sea state determination.
- Monitoring of *coastal areas*, with information on actions played by oceans on land shape, thus supporting analyses on erosion and dilation phenomena.
- Supporting and monitoring to *fish production* activities, providing information and support in large scale fishing and fish protection policies.
- Monitoring of *marine ice*, for climate changes monitoring and navigation issues in Polar Regions.
- *Oil spill* identification and monitoring, with activities related to identification of natural oil spills for exploitation of natural resources, and man-made intentional oil spills caused by illegal damping from marine tanks, as well as disasters relevant to human activities concerning marine tanks and platforms.
- *Search and rescue* activities in the oceans in support of civil protection operations.
- Aid in *ship routing* under ocean navigation and in presence of ice; this activity on global scale would also support any other for localization and tracking of vessels in the oceans.

## 2.3. *Atmosphere*

- Monitoring of the atmosphere *constituents*, including carbon dioxide, methane, ozone and aerosols.
- Monitoring of the *air quality* also in support of forecasts activities.
- Estimation of the UV *radiation* on global scales through determination of aerosol and ozone concentration.
- Monitoring of *greenhouse effect* and related reactive gases.

Three main cross-cutting domains[2] (Emergencies, Security and Climate) are also individuated. Services related to them are obtained from those presented with reference to the Earth domain services once they are integrated with appropriate models and operative issues that may support the monitoring requisites for these services. Major topics of these services are briefly presented hereafter: they are categorized by means of the three key cross-cutting domains along with pertinent subcategories for each domain of application.

- Emergencies. In this case, remote sensing data provide support in pre-crisis, crisis and rescue phases. Remote sensing data can conveniently support guidance of humanitarian aids, identification of fires, monitoring of floods and storms in some cases with the advantage of being independent from the cloud coverage, precise identification of oil spills over the oceans, assessment of damages in earthquakes. More in general the different characteristic of available sensors allows unique support in the monitoring of most of technological accidents.
- Security. Remote sensing data provide valuable information to border controls, support guidance in humanitarian aids missions, contribute on local and worldwide maritime surveillance, and can be conveniently used to support conflicts prevention and early warning activities.
- Climate changes. In this case, remote sensing data role is clear considering the ability to provide global scale monitoring of parameters that are both precursor and consequences of climate

changes on both local and global scales. The support to weather and changing forecast issues is promising and evident.

Major role of the research in the field is to develop new models, techniques and algorithms to extract information from the remotely sensed data[1]. The scientific literature is huge and classify all the research results is almost impossible, and in any case beyond the scope of this chapter. Hereafter, the author only lists some research items personally published. This is done just to show one possible logical scheme to sort the research activities and present some related titles to the interested reader. Models activity mainly involves the scattering mechanisms describing the electromagnetic field scattered or emitted from the scene in terms of the scene parameters. Land[10,11], ocean[12,13] and man-made scattering models are studied[14,15].

Techniques developments operate on the acquired data that usually depend, in a very involved way, on a large set of sensors and scene parameters, trying to isolate contribution of a single physical parameter of the scene holding a clear meaning. Hence, attention is posed on natural surface roughness[10,16-20] and electromagnetic parameters[19-21], ocean state[12] and pollution conditions[13], urban monitoring concerning the buildings materials[22] and shape[23-24].

Algorithms are conceived to reliably, efficiently and effectively estimate those parameters from the acquired actual data. In addition, researchers are sometimes required to conceive advanced remote sensing missions, simulate[25-28] the expected results and propose them to the community. Firms and industries of any size are required to support the initiative by implementing, and providing reliability, to the innovative product proposed by the scientific activities.

Connection expected between public needs and commercial outcomes let GMES be fully integrated within the activities of the European Union towards year 2020.

## 3. Space Segment

Different satellites are available for monitoring the Earth. They can be categorized according to the employed technique as reported hereafter.

                                    *D. Riccio*

Fig. 5. COSMO-SkyMed synthetic aperture radar image of an area around Napoli, Italy (ASI image), wherein urban and natural features are mixed up. Different textures relevant to urban and rural areas are evident. Roads, single buildings, shape and extension of the vegetated areas can be easily recovered in this image. The image is acquired in a side-looking configuration so that the extension of the shadow areas allows estimating the vegetation height.

Tables 1-3 are used to present key parameters of the upcoming generation of sensors. Reference is made to the satellites of the GMES program.

### 3.1. *Synthetic Aperture Radars*

Synthetic Aperture Radars are active sensors, operating at microwave frequencies, for synoptic imagery independent on the weather and day/night conditions, and relevant to the entire Globe monitored in the land and ocean domains[3]. An antenna mounted on the spacecraft orbiting around the Planet along a direction that define the azimuth coordinate, transmits chirp modulated pulses and receives the corresponding echoes reflected by the Earth surface. The raw data acquired by the sensor are processed to obtain resolutions in azimuth and range direction of the order of some meters. Hence, SAR images basically represent the radar reflectivity of the scene at the resolution dictated by the sensor and the processing chain.  However, Synthetic Aperture Radars are suggesting very flexible ways to achieve much more than single images of our Planet. Major source of flexibility are presented in the following items.

- SARs are often arranged in constellation of satellites (e.g., COSMO-SkyMed[8,9] is formed by four satellites) to provide images with a smaller revisit time on any area of the Globe.
- SAR are arranged to operate, in interferometric configurations (e.g, TanDEM-X is formed by two TerraSAR-X[7] satellites arranged in a tandem configuration), with more than one sensor on "slightly"-different "almost"-parallel trajectories. This configuration achieves data able to generate three dimensional profiles of the Earth surface, as well as, by means of repeated passes, Earth surface small movements.
- SAR databases are constantly increasing: the available SARs are continuously acquiring hundreds of data sets each day, so that, on each area of the Earth, the historical series of data is continuously increasing; this clearly supports applications employing repeat pass interferometry, multi-baseline and differential interferometry methods.
- Modern SAR missions are conceived to operate with different operational modes[3,7-9], the standard mode, being the stripmap one

in which a basic compromise is reached between SAR image resolutions and coverage; larger coverage (at the expense of resolution) are obtained in the scansar operational modes[3,7-9]; conversely, smaller resolutions (at the expense of coverage) are obtained in the spotlight operational modes[7-9].

- SAR sensors are now frequently designed to operate in polarimetric configurations[3,7-9]: the transmitted and received electromagnetic fields are both acquired separately with respect to two (in general linear) components. Consequently, from the co-polarization and the cross-polarization components, the scattering matrix can be obtained, thus increasing not only the number of images that can be acquired at each pass, but also opening the study about the scene scattering mechanisms that can help identifying value added quantities such as surfaces soil moisture or roughness.

The choice to operate with lighter satellites is bringing space agencies to plan missions equipped with single band SARs. However, some space agencies are now operating (or are seriously considering to do that in the near future) in a somehow complementary form which is achieved by conceiving SAR operating in different bands; for instance, a very interesting aggregate set of S-band (NovaSAR), C-band (Sentinel-1)[3], and X-band (COSMO-SkyMed, TerraSAR-X)[7-9] spaceborne SAR sensors is going to be available in the near future from the European space agencies.

The future of SAR images provided at the European level is based on the above presented mix of SAR sensors operating in different bands at microwaves, see Table 1. Most of these sensors are aggregated in constellations, the two-satellite ESA Sentinel-1 constellation provides SAR data compatible with the acquired data coming from the ERS-1/2 missions characterizing the '90, and the ASAR images acquired in the first decade of the new millennium during the 50,000 orbits Envisat completed.

Table 1. SAR sensors. Sentinel 1 is shown together with the previous ESA SAR spaceborne sensor, ASAR on board to the Envisat satellite and the two GMES supporting missions COSMO/SkyMed and TerraSAR-X that are already operational.

| Synthetic Aperture Radars | | | | |
|---|---|---|---|---|
| Mission | Sentinel-1 | Envisat/ASAR | COSMO-SkyMed | TerraSAR–X |
| Satellites number | 2 | 1 | 4 | 2 |
| Nominal Height [km] | 693 | 799.8 | 619.6 | 514 |
| Repeat cycle [days] | 12 (1-satellite) 6 (2-satellites) | 35 | 16 | 11 |
| Central frequency [GHz] | 5.405 | 5.331 | 9.6 | 9.65 |
| Year of Launch | 2013-2015 | 2002 (ended 2012) | 2007-2010 | 2007-2009 |
| Swath dimensions along range (min) [km] | 80 (st) 400 (sc) | 100 (st) 400 (sc) | 40 (st) 200 (sc) 10 (sp) | 30 (st) 100 (sc) 10 (sp) |
| Swath dimensions along azimuth [km] | 80-1500 (st) 400-1500 (sc) max 25 min/orbit | 100 (st) 400 (sc) | 40-1600 (st) 100-1600 (sc) 10 (sp) | 50-1500 (st) 100-1500 (sc) 5 (sp) |
| Resolutions azimuth * range (min) [m*m] | 5*5 (st) 40*20 (sc) | 30*30 (st) 150*150 (sc) | 3*3 (st) 30*30 (sc) 1*1 (sp) | 3* 3 (st) 16*16 (sc) 1*1 (sp) |
| Polarizations | HH, VV<br><br>dual pol (HH-HV,VV-VH) | HH, VV<br><br>dual pol (HH-HV,VV-VH) | HH, VV, HV, VH,<br><br>ping pong (HH-VV, HH-HV, VV-VH) | HH, VV<br><br>dual pol (HH-VV, HH-HV, VV-VH) quad pol |
| Incidence angles | 20°-47° (st) 20°-47° (sc) | 15°-45° | ≅20°-60° (st) ≅20°-60° (sc) ≅20°-60° (sp) | 20°-45° (st) 20°-45° (sc) 20°-55° (sp) |

Notes: Parameters may vary from those reported above as function of the specific SAR product. Operational mode: stripmap (st), scansar (sc) and spotlight (sp). Some sensors may have different operational modes or different settings for each operational mode: to each setting different parameters are referred to. Those reported in the table provides a first idea on the sensors characteristics, the more interesting or limit values being reported, not the whole set. Experimental configurations are not reported here. The interested reader is invited to refer to the technical literature[3,7-9] of each sensor for full presentation of the different sensor settings. Dual pol and ping pong polarization modes may vary for each sensor. The two TerraSAR-X satellites jointly operate in the TanDEM-X mission to acquire DEM of the scene.

## 3.2. *Optical Sensors*

At frequencies around those typical of the human view, the optical sensors are able to measure the solar energy reflected by the Earth surface[4]. They can be further categorized in view of their resolution (the minimum distance at which two objects can be separated) and coverage (the maximum area imaged around a generic fixed point of reference), as follows.

- *Low resolution* (of the order of hundreds of meters) sensors for wide area applications including continental and national scales. They are mainly applied for forest monitoring, desertification assessment, estimation of water resources, agriculture applications, ocean monitoring, coastal monitoring.
- *Medium resolution* (of the order tens of meters) sensors that usually also include multi- or hyper-spectral monitoring with the ability to operate also in panchromatic operational modes thus obtaining better geometrical resolutions at the expenses of frequency selective information. They are mainly employed for regional scale monitoring of land and ocean as well as for security applications.
- *High resolutions* (of the order of meters) sensors for local areas to monitor specific applications or as a general support to security and emergencies applications.

The future of optical high-resolution images provided at the European level is based on the two-satellite ESA Sentinel-2 constellation providing multispectral data relevant to bands of the electromagnetic spectrum corresponding to the visible, near and short wave infrared allowing vegetation, snow, ice, clouds monitoring, see Table 2.

Table 2. Sentinel-2 Multispectral Spectral Instrument.[4]

| Bands | | | | |
|---|---|---|---|---|
| Number | Wavenumbers [nm] | Resolution [m] | Type | Use |
| 1 | 433-453 | 60 | V | aerosols corrections |
| 2 | 457.5-522.5 | 10 | V | blue image |
| 3 | 542.5-575.5 | 10 | V | green image |
| 4 | 650-680 | 10 | V | red image |
| 5 | 697.5-712.5 | 20 | V | vegetation characteristics |
| 6 | 732.5-747.5 | 20 | V | vegetation characteristics |
| 7 | 773-793 | 20 | V | vegetation characteristics |
| 8 | 784.5-899.5 | 20 | NIR | near infrared image |
| 8b | 855-875 | 20 | NIR | vegetation |
| 9 | 935-955 | 60 | NIR | water vapour |
| 10 | 1365-1395 | 60 | NIR | cirrus detection |
| 11 | 1565-1655 | 20 | SWIR | snow, ice, cloud, detection |
| 12 | 2100-2280 | 20 | SWIR | vegetation moisture |

| Parameters | |
|---|---|
| Satellites number | 2 |
| Nominal Height [km] | 786 |
| Revolutions per day | 14 |
| Revisit time [days] | 5 (two-satellites) <br> 10 (one-satellite) |
| Launch date [year] | 2014 |
| Swath dimensions across track [km] | 290 |

Notes: V-Visible; NIR-Near Infra-Red; SWIR-Short Wave Infra-Red

## 3.2. Radiometers

Radiometers are passive sensors able to measure the energy emitted from the Earth surface[5]. The main source of energy for these remote sensors is provided by the sun. These passive sensors are able to operate in different bands. They are mainly employed to monitor:

- the thermal emission from land and ocean;

- the dielectric constants of the surfaces, in particular if operated at microwaves.
- The future of Radiometers provided at the European level within the GMES program is based on the two-satellite ESA Sentinel-3 providing the Sea and Land Surface Temperature Radiometer which is expected to extend the remote sensing dataset previously acquired within the Envisat Advanced Along Track Scanning Radiometer, see Table 3.

### 3.3. *Spectrometers*

Spectrometers are employed on the Earth to monitor specific chemistry factors involved in the atmosphere and concerning its composition and quality[4].

The future of Spectrometers data provided at the European level is based on the two-satellite ESA Sentinel-3 including the Sea and Land Surface Temperature Radiometer, see Table 3, which is expected to extend the remote sensing dataset previously acquired within the Envisat Medium Resolution Imaging Spectrometer.

### 3.4. *Altimeters*

Altimeters are active sensors, mostly applied at optical (Lidar) or microwave (radar) frequencies[3]. They are mainly employed to monitor the following features:

- Local or regional features on solid Earth
- sea level at global scale
- marine ice position and composition.

The future of Altimeters provided at the European level is based on the two-satellite ESA Sentinel-3, see Table 3, including Ku and C-band altimeters that are expected to extend the remote sensing dataset previously acquired within the Envisat Earth Explorer Cryo-Sat mission.

Table 3. Sentinel-3 Instruments.

| Instruments | | | | |
|---|---|---|---|---|
| SLSTR<br>Sea and Land Surface<br>Temperature Radiometer | | OLCI<br>Ocean and Land Colour<br>Instrument Spectrometer | | Topography System |
| Central wavenumber [nm] | Resolution [m] | Central wavenumber [nm] | Resolution [m] 300-1000 | Synthetic Aperture Radar Altimeter<br>C- band (5.41 GHz)<br>Ku- band (13.575 GHz) |
| 555 | 500 | 400 | | |
| 659 | 500 | 412.5 | Depending | |
| 865 | 500 | 442.5 | on the | |
| 1375 | 500 | 510 | specific | |
| 1610 | 500 | 560 | product | |
| 2250 | 500 | 620 | | Supported by corrections provided by |
| 3740 | 1000 | 665 | | |
| 10950 | 1000 | 673.75 | | Microwave Radiometer for atmospheric effects |
| 12000 | 1000 | 681.25 | | |
| | | 708.75 | | DORIS/GPS/Laser retroreflector for orbit effects |
| | | 753.75 | | |
| | | 761.25 | | |
| | | 764.375 | | |
| | | 767.50 | | |
| | | 778.75 | | |
| | | 865 | | |
| | | 885 | | |
| | | 900 | | |
| | | 940 | | |
| | | 1020 | | |

| Key Parameters | |
|---|---|
| Satellites number | 2 |
| Nominal Height [km] | 814.5 |
| Revolutions per day | 14 |
| Revisit time [days] | 1-3 days (SLSTR, OLCI) |
| Launch date [year] | 2015 |
| Swath dimensions across track [km] | 1440 km (OLCI) |

Notes: Parameters

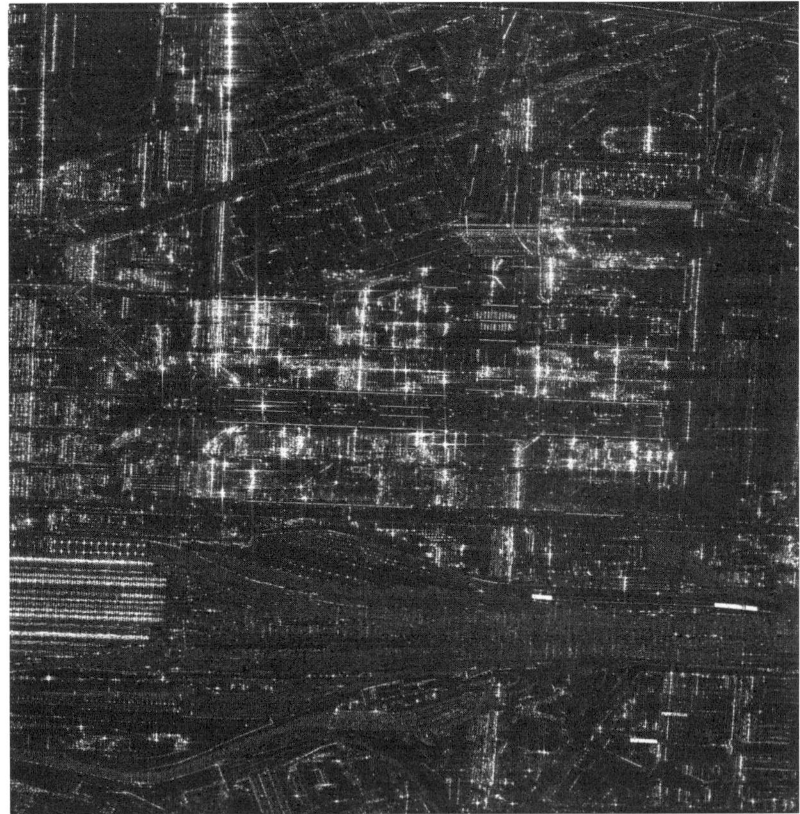

Fig. 6. COSMO/Skymed synthetic aperture radar image of the central area of Napoli (ASI Image). The central area refers to the Napoli business district characterized with regularly spaced and oriented buildings with maximum height of approximately 100 meters. Central station platforms roofs (bottom-left) and railways (bottom) are evident in the figure. A train, producing a very bright line in the bottom right part of the image is present. Brighter pixels in the image are due to the multiple bouncing effects typical of the dihedral formed by the buildings walls and the surrounding terrain. Buildings in the business districts leads to multiple bounce reflection effects generating bright lines aligned to the line of flight of the sensor.

## 3.5. *Scatterometers*

Scatterometers are active sensors, mostly working at microwave frequencies that acquire precise information on the scene reflectivity at low spatial resolutions. Scatterometers are mainly employed to:

- Support precise measurements of altimeter data
- Estimate wind patterns over the oceans. This is obtained for measurements at the microwaves considering that the ocean

surface at the scale of gravity-capillary waves is roughened by the local winds.

## 4. Conclusions

The above reported material shows how the remote sensing arena is entering a new area in the Earth monitoring providing a new generation of spaceborne sensors. But, not only are the sensors increasing their performances: more importantly, the data distribution policy is often going to change in favor of a larger use of the acquired public remote sensing data. As a matter of fact, a significant amount of acquired data by different sensors at different frequencies is going to be freely distributed and available. For instance, data from the Sentinel mission will be opened to public, scientific and commercial entities, and data will be obtained online and free of charge. Moreover, also the support to possible users is going to become more reliable and friendly. As a matter of fact, software utilities, to import the data and support applicative, issues are being available to the interested technical staff that is in charge to use remote sensing data.

## Acknowledgments

This work was partly supported by the Italian Space Agency within the projects "Buildings Feature Extraction from Single SAR Images: Application to COSMO-SkyMed High Resolution SAR Images", and "Exploitation of Fractal Scattering Models for COSMO-SkyMed Images Interpretation", leaded by the author of this Chapter.

The author would like to thank the space agencies (NASA, ESA, and ASI) that provided the images used in this paper.

## References

1. P. Imperatore and D. Riccio, "*Geoscience and Remote Sensing, New Achievements*", IN-TECH eds., Vienna (Austria), 1-508 (2010).
2. J. Aschbacher and M. Milagro-Pèrez, *Remote Sensing of Environment*, vol. **120**, 3-8 (2012).
3. R. Torres, P. Snoeij, D. Geudtner, D. Bibby, M. Davidson, E. Attema, P. Potin, B. Rommen, N. Floury, M. Brown, I. N. Traver, P. Deghaye, B. Duesmann, B. Rosich, N. Miranda, C. Bruno, M. L'Abbate, R. Croci, A. Pietropaolo, M. Huchler and F. Rostan, *Remote Sensing of Environment*, vol. **120**, 9-24 (2012).
4. M. Drusch, U. Del Bello, S. Carlier, O. Colin, V. Fernandez, F. Gascon, B. Hoersch, C. Isola, P. Labertini, P. Martimort, A. Meygret, F. Spoto, O. Sy, F. Marchese and P.

Bargellini, *"Sentinel-2: ESA's Optical High-Resolution Mission for GMES Operational Services"*, vol. **120**, 25-36 (2012).

5.  C. Donlon, B. Berruti, A. Buongiorno, M.-H. Ferreira, P. Femenias, J. Frerick, P. Goryl, U. Klein, H. Laur, C. Mavrocordatos, J. Nieke, H. Rebhan, B. Seitz, J. Stroede and R. Sciarra, *Remote Sensing of Environment,* vol. **120**, 37-57, (2012).

6.  P. Ingmann, B. Veihelmann, J. Langen, D. Lamarre, H. Stark and G. Courreges-Lacoste, *Remote Sensing of Environment*, vol. **120**, pp. 58-69 (2012).

7.  W. Pitz and D. Miller, *IEEE Transactions on Geoscience and Remote Sensing*, **48**, 615-622 (2010).

8.  ASI, "COSMO-SkyMed SAR Products Handbook" (2009).

9.  ASI, "COSMO-SkyMed System Description & User Guide" (2007).

10. G. Franceschetti and D. Riccio, *Scattering, Natural Surfaces and Fractals*, Academic Press eds., Burlington, MA (USA), 1-304 (2007).

11. P. Imperatore, A. Iodice and D. Riccio, *IEEE Transactions on Geoscience and Remote Sensing*, **GE-47**, 1634-1643 (2009).

12. G. Franceschetti, M. Migliaccio and D. Riccio, *IEEE Transactions on Geoscience and Remote Sensing*, **GE-36**, 84-100 (1998).

13. G. Franceschetti, A. Iodice, D. Riccio, G. Ruello and R. Siviero, *IEEE Transactions on Geoscience and Remote Sensing*, **GE-40**, 1935-1949, (2002).

14. G. Franceschetti, A. Iodice and D. Riccio, *IEEE Transactions on Geoscience and Remote Sensing*, **GE-40**, 1787-1801, (2002).

15. G. Franceschetti, A. Iodice, D. Riccio and G. Ruello, *IEEE Transactions on Geoscience and Remote Sensing*, **GE-41**, 1986-1995 (2003).

16. G. Di Martino, A. Iodice, D. Riccio and G. Ruello, *IEEE Transactions on Geoscience and Remote Sensing*, **GE-48**, 3280-3289 (2010).

17. G. Di Martino, D. Riccio and I. Zinno, *IEEE Transactions on Geoscience and Remote Sensing*, **GE-50**, 630-644 (2012).

18. G. Di Martino, A. Iodice, D. Riccio and G. Ruello, *IEEE Transactions on Geoscience and Remote Sensing*, GE-45, 1559-1570 (2007).

19. G. Franceschetti, A. Iodice, S. Maddaluno and D. Riccio, *IEEE Transactions on Geoscience and Remote Sensing*, **GE-38**, 641-650 (2000).

20. A. Iodice, A. Natale and D. Riccio, *IEEE Transactions on Geoscience and Remote Sensing*, **GE-49**, 2531-2547 (2011).

21. E. Ceraldi, G. Franceschetti, A. Iodice, D. Riccio and G. Ruello, *IEEE Transactions on Geoscience and Remote Sensing*, **GE-43**, 295-305 (2005).

22. R. Guida, A. Iodice, D. Riccio and U. Stilla, *IEEE Journal of Selected Topics in Applied Earth Observations and Remote Sensing,* **JSTARS-1**, 107-119 (2008).

23. R. Guida, A. Iodice and D. Riccio, *IEEE Transactions on Geoscience and Remote Sensing*, **GE-48**, 930-938 (2010).

24. R. Guida, A. Iodice and D. Riccio, *IEEE Transactions on Geoscience and Remote Sensing*, **GE-48**, 2967-2979 (2010).

25. G. Franceschetti, M. Migliaccio, D. Riccio and G. Schirinzi, *IEEE Transactions on Geoscience and Remote Sensing*, **GE-30**, 110-123 (1992).

26. G. Franceschetti, A. Iodice, M. Migliaccio and D. Riccio, IEEE Transactions on Geoscience and Remote Sensing, **GE-36**, 950-962 (1998).

27. S. Cimmino, G. Franceschetti, A. Iodice, D. Riccio and G. Ruello, *IEEE Transactions on Geoscience and Remote Sensing*, **GE-41**, 2329-2337 (2003).

28. G. Franceschetti, R. Guida, A. Iodice, D. Riccio and G. Ruello, *IEEE Transactions on Geoscience and Remote Sensing*, **GE-42**, 2385-2396 (2004).

# PHOTONIC TECHNOLOGIES FOR THE SAFEGUARDING OF CULTURAL ASSETS


Costanza Cucci[1]* and Vivi Tornari[2]

[1]*Institute of Applied Physics "Nello Carrara"*
*Italian National Research Council (IFAC-CNR)*
*Via Madonna del Piano, 10 I50019 Sesto Fiorentino (Florence) Italy*

[2]*Foundation for Research and Technology-Hellas (FORTH),*
*Institute of Electronic Structure and Laser (IESL),*
*N. Plastira 100, Voutes, 71110, Heraklion, Crete, Greece*
**E-mail: c.cucci@ifac.cnr.it*



Safety and security of cultural heritage entail a huge variety of issues, all inherent to the protection of cultural assets from risks of damages of both anthropic and natural origin. Due to the high heterogeneity of cultural assets, which includes movable and immovable cultural properties (paintings, art objects, monuments, archeological sites, etc.), the safeguard of this patrimony involves diverse aspects and different classes of problems that need advanced technological solutions. Photonic technologies may offer these innovative solutions and they are increasingly used in the cultural heritage field to address several issues of safety and security. In this chapter two examples of implementation of photonics for this applicative area are illustrated. The first example deals with applications of optical sensors and optical fiber-based systems for monitoring conditions of artworks on exhibits while the second example is from the field of laser-technology and deals with the application of coherent interferometry to assess invisible impact caused on transit of movable artworks to exhibitions by repeated conditions of alteration and handling.


## 1. Introduction

In the Cultural Heritage (CH) field, keywords like "safety" and "security" are not referable to a specific applicative area; rather, they range over a widespread and heterogeneous class of issues, such as the prevention of natural degradation in cultural items, the risk assessment and the management of momentous events (fire, floods, etc.) in museums and archives, as well as all topics related to protecting against anthropic activities that can threaten artworks (e.g. the negative consequences of mass tourism on museum objects, the risks related to the handling and transportation of movable assets, problems of fraud, vandalism, etc.).

Within such a complex context, it is not surprising that photonic technologies play a key role in addressing safety and security issues in the CH field, by offering huge potential for the development of systems and devices tailored to the solution of specific problems related to the safeguarding of cultural assets.

Without claiming to be exhaustive, we will illustrate in this chapter two main representative topics, both of which are intrinsic to different concepts of the use of photonics in the area of the security and safety of cultural assets.

The first topic, which is presented in Sec. 2, focuses on sensoristic applications of photonic technologies in museums and galleries. Several examples of optical sensors and optical fiber-based systems intended for monitoring environmental conditions and controlling the impact on artefacts on display are reviewed. Particular attention is dedicated to the latest generation of optical sensors, early-warning systems and optical fiber-based devices designed to respond to specific requirements from the perspective of the preventive conservation of artworks.

The second representative topics is dedicated to laser technologies applied to the safeguarding and control of the integrity of artifacts. In Sec. 3, a novel cutting-edge technology based on laser interferometry is discussed in detail and proposed as a prototype methodology for the assessment of impact suffered over time by movable artworks that are subjected to repeated handling, moving, and transportation. This methodology, which was developed within the framework of an EC-funded Project (MULTIENCODE 006427), is one of the most up-to-date results of research in the field of photonic applications for preventive conservation.

## 2. Optical Sensors and Optical Fibers-based Devices for Applications in Museums

### 2.1. *Introduction*

Artworks on display in museums and galleries may be threatened by different environmental factors  - light, UV radiation, temperature and relative humidity, pollutants, etc. -  which, to different extents, cause a progressive and irreversible decay in their constitutive materials.

Consequently, continuous monitoring of environmental conditions in museums is considered to be a priority requirement for a correct conservation policy[1,2]. Guidelines accepted in the most important museums  provide the recommended limits for the various environmental factors, taking  into account the various categories of artifacts  and the typologies of the specific materials to be preserved[3,4,5]. The conventional approach to the problem is to set up a monitoring system based on the use of multiple devices and/or data-loggers that measure each parameter separately, at the desired acquisition rates. Although nowadays many low-cost and user-friendly compact devices for monitoring several environmental factors are available on the market, their maintenance and management, together with the overall data analysis and interpretation, still require trained staff and may be costly and time-consuming. Very recent studies on this topic have proposed resorting to wireless sensor networks in order to provide a capillary and automated monitoring of environmental conditions[6,7]. However, although sensors networks appear to be very promising for museum applications, this  is still a subject of on-going research, the results of which are not yet ready for widespread applications in the control of museum microclimates. Along with these studies, in recent decades another important branch of research has flourished that is devoted to the design of a new generation of sensors, dosimeters, and other devices tailored for applications in the CH field.

These new tools include different kinds of passive sensors, smart devices, and low-cost methodologies for preventive conservation, which are designed to make widespread monitoring easier and to prevent the deterioration of objects on display[8]. Among these new tools, a foremost position is occupied by so-called "impact-sensors", which have been specifically developed to provide an assessment of the cooperative effects of environmental agents. Indeed, as reported by several studies in the literature, many deterioration phenomena affecting artistic materials (discoloration, yellowing, embrittlement, etc.) are caused, or accelerated, by the synergistic action of environmental factors, which do not act only individually, but also give rise to cooperative phenomena[9,10]. Typical examples are the various photo-chemical reactions caused by the combination of light and pollutants. The class of impact sensors includes different concepts of devices based on different working principles.

Nevertheless, most of them have a common denominator in their implementation, namely the use of photonic technologies, as will be discussed in Sec. 2.2. Indeed, the high versatility and modularity of opto-electronical components make possible the design of a wide range of customized systems that are tailored to very specific applications.

Another noteworthy category of smart devices for museum applications includes optical fibers-based systems. Also in this case, the flexibility and intrinsic compliance of optical fibers has been exploited in the design of customized systems that are aimed at fulfilling specific needs related to the safeguarding of cultural items. Examples of optical fiber-based systems are reported in Sec. 2.3.

## 2.2. *Colorimetric Passive Sensors for Museum Environments*

"Impact sensors" are designed to mimic, in a very simplified way, certain reactions induced in material by the exposure to environmental factors that are responsible for alterations observed in artworks on display.

These sensors, which are passive indicators, are typically based on disposable materials which, upon the action of one or more environmental agents, exhibit a behavior similar to the objects to be surveyed. These passive sensors are designed to react with a detectable change in some measurable parameter that is affected by the agent to be monitored. The measurement of this change is used to provide an assessment of the impact undergone by an object exposed under the same conditions. Impact sensors act, in fact, like dosimeters: they are usually disposable, low-cost, and aesthetically non-distracting. Depending on their characteristics and design, impact sensors have been used in the CH field either to monitor the effects of specific factors (e.g. light, humidity, pollutants, etc.), or to assess cumulative environmental effects. It must be emphasized that, no matter what type of impact sensor is considered, these indicators are all based on a simulation principle. However, they may have different working principles. In some cases, impact sensors are made in order to reproduce the same reaction observed in the object under control, whereas in other cases these sensors are designed to be responsive to some factor that affects the object, but involves a more rapid reaction time. The latter is the case of early-warning systems, which are aimed at preventively signaling possible critical situations.

Examples of the first type of sensors were successfully used to investigate the susceptibility of several pigments and artistic materials to major environmental agents (light, temperature, relative humidity and selected pollutants) within the framework of an EC-funded Research project[11,12]. This pioneer study introduced for the first time the idea of evaluating the impact of the microenvironment on paintings on display in museums by using a set of test panels or 'mock paintings', displayed close to the artworks to be monitored. These test panels consisted of several strips obtained by using different artistic materials of primary interest (pigments and binding media). The strips were prepared by following the traditional artists' formulas. Several replicas of the same mock panel were made and then treated using artificial ageing tests in climate chambers under different conditions: varying light, temperature, relative humidity, and pollutant (e.g $NOx$ concentrations).

Subsequently, the artificially-aged samples were analyzed, and the results were compared with those obtained from p the test panels displayed in various museum sites. The results obtained showed that some pigments exhibited a color variation due to the action of selected environmental agents, whereas other pigments remained unchanged.

Moreover, the comparison between the laboratory tests and the in-field exposures pointed out that a given color variation could be caused by various agents, and that the final alterations registered in field could be attributed to the cumulative action of different factors. These types of impact sensors were found to be a key tool in studying degradation processes in a museum environment, where a number of different materials coexist and respond differently to the same environmental conditions[13].

Another important class of passive sensors for applications in museums is that of the "early-warning systems". These are passive indicators designed to exhibit a rapidly detectable reaction so as to give a warning signal before the display conditions become critical for the artworks. During recent years, various examples of early-warning sensors have been proposed that are aimed at monitoring different applications in a museum environment[14-17]. Among these, a very noticeable example is a light-dosimeter designed for monitoring highly photosensitive artworks. It was developed within the framework of the EC-funded LiDo project, and was then marketed under the name of LightCheck®Ultra[18,19]. This particular passive sensor consisted of a photosensitive dye/polymer coating applied to an inert support. Upon exposure to the visible light, this indicator exhibited a progressive color change that spanned different chromatic phases, from an intense blue, through purple and pink as far as white. The indicator was equipped with a color reference scale that could be used to correlate the color observed on the dosimeter with the dose of light received. Thanks to the high sensitivity of its coating, this indicator responded to very low levels of light exposures, and could be proposed as a unique tool for monitoring the most photosensitive category of artifacts.

It should be noted that all the examples described above may be classified as passive optical sensors, and are based on a colorimetric principle. These sensors are designed to react with a detectable color change upon exposure to one or more environmental agents to be monitored.

It is worth recalling that colorimetric analysis is a well-established tool that is widely used and for manifold purposes in the conservation field. Indeed, since some of the most evident and dramatic deterioration phenomena in artworks are accompanied by chromatic alterations, colorimetry was adopted primarily as a methodology for quantifying and assessing the color alterations of polychrome surfaces[20-23]. Afterwards, with the use of colorimetric impact sensors an innovative concept was introduced in conservation science, namely the exploitation of the degradation effect itself (i.e. the color change) as parameter to assess the overall impact on the object to be protected. This new concept marked the beginning of a new research area in the photonics for optical sensors customized for applications in a museum environment.

### 2.3. *Optical Fibers-based Systems for the Safety and Control of Cultural Items*

Thanks to their high versatility, optical fiber technologies have attained wide diffusion in the field of conservation. Optical fibers are based on transparent and dielectric materials, and this makes them particularly suitable for manifold applications in museum environments. Transparency makes it possible to satisfy many aesthetic requirements, and dielectricity provides intrinsic safety against risks of spikes, electric discharges, explosions, and fires.

One of the foremost uses of optical fibers in museums involves the lighting design and illumination of display environments, where compromises between different exigencies - such as visual comfort for the public and the protection of artifacts from excessive light exposures - have to be found. Since optical fibers make it possible to remotely guide the light from the source, innovative lighting systems are proposed that enable an appropriate illumination even in areas that are hard to reach[24]. Moreover, by means of optical fibers, every undesired component of the radiation emitted by light sources (namely, UV and IR components) can be easily filtered out, so as to avoid or reduce the damaging action of light on artworks on exhibit. For example, optical fibers offer suitable solutions for a safe illumination of confined environments, such as showcases, in which internal heating needs to be avoided in order to correctly preserve the objects on display[25].

Optical fibers technology is also extensively used in sensoristic applications so as to ensure the care and safeguarding of cultural items.

A variety of innovative sensors and customized devices based on the use of optical fibers have been developed in order to respond to different and specific conservation needs. Prototypes of optical-fibers sensors have been proposed for various applications in indoor environmental control: for example, the detection of selected pollutants (POF type sensors)[26], lighting control[27,28] and the monitoring, both in-situ and in real time, of temperature fluctuations[29]. Fiber Bragg Gratings (FBG) have been proposed for the control of stress and strain in wooden artworks[30], and are widely used in structural health monitoring, which is an area of great interest also for the cultural heritage field[31].

To conclude this brief overview, it is worthwhile examining the importance of optical fibers for the implementation of several optical techniques - such as Fiber Optics Reflectance Spectroscopy (FORS), Raman, fluorescence, etc. - that are widely used for performing non-invasive diagnostics on various categories of artworks. Optical fiber modules and accessories are used, in fact, to equip different types of portable optical instrumentation which are used to perform in-situ, no-sampling measurements (Fig. 1), thus making possible the characterization and identification of materials under conditions of total respect for the integrity of the object under examination[32,33].

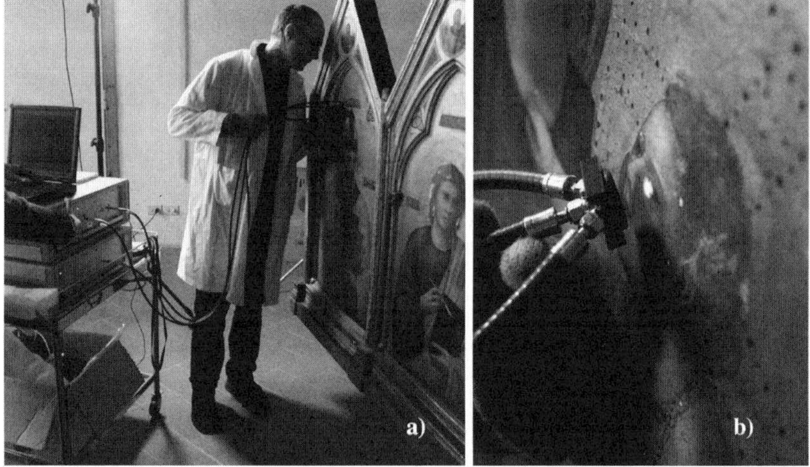


Fig. 1. (a) Fiber Optics Reflectance Spectroscopy (FORS) in-situ measurement campaign on the artwork "Polittico di Badia" (ca. 1300) by Giotto; Uffizi Gallery, Florence. (b) The non-contact probe-head connected to the optical-fibers bundle.

## 3.　The Use of Laser Interferometry in the Direct Impact Assessment of Museum Objects in Transit

### 3.1. *Introduction*

#### 3.1.1. *Impact in Structural Documentation Terms*

Direct impact assessment of an artwork under examination is provided by direct-recording laser interferometry of the surface micro-displacements. Interferometric measurements make it possible to freely select a range of spatial resolutions for the application request.

The need to directly assess the artwork is associated with the problems encountered locally in the distant past, but which are projected and amplified worldwide with the mass cultural tourism that has generated an increased demand for the loan of objects for travelling exhibitions. Objects of high aesthetic, cultural and historical value are constantly exchanged between museums and galleries. At this point, a clear definition of the term "impact" for movable works of art is suggested. As is well known, works of art and museum objects are susceptible to deterioration due to the fundamental operating system of museums worldwide, which is dominated by the unavoidable necessity of their display. The operation involves repeated handling and conservation treatments, various means of transportation, not to mention sudden environmental and climatic changes[2,34]. In each of the above-mentioned procedures, art objects suffer the impact of deterioration in the form of chemical, physical and structural alterations. Photographic records, coupled with routine condition checks, prove that short-term damage is expected to an acceptable degree; otherwise, insurance premiums would have to rise to prohibitive levels in response to claims. Instead, long-term damage involves a slow but steady deterioration process.

Existing or short-term damages and induced defects affect the structural integrity in the long term, and well before visible signs of decay occur. Thus, by becoming centres of deterioration further invisible alterations accumulate causing new visible defects and the propagation of visible damage[35,36].

The notion of main importance here is that the deterioration process is identifiable when the data on damage is monitored repeatedly in order to be compared with previous data[37,38]. A deterioration-monitoring process makes possible the early identification of alterations and thus causes preventive measures and safety rules to be applied and observed in the early stages. The development of methods and instruments for evaluating damage alterations and for responding to decisions regarding crucial physicochemical and mechanical preventive deterioration is of critical importance in preserving the cultural heritage. It makes possible to guarantee proper treatments, safer interventions, safer microclimate maintenance, (more) secure handling and packaging, and fighting actions against fraud.

In this short review, a direct monitoring system is presented for directly assessing the impact of damage on the surface of an artwork.

The development of an experimental methodology, procedure and system aimed at a systematic assessment of impact over time in movable artworks implies a procedure through which an artwork is forced to provide spatial signals in an initial position, with which spatial signals are later compared. To acquire the spatial response in a non-destructive and non-contact procedure, an implementation of spatially-sensitive probes of coherent interferometry has been exploited[39-42]. It has been affirmed that the probe beam obtained in the form of interference fringe-patterns under parametrical investigation of the experimental boundary conditions is a rich self-encoded signal of the artwork condition, and for this reason alterations for a comparison over time between "before and after" impact relations are meaningful. Multiple benefits derive for artwork management in cases in which routine impact monitoring is required.

### 3.1.2. *Laser-Coherent Interferometry in Structural Documentation*

Laser holographic interferometry is a coherent, high-resolution, quantitative full-field technique for the non-contact and non-destructive detection of invisible subsurface or bulk defects using remote surface monitoring[43-45]. It makes use of the principle of wave superposition in a way that the results of their combination have meaningful properties in diagnostics[46-50]. Two waves with the same frequency combine (here, laser light waves), resulting in a fringe pattern that is determined by the phase difference between the two waves. If waves combine in phase, they result in constructive interference that is seen as bright fringe; if waves are out of phase, they result in destructive interference that is seen as dark fringe. Each pair of fringes measured at bright $N^{th}$ fringes is considered to be a fringe-pattern corresponding to a phase change equal to $N\lambda/2$. If exists a subsurface defect causing at neighbouring points a mechanical instability preventing the surface from homogeneous displacement the equidistant intensity is lost and the pattern becomes distorted. The fringe patterns detect impact and deterioration rate directly from the surface and could be considered as direct impact sensors. Impact Assessment is here examined on a multiple spatial-frequency measuring procedure introducing direct-full field recording principle as direct-impact sensors for impact monitoring.

## 3.2. *Impact Assessment Procedure*

Impact is defined and considered to be any slow or rapid distorting effect that causes some sort of alteration in the physicochemical condition of an artwork. When the physicochemical alterations reach a magnitude that can affect irreversibly the structural and mechanical properties of the artwork in question a disintegration impact is identified. Several decay processes then begin. With time, an artwork affected mechanically loses its elasticity and its stability properties resulting in higher ability of movement among the constituent materials and parts, as layers and support, making it possible for microstructural deformation to establish itself and to grow. A mechanically affected construction is destabilised, and cycles of dimensional-changes increase with time until its elasticity properties are violated and final collapse, breaking the body, according to Hooke's law of elasticity, is established. The natural ageing process of an artwork is smooth, long and continuous, and the magnitude of dimensional change of decay effects is too small to be detected until the damage and the onset of impact become visible.

An alteration in the spatial coordinates of an artwork is a measurable and direct microscopic result of natural or induced decay effects, even if mechanical instability is not visibly observed. Changes measured in spatial coordinates indicate dimensional alteration that is negligible compared to the macroscopic size of the artwork. Moreover, the weak areas of the object that are primarily affected and become unstable generate own fringe pattern whereas are invisible to conventional investigation, e.g. x-rays. After spatial information is retrieved, through monitoring spatial responses over time changes can be assigned to prevent artwork from irreversible visible signs of deterioration.

Within this context, coherent interferometry techniques provide sensitive methods to capture 2D-3D spatial displacement. Dimensional changes are detected and measured as optical path changes, with the laser light wavelength serving as the unit of measurement. A variety of optical arrangements can be manipulated in order to adapt to the magnitude of the expected impact and to experimental constraints that are encountered in the conservation field[51-53]. The technology uses non-penetrating irradiation that relies on a reflection of laser beams diffused from the artwork surface that do not require any contact with delicate surfaces nor with preparation of the surface[54-56]. The levels of irradiation and exposure used are safe for varnishes and pigments[57-59]. In this context, the technology involved can be classified as non-destructive, non-contact and non-invasive, and allows the new application to artworks to flourish in routine manner[60]. Structural alterations are self-encoded signatures of the artwork which can witness impact and originality. By generating the process in order to monitor the structural alteration over time within strict boundary conditions, a novel application for impact assessment is introduced.


### 3.3. *Interferometry Principle*

Any spatial alteration of surface points in time, $(t_1)$ before and $(t_2)$ after displacement, acquiring two different spatial positions, $(P_1)$ and $(P_2)$, generate in intensity $(I)$ a phase difference $\Delta\varphi$ that is due to a change in the surface reflected optical path, (such as $I_1$ before and $I_2$ after displacement). The relation is described in Eq. (1).

$$I^2 = (I_2 - I_1)^2 = 4I_1I_2[1 - \cos(\Delta\varphi)] \qquad (1)$$

The principle of wave superposition is the basis in the procedure implemented in the impact monitoring. A surface that is double exposed before and after a displacement results in a path change relevant to the observation system, and the signal is significantly modulated in intensity. Intensity modulation is captured in an optical or digital photosensitive medium, thus providing a visual interference pattern that reveals a deformation in the reflecting surface as seen in Fig. 2. Micrometric displacement is recorded and studied by laser coherent interferometry.

The spatial sensitivity of the detection system depends on the optical geometry. It is assumed that the ideal resolution, depended on the application constrains and requirements, cannot be found in a single optical geometry. Hence, flexible geometry adjustments provided from hybrid of optical configurations is required. An experimental optical geometry implementing a novel hybrid optical design was developed in order to adjust to the broad range of spatial resolution and diverse needs.

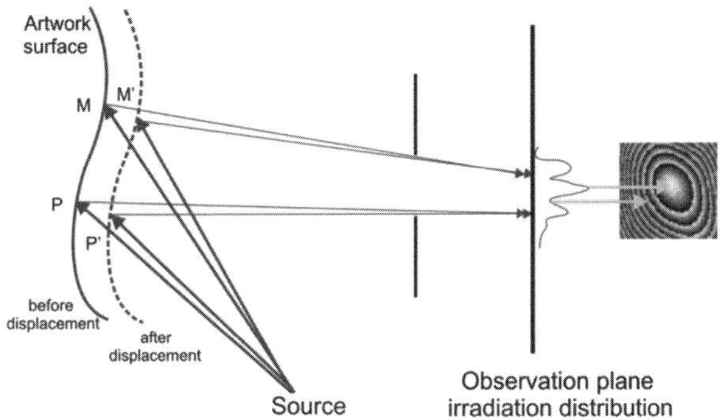


Fig. 2. Interference by means of a coherent laser beam reflected twice in surface displacement.

The optical geometries implemented in the hybrid optical system for impact assessment are shown in Fig. 3. An amplitude division interferometer is equipped with a) a shearing element to generate a mutual coherent wave generating a sheared wave front with interference field properties of high stability and decreased spatial resolution <100 lines/mm[61]; b) an off-axis interference geometry with the addition of a reference beam in front of a high resolution CCD (1200x1200 pixels), generating an interference intensity distribution field susceptible to low vibrations and moderate spatial resolution >100 lines/mm[62]; c) a photorefractive crystal used as a high-density-content photosensitive medium to reach spatial frequencies >1000 lines/mm. The optical arrangements operate with internal algorithms of phase-shifted interferometry[63]. The optical possibilities of the hybrid configuration reveals spatial changes hosting a variety of effects suited to perform examination in variety of environments and complexity of artworks, as it is required in the detection of impact assessment. The geometries make possible an extended field of view ($\approx$ 50 cm diameter) allowing an instantaneous study of multiple object points. The detection is named "full-field", as opposed to point-by-point scanning systems.

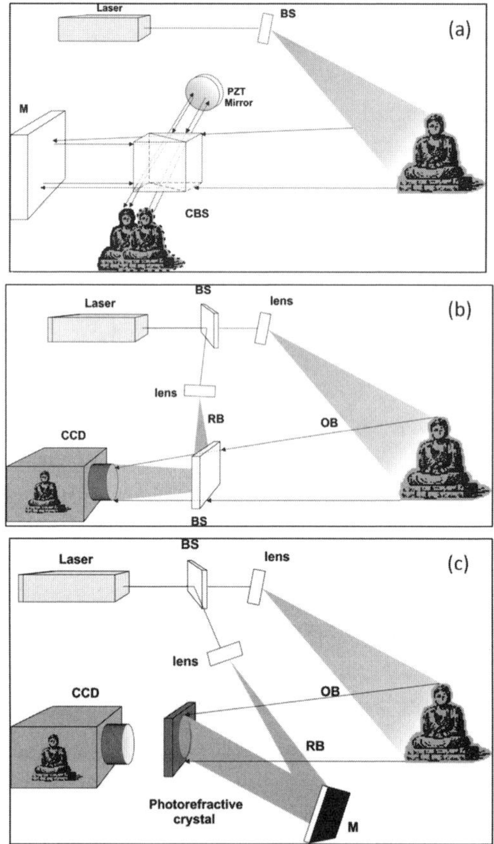


Fig. 3. Coherent interferometry makes possible to record geometries that have various spatial frequencies.

### 3.4. *Direct Impact Assessment In Brief*

Impact assessment involves an investigation of initial position termed "reference position", and stores current and future displaced positions in a database as long as routine updates of artwork conditions are carried out. The procedure for obtaining either the reference or the next states is very similar to double-exposure or sequential recording procedures[62].

Spatial alterations are detectable and measurable over time under specific and confidential strict boundary conditions stored in a database.

The post-processing analysis is based on the artwork's overall and localized spatial responses examination. Update and comparison of data result to a determination of positive or a negative impact in respect to previous condition. By controlled monitoring of reference fringe patterns in respect to subsequent alterations, each artwork obtains a unique and self-coded "signature". Using same data for impact assessment and anti-fraud assessment the application offers a multifunctional anti-fraud technology.

### 3.5. *Characteristic Examples*

#### 3.5.1. *Wooden Panel Paintings: Impact with Ageing*

A model with layers of animal glue, canvas, gesso and animal glue was prepared on soft wood with dimensions 15 cm x 20 cm. It was initially examined for spatial signatures, and then aged for 60-years (protocol)[a].

The sample was re-examined for signature changes. An example is shown in Figs. 4 (a) and (b) using visual raw-data instead of intensity profile.

---

[a] By Dr Eleni Kouloumpi, a conservation scientist at the National Gallery of Athens conservation labs, EPMAS-NGA.

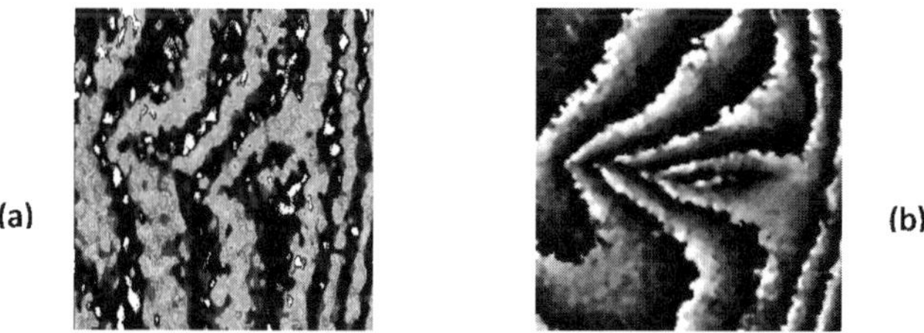



Fig. 4. Impact reveals spatial signatures of model, at (a) before and (b) after ageing. Note: in the (a) signature evidence of a crack-to-be, in (b) signature of the crack.

### 3.5.2. *Model Canvas Paintings: Accidental Impact*

A model canvas, 35 x 35 cm, was constructed on a stretcher with expandable corners, using a controllable screw system and heavily sized with animal glue, a coat of chalk and glue, and various low-tack materials in order to produce poor adhesion. The artwork, which was highly susceptible to vibrations and was examined by low resolution, was hit with a hammer in order to generate an impact point of approximately 20 mm in diameter. The measurements performed show that the strain levels adjacent to this impact point were initially very high, but that the strain was reduced dramatically with time and achieved a stable level after several days. Results from the impact damage caused by a hammer are shown in Fig. 5 (a) as the reference phase map, while the high strain concentrations around the impact point 300 s. after impact are shown in Fig. 5 (b). The best visualisation was achieved after a post-processing of the measurement between unwrapped phase maps.

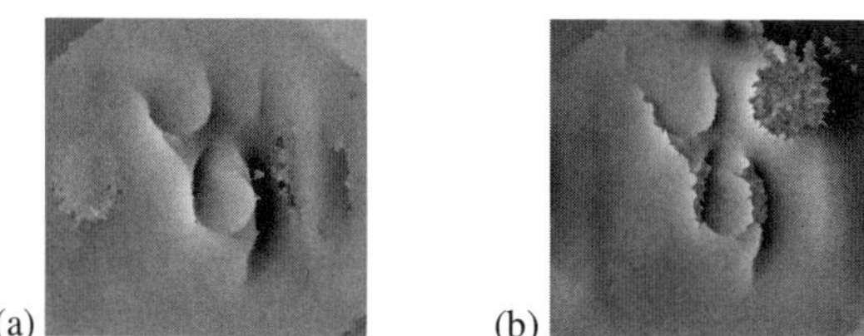



Fig. 5. The strain map (a) before impact, (b) after impact. Note: in (a) initial signatures retrieved and in (b) new signature is added.

### 3.5.3. Short Term vs. Long Term Impact

Models of icons used as samples to visualize and assess the slow but steady impact of the decay processes. Icons were examined with high resolution arrangement both before and after the ageing process, with an aging protocol that aimed to generate a natural decay of about 60 years.

Even the long-term slowly-affecting natural ageing imposes impact as was detected and assessed. Results confirm the feasibility of using spatial information concept and procedure as direct impact assessment sensors.

## 4. Conclusions

It has been shown that impact can be retrieved directly from the surface of an artwork through changes in its spatial coordinates over time, provided that the routine investigation scheme initially applied can be reproduced in the coherent interferometry geometries. A standard investigation protocol and comparison routine was applied by means of a hybrid optical system for Impact Detection, which incorporates tailored geometries. These have significant advantages, such as immunity to vibration, real-time monitoring, high information content, and can be selected according to constraints that are imposed in different case-study.

The custom-made hybrid optical system was envisaged and implemented in order to assess impact, originality and deterioration of artworks in transit or in exhibition. It is recently exploited in environmental and microclimate impact assessment. The Impact Detection System (IDS) and Impact Assessment Procedure (IAP) are the actual results of R&D project that are now being exploited in diverse fields and applications.

## Acknowledgments

Documentation on FORS measurements was provided by Dr Marcello Picollo (IFAC-CNR, Italy), who is gratefully acknowledged also for fruitful discussions. Research on Laser interferometry was primarily funded by the 6th FP EC project MULTIENCODE and partially by IESL-FORTH and partners ITO, CSL-ULG, OPTRION, EPMAS-NGA and TATE. The MULTIENCODE consortium is acknowledged for creative cooperation and results. The project output is being exploited in the Climate for Culture (7th FP), CHARISMA (7th FP), SYDDARTA (7th FP,) and in 7th FP aeronautics project (ULG).

## References


1. M. Cassar, Environmental Management: *Guidelines for Museums and Galleries* Routledge Eds. (1994).

2. G. Thomson, *The museum environment,* 2nd Ed., Butterworth Heinemann Eds., Elsevier, Oxford (1994).

3. J. Tétreault, *Airborne Pollutants in Museums, Galleries and Archives: Risk Assessment, Control Strategies and Preservation Management* (Canadian Conservation Institute, Ottawa (2003).

4. G. Pavlogeorgatos, *Building and Environment,* **38**, p. 1457 (2003).

5. *Guidelines on Technical-Scientific Criteria for Application of Standards in Museums,* Law by Decree n. 112/1998, Ministerial Decree 10 May 2001, *Italian Republic's Official Gazette (19 October 2001) n. 244 (in Italian.)*

6. L. M. Pestana Leão de Brito and L. M. Rodríguez Peralta in: *Cooperative Design, Visualization, and Engineering Lecture Notes in Computer Science*, Procedings, of the 5th Int. Conference CDVE 2008, Spain, Vol. **5220**, Ed. Y. Luo, p. 107, Springer (2008).

7. C.T. Angeles, M.C.R. Talampas, L.G. Sison and M. N. Soriano, in: *Networking, Sensing and Control, ICNSC 2008. IEEE International Conference*, Hainan, China, p. 1263 (2008).

8. M. Bacci, C. Cucci, A. A. Mencaglia and A. G. Mignani, *Sensors*, **8**, p. 1984, (2008).

9. M. Odlyha, J. J. Boon, O. van den Brink and M. Bacci, *Journal of Thermal Analysis and Calorimetry,***49**, p. 1571 (1997).

10. R. L Feller, in: *Contributions to Conservation Science: A Collection of Robert Feller's Published Studies on Artists' Paints, Paper and Varnishes*, Eds. P.M. Whitmore, Carnegie Mellon University Press, Pittsburgh, p. 107 (2002).

11. ERA (Environmental Research for Art Conservation) EC funded project. Contract n. EV5V-CT-94-0548.

12. M. Bacci, M. Picollo, S. Porcinai and B. Radicati, *Environ. Sci. Technol.*, **34 (13)**, p. 2859 (2000).

13. M. Odlyha, C. Theodorakopoulos, D. Thicket, M. Ryhl-Svendsen, J. M. Slater and R. Campana, in: *Museum Micrioclimates the Proc. Conference Copenhagen 19-23*

*Nov. 2007*; Eds T. Padfield, K. Borchersen, National Museum of Denmark Copenhague, p. 73 (2007).

14. M. Bacci, C. Cucci, A.-L. Dupont, B. Lavédrine, M. Picollo and S. Porcinai, *Environmental Science and Technology*, **37**, p. 5687 (2003).

15. P. Mottner, in: *Strategies for saving our cultural heritage. Proceed. Int. Conference on Conservation Strategies for Saving Indoor Metallic Collections* (CSSIM - ICOM-LIC Meeting, Cairo, 25. February - 1 . March 2007), Eds. V. Argyropoulos, A. Hein, M.A. Harith, (2007).

16. E. Dahlin, T. Grontoft, S. Rentmeister, C. Calnan, J. Czop, K. Hallett, D. Howell, C. Pitzen and A. Sommer Larsen, in: *Proceed. of ICOM Committee for Conservation, 14th Triennial Meeting, The Hague, 12-16 September 2005, Preprints Vol II*, James and James, p. 617, London (2005).

17. N. Carmona, E. Herrero, J. Llopis and M. A. Villegas, *Sensors and Actuators B*, **126,** p. 455 (2007).

18. EC funded project Contract n°: EVK4-CT-2000-00016, http://www.lido.fhg.de.

19. A.-L. Dupont, C. Cucci, C. Loisel, M. Bacci and B. Lavédrine, *Studies in Conservation*, **53(2),** p. 49 (2008).

20. M. Bacci, F. Baldini,R. Carlà and R. Linari, *Appl. Spectrosc.*, **45**, p. 26 (1991).

21. M. Bacci, A. Casini, C. Cucci, M. Picollo, B. Radicati and M. Vervat, *J.of Cultural Heritage*, **4(4)**, p. 329 (2003).

22. L. Bullock, *The National Gallery Technical Bulletin*, **2(1)**, p. 48 (1978).

23. M. Bacci, M. Picollo, S. Porcinai and B. Radicati, *Techne*, **5**, p. 28 (1997).

24. B. Cook, *Engineering Science and Education Journal*, **9(5),** p. 207 (2000).

25. D. Camuffo, G. Sturaro and A. Valentino, *Thermochimica Acta* **365**, p. 65 (2000).

26. E. Angelini, S. Grassini, D. Mombello, A. Neri, M. Parvis and G. Perrone, *Applied Physics A: Materials Science & Processing*, **100(3)**, p. 975 (2010).

27. M. Bacci, C. Cucci, A. A. Mencaglia, S. Porcinai and A. G. Mignani in: *Optical Metrology for Arts and Multimedia*; Ed. R. Salimbeni, Proc. SPIE, **5146**, p. 170, (2003).

28. A. G. Mignani, M. Bacci, A. A. Mencaglia and F. Senesi, *IEEE Sensors J.*, **3(1)**, p.108 (2003).

29. A. G. Mignani, R. Falciai and C. Trono in: *Optical Sensors and Microsystems. New concepts, Materials Technologies,* Vol. II, Eds. S. Martellucci, A. N. Chester and A. G. Mignani, (Kluwer Academic Pubs. USA, p. 253 (2000).

30. R. Falciai, C. Trono, G. Lanterna and C. Castelli, *J.of Cultural Heritage*, **4,** p. 285 (2003).

31. P. Antunes, H. Lima, N. Alberto, L. Bilro, P. Pinto, A. Costa, H. Rodrigues, J. L. Pinto, R. Nogueira and H. Varum, et al. in *New Developments in Sensing Technology for Structural Health Monitoring. Lecture Notes in Electrical Engineering,* Vol. 96, p. 253 (2011).

32. M. Bacci, *Sensors and Actuators B,* **29(1)**, p. 190 (1995).

33. M. Leona and J. Winter, *Studies in Conservation*, **46( 3)**, p. 153 (2001).

34. V. Bucur in *Springer Series in Wood Science*, Springer, Berlin, (2003).

35. V. Tornari, A. Bonarou, V. Zafiropoulos, C. Fotakis, N. Smyrnakis, S. Stassinopoulos in: *Proceedings of the 5th International Conference on Lasers in the Conservation of Artworks (LACONA V)* Eds. K. Dickmann, C. Fotakis, J. F. Asmus, Springer Proceedings in Physics 100, p. 513 (2005).

36. K. D. Hinsch, G. Gulker and H. Helmers. *Opt. Laser Eng.*, **45**, p.578 (2007) .

37. V. Tornari, A. Bonarou, V. Zafiropulos, L. Antonucci, S. Georgiou, C. Fotakis in: *ROMOPTO Conference*, Bucharest, Ed. V. I. Vlad, Proc. SPIE, **vol. 4430**: p. 153 (2000).

38. V. Tornari, E. Bernikola, W. Osten, R.M. Grooves, M. Georges, G.M. Hustinx, E. Kouloumpi, M. Doulgeridis, S. Hackney and T. Green, *Coherent metrology in impact assessment of movable artwork: the MULTIENCODE project* in CHRESP 8th European Commision Conference on Sustaining Europe's Cultural Heritage, Lijubliana Slovenia 10-12 November, 2008.

39. V. Tornari, *Analytical and Bioanalytical Chemistry* **387(3)**, p. 761, (2007).

40. C. Thizy, M. P. Georges, E. Kouloumpi, T. Green, S. Hackney and V. Tornari in: *Controlling Light with Light: Photorefractive Effects", Photosensitivity, Fiber Gratings, Photonic Materials and More*, OSA Technical Digest (CD) (Optical Society of America, paper MB49 (2007).

41. R. M. Groves, W. Osten, S. Hckney, E. Kouloumpi, V. Tornari, in: *LACONA VII, Proceedings of the International Conference on Lasers in Conservation of Artworks* Madrid Spain, Eds. M. Castillejo, P. Moreno, M. Oujja, R. Radvan, J. Ruiz , CRC Press, September 2007.

42. V. Tornari, E. Bernikola, W. Osten, R. M. Grooves, G. Marc, G. M. Hustinx, E. Kouloumpi and S. Hackney in: *Multifunctional Encoding System for Assessment of Movable Cultural Heritage, ICEM 13*, International Conference on Experimental Mechanics, Alexandroupolis, Greece, Abstract Conference Proceedings, Ed. E.E Gdoutos, Springer, pp 715, July 2007.

43. C. M. Vest, *Holographic Interferometry,*Wiley, New York (1979).

44. R. K. Erf, *Holographic nondestructive testing.* London, U.K., Academic Press (1974).

45. W.E. Kock, *Engineering applications of lasers and holography*, New York, U.S.A., Plenum Press (1975).

46. S. Amadesi, A. D'Altorio and D. Paoletti *Optical Engineering,* **22**, p. 660 (1983).

47. V. Tornari, V. Zafiropoulos, A. Bonarou, N. A. Vainos and C. Fotakis, *J. Opt. Las. Eng.* **34**, p. 309 (2000).

48. U. Mieth, W. Osten and W. Juptner in: *Fringe 2001 - The 4th International Workshop on Automatic Processing of Fringe Patterns*, Elsevier, p. 163, (2001).

49. D. Paoletti and G. Schirripa Spagnolo in: *Interferometric methods for artwork diagnostic*, Progress in Optics XXXV, p. 197 (1996).

50. P.M. Boone and V. B. Markov, *Studies in Conservation*, **40**, p.103 (1995).

51. K. A Haines, B.P. Hildebrand, *Appl. Opt.*, **5**, p. 595 (1966).

52. P.K. Rastogi, *Applied Optics*, **23(6)**, p. 924 (1984).

53. M. Takeda., H. Ina and S. Kobayashi, *J. Opt. Soc. Am.,* **72**: p. 156 (1982).

54. P. M. Boone, *Optical Engineering,.* **21(3),** p. 407 (1982).

55. G. Schirripa Spagnolo and  D.Paoletti, *Optical Review*, **Vol.1,.2** p. 284 (1994).

56. M. Kalms, W. Osten, W. Jüptner, W. Bisle, D. Scherling and G. Tober in: *Proceedings of SPIE*, **3824,** p.280 (1999).

57. V. Tornari, A. Bonarou, V. Zafiropoulos, L. Antonucci and C. Fotakis in: *ROMOPTO 2000: Sixth Conference on Optics*, Ed. V.I. Vlad, Proceedings of SPIE **4430**: p. 153 (2001).

58. S. Georgiou, V. Zafiropoulos, V. Tornari and C. Fotakis, *Laser Physics*, **8**, p. 307 (1998).

59. S. Georgiou, V. Zafiropoulos, D. Anglos, C. Balas, V. Tornari and C. Fotakis *Applied Surface Science*, **127–129,** p. 738–745 (1998),

60. V. Tornari, E. Tsiranidou, Y. Orphanos, C. Falldorf, R. Klattenhof, E. Esposito, A. Agnani, R. Dabu, A. Stratan, A. Anastassopoulos, D. Schipper, J. Hasperhoven, M. Stefanaggi, H. Bonnici and D. Ursu in: *Proceedings of the 6^{th} International Conference on Lasers in the Conservation of Artworks (LACONA VI), Vienna 21-25 Sept.* Springer Proceedings in Physics (2007).

61. L. Yang, F. Chen;W. Steinchen; M. Y. Hung, *Journal of Holography and Speckle, American Scientific Publishers*, **Vol. 1( 2)**, pp. 69-79(11) (2004).

62. M. P. Georges and P.C. Lemaire *Applied Physics B, Lasers and Optics,* **68(5)**, p.1073 (1999).

63. R. Jones. and C. Wykes, *Holographic and Speckle Interferometry.* 2^{nd} Ed. Cambridge University press (1989).

# RAMAN BASED DISTRIBUTED OPTICAL FIBER TEMPERATURE SENSORS: INDUSTRIAL APPLICATIONS AND FUTURE DEVELOPMENTS


Fabrizio Di Pasquale[1], Marcelo A. Soto[2] and Gabriele Bolognini[3,*]

$^a$*T$^l$eCIP Institute, Scuola Superiore Sant'Anna, via G. Moruzzi 1, 56124*
*Pisa, Italy*
*[2]EPFL Swiss Federal Institute of Technology, Institute of Electrical Engineering*
*SCI STI LT, Station 11, CH-1015 Lausanne, Switzerland*
*[3]Istituto per la Microelettronica e Microsistemi, CNR*
*via P. Gobetti 101, 40129 Bologna, Italy*
*[*]E-mail: bolognini@bo.imm.cnr.it*



This book chapter describes Raman Based Distributed Optical Fiber Sensors and their main industrial applications. In particular, after a brief introduction on the basic physical mechanisms, and a theoretical description of distributed optical fiber sensors based on spontaneous Raman scattering, we have focused our attention on the most attractive applications in terms of market opportunity: fire detection in tunnels and industrial plants and leakage detection in pipelines. We show that in order to achieve long sensing distances with high temperature resolution and good reliability, most standard Raman based distributed temperature sensors make use of graded-index multi-mode fibers, in double-ended or looped configuration. Advanced pulse coding techniques can however be effectively applied to make Raman sensors based on standard single mode fibers competitive with respect to other distributed optical fiber sensing technologies, such as Brillouin-based optical fiber sensors. Future developments of Raman based distributed temperature sensors will likely exploit standard single mode fibers, achieving meter-scale spatial resolutions over ultra-long distances, with high temperature resolution, fast response time, high robustness and reliability.

## 1. Introduction

Raman Based Distributed Temperature Sensors (RDTS) have been intensively studied in the last years[1-3], due to their spatially-resolved measurement capabilities over tens of kilometers of optical fiber with meter scale spatial resolution, and their consequent great advantages over conventional electrical sensors. RDTS have found successful applications in many different industrial areas, such as fire detection, power cable monitoring and leakage detection in oil and gas industry.

RDTS can provide industry with a sensing system characterized by several features which can constitute a significant competitive advantage in real-field applications, especially under extreme environmental conditions. Actually, since the sensing element is given by dielectric silica optical fibers that are interrogated through light propagation, RDTS can be safely and conveniently installed in all environments that are not compatible with standard sensors which exploit conductors (e.g. copper wires) or electrical measurements. Examples of such environments are given by temperature sensing in proximity of high voltages (such as in power plants or in distribution stations), along pipelines, plants or tanks carrying explosive gases, liquids or solids, in present of magnetic fields, and, in general, in hazardous environments or whenever issues associated to Electromagnetic Interference (EMI) are encountered.

As it will be detailed in the next section, most RDTS systems are based on Optical Time-Domain Reflectometry (OTDR) measurements of the temperature-dependent backscattered Raman Anti-Stokes (AS) signal, which is then normalized to either the Raman Stokes (S) or Rayleigh backscattered component[3-5] to compensate for spurious losses along the sensing fiber. The most common solution, called Single-Ended RDTS, interrogates the fiber from one end only, allowing in principle for extended sensing distances[5]; however single-ended RDTSs are intrinsically affected by issues arising from the different fiber losses at the AS and S wavelengths and their time-dependent variations (Wavelength-Dependent Losses, WDL)[6]; in particular their slow variation with time can cause distortion in the measured temperature traces which are particularly relevant in harsh application environments, such as in nuclear plants monitoring, where the presence of ionizing radiation effectively increases the optical fiber attenuation with time, inducing significant WDL[6,7]. Fiber losses can also be strongly affected in hot and humid environments, such as for example in geothermal wells applications, where the differential wavelength-dependent attenuation of the fiber typically varies with time as a consequence of high temperature and hydrogen concentration[8]. For all these reasons, a double-ended interrogation scheme (also called loop configuration technique)[6] is usually employed, where AS and S traces are alternately acquired in forward and backward directions and then properly averaged[6]; such scheme, although potentially less attractive in terms of maximum sensing distance, provides inherent correction of WDL and local losses, leading to highly reliable systems for industrial applications.

Raman DTS systems based on Multi-Mode (MM) Graded-Index Fibers are finding wide applications in strategic industrial sectors like transportation and energy. MM fibers are preferred to Single Mode (SM) ones due to the higher allowed peak power and higher backscattering coefficient, leading to longer sensing distances (up to a few tens of kilometers). However intermodal dispersion in MM fibers induces pulse broadening along propagation, ultimately limiting the achievable spatial resolution. Advanced techniques based on pulse coding are under development to enhance the sensing capabilities on SM fibers, providing then attractive solutions with sensing distances up to several tens of kilometers and meter scale spatial resolutions.

Note that industrial applications which only require distributed temperature measurements, can strongly benefit from the availability of long range RDTS systems; in such applications RDTS can also strongly compete with Brillouin Optical Time-Domain Analysis (BOTDA) sensors, which exploit a basic physical mechanism intrinsically affected by cross-sensitivity between strain and temperature.

In this work we will first provide a brief description of the physical mechanisms and introduce a theoretical model governing RDTS systems[9]; we will then describe their main industrial applications, concluding with an outlook on advanced techniques to improve the RDTS performance[10].

## 2. Physical Mechanism and Theory

The physical mechanism which is exploited to infer the temperature of the fiber (and of surrounding environment) through light wave measurements is the so-called Raman effect, an optical scattering phenomenon discovered in liquids in 1922 by C.V. Raman, and involving the generation of new light spectral components in presence of a strong light propagating through a gas, liquid or solid. It can be seen as an inelastic scattering between light (photons) and molecular vibrations (phonons) of the silica glass, hence involving a decrease (or increase) of phonon energy (i.e. optical frequency) after the photon-phonon interaction. More specifically, if some energy is transferred *from* the incoming *photons to* the glass *phonons*, then the emerging photons will possess less energy than the incoming ones, resulting in the generation of a new light component with a lower frequency (longer wavelength), which is known as Stokes component; if, on the contrary, some energy is transferred *from* the glass *phonons to* the incoming *photons*, then the emerging photons will possess more energy, resulting in the generation of a new light component at a higher frequency (smaller wavelength), which is called anti-Stokes component. As depicted in Fig. 1, the generated spectral components (Stokes and anti-Stokes) will be specularly separated from the incoming light (also named optical pump), with a frequency difference depending on the specific material; more specifically, their intensity will be maximum at a given frequency shift value, known as the Raman frequency shift (in particular, the Raman frequency shift in silica glass has a value of ~13.2 THz). Since molecular vibrations are strongly connected to temperature, Spontaneous Raman Scattering (SpRS) is inherently temperature-dependent, and is then fruitfully employed in silica glass fibers[11] to provide temperature sensing.

                              *G. Bolognini et al.*

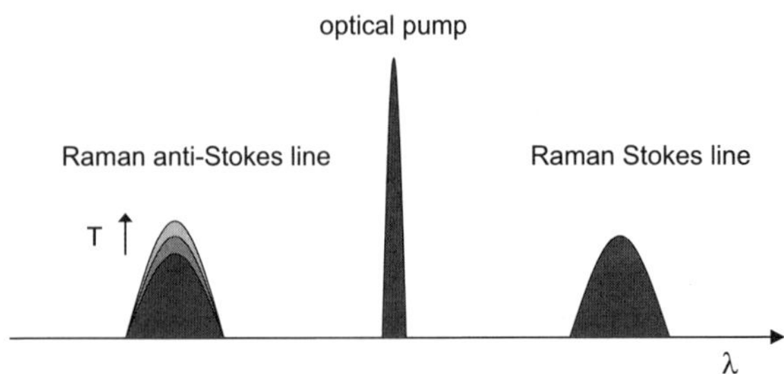


Fig. 1. Raman scattering in glass involves the generation of a (strongly temperature-dependent) anti-Stokes light component, and a (slightly temperature-dependent) Stokes component.

The intensity of the anti-Stokes light exhibits a strong dependence on the temperature, while the intensity of the Stokes light is found to be only slightly temperature dependent[5]. In order to provide temperature measurements that are less dependent on fluctuations of fiber loss, power or other effects (details are given in the discussion below), usually the power ratio of the anti-Stokes over the Stokes light is used (also the ratio of anti-Stokes over Rayleigh scattering is sometimes employed).

It can be seen that such a ratio *R(T)* has a dependence on temperature *T* at a specific scattering point which can be expressed as:

$$R(T) \equiv \frac{P_{AS}}{P_S} \propto \left( \frac{\lambda_S}{\lambda_{AS}} \right)^4 \exp\left( -\frac{h\Delta v}{k_B T} \right) \qquad (1)$$

where $P_S$ ($P_{AS}$) and $\lambda_S$ ($\lambda_{AS}$) are the power and wavelength for the Stokes (anti-Stokes) light respectively, $h$ is the Planck's constant, $\Delta v$ is the Raman frequency shift, and $k_B$ is the Boltzmann constant.

The temperature sensitivity of the AS/S ratio in silica glass, i.e. the temperature derivative of *R(T)*, is actually an important parameter in RDTS; a relative sensitivity of 0.8 %/°C is commonly observed around room temperature (25 °C), a value that slightly changes to 1.2 %/°C at about -50 °C[2]. Although the absolute power of the AS Raman component is more than 30 dB weaker than the Rayleigh backscattered component, its temperature dependence is high enough to provide a mechanism to perform distributed temperature sensing for real applications using an optical fiber[1-3].

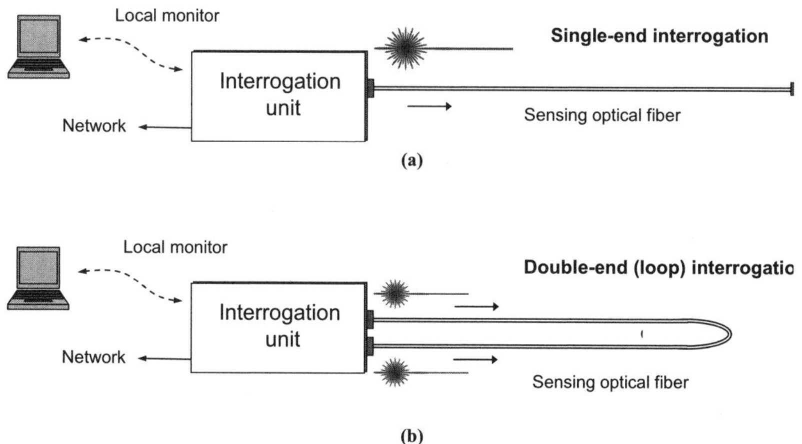


Fig. 2. (a) Basic single-end interrogation scheme and (b) double-end (or loop) interrogation scheme in ODTR-based Raman DTS.

In order to obtain the spatial information in RDTS systems, OTDR techniques are usually exploited[4]. This method is often called Raman-OTDR, and in its basic configuration, involving interrogation of a single fiber-end, it essentially consists in launching short laser pulses into the sensing fiber and detecting the backscattered spontaneous Raman signals (filtered Stokes and anti-Stokes components) with high temporal resolution[2,3] within the DTS interrogation unit, as depicted in Fig. 2a.

Unfortunately, the AS Raman signal reaching the receiver inside the DTS interrogation unit does not only depend on the fiber temperature, but also on the fiber attenuation and local losses (such as splices, connectors and bending losses)[3]. This dependence can actually induce significant errors in the temperature measurements, since variations of the local fiber losses and attenuation coefficient at the AS wavelength can be easily misinterpreted as temperature variations. To overcome this problem, as mentioned above, the AS Raman-OTDR trace has to be normalized by a temperature-independent OTDR signal, such as the Raman Stokes[3] or Rayleigh[5] intensity.

When the backscattered Raman Stokes light from OTDR is used to normalize the AS trace[3,6], we can obtain the following temperature-dependent ratio at the receiver representing the response from the whole sensing fiber:

$$R(z) = \frac{P_{AS}}{P_S} = C_R(z)\exp[\frac{-h\Delta v}{K_B T(z)}]\exp\{\int_0^z -[\alpha_{AS}(\xi) - \alpha_s(\xi)]d\xi\} \qquad (2)$$

In Equation (2), $\Delta v$ represents the frequency separation between anti-Stokes and pump signal, $h$ is the Planck constant, $k_B$ is the Boltzmann constant, $T(z)$ is the local fiber temperature, $\alpha_{AS}$ and $\alpha_S$ are the attenuation coefficient at the anti-Stokes and Stokes wavelengths, and $C_R(z)$ is a constant that takes into account the differences in the Raman capture factor and the response of the receiver for the different wavelengths. Note that the ratio $R(z)$ depends on the differential WDL of the fiber. If this factor is properly characterized as a function of the distance, it can be corrected from Eq. (2).

Unfortunately, the WDLs are not constant in time, and might change depending on external environmental conditions[6-8]. Hence, the ratio AS/S (or AS/Rayleigh) is expected to change during the sensor lifetime due to ageing process or external harsh environmental conditions, leading then to possible errors in the temperature estimation[6,12]. Such errors actually consist in the gradual deviation of the temperature trace with respect to the real value, effect that increases with the distance given the exponential factor in Eq. (2), and turns to be a critical factor at long sensing distances. As a consequence, the single-ended scheme described by Eq. (2) turns out to be not optimal for many cases since it requires a periodic sensor calibration, which is however not possible in many in-field applications[6].

In order to correct the above-mentioned problem, measurements can be performed in a double-ended configuration[6], as depicted in the scheme of Fig. 2b. In such a scheme, pump pulses are alternatively sent into the sensing fiber from both fiber ends, so that the AS signal and either the Stokes or the Rayleigh component can be measured in both forward and backward directions. Then, the geometric mean is taken between traces acquired in the two opposite directions. Even though the RDTS system requires access to both fiber-ends, this configuration has been demonstrated to provide a simple and effective solution to compensate differential attenuation issues[6]. The ratio between the anti-Stokes to

Stokes power can then be obtained in both forward ($R^{For}(z)$) and backward ($R^{Back}(z)$) directions according to:

$$R^{For}(z) = \frac{P_{AS}^{For}}{P_{S}^{For}} = \frac{C_{AS}^{For}}{C_{S}^{For}} \exp\{-\int_{0}^{z}[\alpha_{AS}(\xi) - \alpha_{s}(\xi)]d\xi\}\exp[\frac{-h\Delta v}{K_{B}T(z)}] \tag{3}$$

$$R^{Back}(z) = \frac{P_{AS}^{Back}}{P_{S}^{Back}} = \frac{C_{AS}^{Back}}{C_{S}^{Back}} \exp\{-\int_{z}^{L}[\alpha_{AS}(\xi) - \alpha_{s}(\xi)]d\xi\}\exp[\frac{-h\Delta v}{K_{B}T(z)}] \tag{4}$$

The ratio in loop configuration $R^{Loop}(z)$ is obtained as the geometric mean of these two ratio, giving rise to:

$$R^{Loop}(z) = \sqrt{R^{For}(z)R^{Back}(z)} =$$
$$= [\frac{C_{AS}^{For}C_{AS}^{Back}}{C_{S}^{For}C_{S}^{Back}}]^{1/2} \exp[-(\alpha_{AS} - \alpha_{S})\frac{L}{2}]\exp[\frac{-h\Delta v}{K_{B}T(z)}] \tag{5}$$

The temperature profile can then be estimated using a reference ratio $R_{REF}^{Loop}(z)$, which is obtained in loop configuration at a given temperature $T_{ref}(z)$, according to:

$$T(z) = \{\frac{1}{T_{ref}(z)} - \frac{K_{B}}{h\Delta v}\ln[\frac{R^{Loop}(z,T)}{R_{REF}^{Loop}(z,T_{Ref})}]\}^{-1} \tag{6}$$

Looking at this expression we can clearly see that using a loop configuration cancels out all loss factors dependent on the fiber position $z$. This feature leads to a self-calibrated temperature measurement which is robust to loss variations occurring during the sensor lifetime.

Note however that although the use of traces in forward and backward directions compensates the effects of WDL on the temperature profile (i.e. possible distortions in the trace), differential losses can induce a significant variation in the temperature sensitivity and an offset with respect to the real temperature[13]. To correct such effects, an internal calibration fiber spool at a known temperature value is typically used in RDTS systems, providing a simple and effective method to auto-compensate possible temperature errors[6,14].

## 3. Industrial Applications

Among the several industrial applications of RDTS systems we will focus our attention on the ones that are considered most attractive in terms of market dimensions and opportunities, namely fire detection (such as in tunnels and industrial plants) and leakage detection (such as in oil, gas, steam and hot water pipelines).

### 3.1. *Fire Detection*

Safety systems in modern transportation and power generation infrastructures require highly reliable and extremely fast heat detection systems. Among the several applications requiring such fire detection systems with a continuous measurement capability over long distances, the most relevant ones are fire alarms in tunnels (railways, highways, underground lines), cable trays and conveyer belts. Other applications which can also benefit from this distributed fire detection technology are warehouses, car parking and large hotels; however, a mass-scale exploitation of RDTS systems in this market sector would require an effective cost reduction of the reading unit, potentially achievable by developing an integrated RDTS sensor chip, fully based on a C-MOS compatible platform employing innovative solutions for multimode Si-based technology[15].

The proposed RDTS solutions for fire detection, especially in hazardous environments, need to be robust, reliable, immune to electromagnetic interference, and to false alarms; RDTS systems operating in loop configurations and with internal calibration fiber spools are particularly suitable to satisfy all such requirements.

RDTS systems often require special fiber cabling protecting the optical fiber itself, which is typically designed for the specific application; in particular, fiber cables for fire detection in harsh environments should be robust, moisture-, corrosion- and rodent-resistant, and be compliant with Low-Smoke-Zero-Halogen Specifications (LSZH). A local monitor and keyboard, and, above all, integration with (often proprietary) communication networks are other key issues for a DTS interrogation unit, as also depicted in Fig. 2, in order to provide automatic activation of protective systems, based on specific info on the different zones under monitoring. In each zone, several specific threshold values are commonly defined in order to quickly detect fire without false alarms; in particular static maximum temperature values and spatial and temporal temperature gradients can be defined in each zone to provide key information on the affected area.

In order to provide a real example of industrial application of a Raman DTS system, we will describe in the following paragraphs the details of the in-field deployment of an OTDR-based RDTS, that has been recently carried out in collaboration with the Italian Railway Network (RFI), for fire detection in railway tunnels[16].

Fig. 3 describes the specific field-trial cable installation scheme along a tunnel (1.85 km length) on the west coast of Italy; the cable goes from the control room where the DTS is located to the tunnel passing through an underground route and an air-route (exposed to sunrays and with notable temperature excursions within a day-night cycle) for a total path of ~2 km. The multi-fiber cable, specially designed for this application, includes 8 multi-mode graded index fibers (50/125) which have been properly fusion-spliced at both fiber-ends, as shown in Fig. 3, in order to create a multiple round-trip path simulating a 16 km long tunnel (Fig. 3 also shows some pictures of the cable, cable termination and splice protection at the tunnel exit).

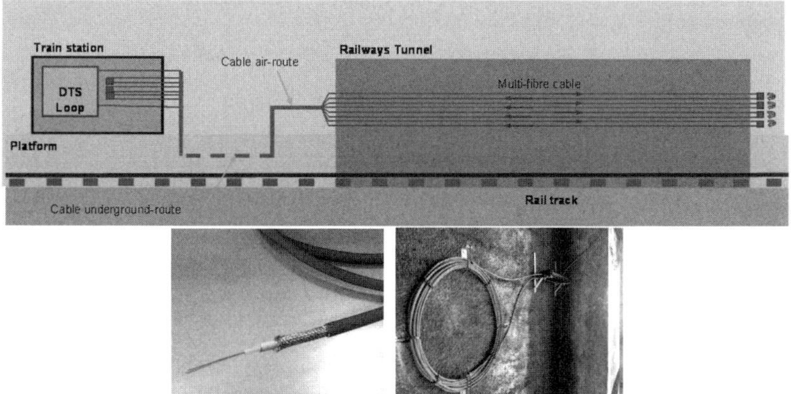


Fig. 3. Field trial scheme and details of the cable and installation.

The RDTS system operates at 1550 nm in loop configuration, with a maximum sensing distance up to 30 km.

Fig. 4b reports in detail a single round-trip temperature profile along the tunnel and within the route connecting the railway station, where the RDTS is located, and the tunnel; the measurement data confirm the reliability of the RDTS measurement system, showing good symmetry in the forward and backward directions, also pointing out the presence of high temperature values in the air route due to direct sun exposure and almost constant temperature along the tunnel, only slightly affected near the tunnel entrances.

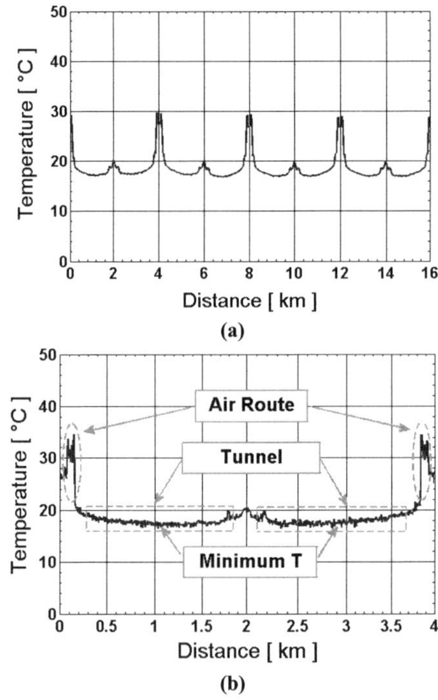


Fig. 4. Temperature profiles (a) along the multi-fiber cable and (b) for one single round trip.

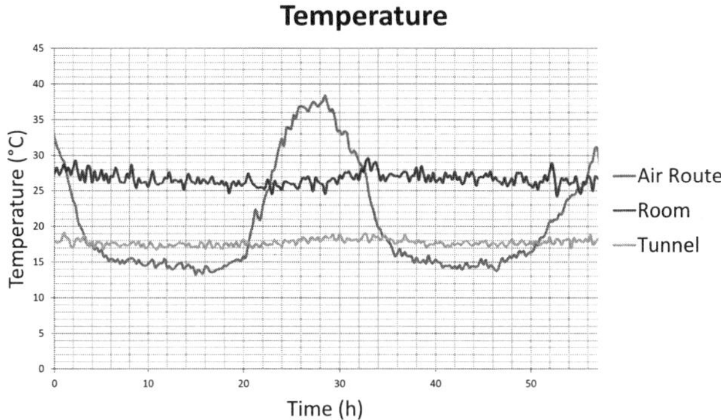


Fig. 5. Temperature monitoring in three locations along the optical fiber (57 hours span).

Fig. 5 shows the temperature variation versus time (with a span of 57 hours) in three different locations along the fiber cable, i.e. inside the control room (red line), in the air route section (blue line) and within the tunnel (green line). It is evident from Fig. 5 that the temperature inside the tunnel is almost constant during standard operating conditions, while significant temperature variations can be observed along the air route.

The DTS specifications indicate the ability to detect temperature variations (over a sensing distance up to 30 km) with a response time down to some seconds, and with temperature and spatial resolutions of 1 °C and 2 meter respectively. We analyzed the DTS performance up 25 km by adding a 9-km MM-fiber spool before the DTS to the setup of Fig. 3. Fig. 6 reports a typical temperature resolution of the RDTS system working in loop configuration, with different measurement times.

The presence of trains stopping inside the tunnel has been clearly identified by RDTS due to the local fiber temperature increase due to heat generation from electric train engines. Specific fire detection field trials are under investigation with the help of fire brigades in order to further validate the technology and identify the best fiber cable specifications for this specific application.

## 3.2. *Leakage Detection in Pipelines*

We are today experiencing a paradigm change in our energy system, mainly driven by important economical and environmental factors such as the need of renewable energy resources, demographic and climate changes. The energy system is becoming more decentralized with the introduction of renewable energy resources, which often lead to a mismatch between energy load and generation. Transmission and distribution grids connecting conventional and renewable energy resources require an intelligent monitoring system; Information and Communication Technology (ICT), together with distributed optical fiber sensors, can provide the right control system to match the power demanded and delivered and to enhance the modern energy system efficiency. Optical fiber sensors, and in particular distributed sensors, can play a key role in this field, for example for monitoring reservoirs, wellbores, pipelines and power cables. Monitoring the temperature profile along geothermal wells, as well as along more traditional oil, and gas pipelines, can then help in optimizing the energy production process, from the initial well assessment to production monitoring and optimization of service life.

Also monitoring the integrity of large pipelines, that are fundamental components of the energy system for oil, gas, steam and hot water distribution, is becoming a major goal of modern management and control systems. The main threats to such distribution systems are connected to landslides, earthquakes and harsh seasonal climate changes which can potentially induce structural problems with consequent leaks.

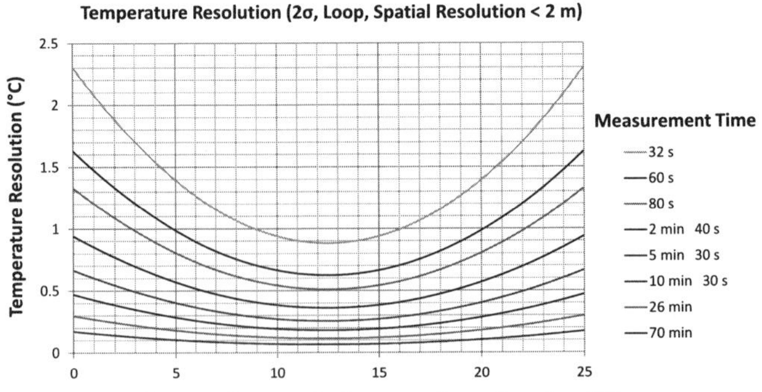


Fig. 6. Temperature resolution vs distance in loop configuration for different measurement times.

Also third-party intrusion is seen as a growing threat in many geographical areas. Distributed optical fiber sensors can be employed for continuously monitoring the entire pipeline, using dedicated fiber optic sensing cables which are able to detect very small variations in both temperature and strain with meter scale spatial resolution along the entire pipeline, providing early warning of potential failures, avoiding then the huge operation cost and environmental impact associated to pipeline failures[17]. Although many distributed optical fiber sensor schemes can potentially be used for simultaneous leakage detection, ground movement and intrusion detection (e.g. those based on Brillouin scattering), in this work we will focus on the application of Raman DTS.

Actually, by measuring the temperature profile only, without cross-sensitivity between strain and temperature, RDTS can be very effective to detect leakages of oil, gas, steam and hot water as well as for intrusion detection. The features of the technology are discussed in this section by analyzing specifications and issues in a real field-trial industrial application of RDTS for leakage and intrusion detection in a buried hot water pipeline in northern Italy.

Fig. 7 shows the set-up which has been used for in-field assessment of the RDTS technology, provided by Fibersens S.r.l., for leakage and

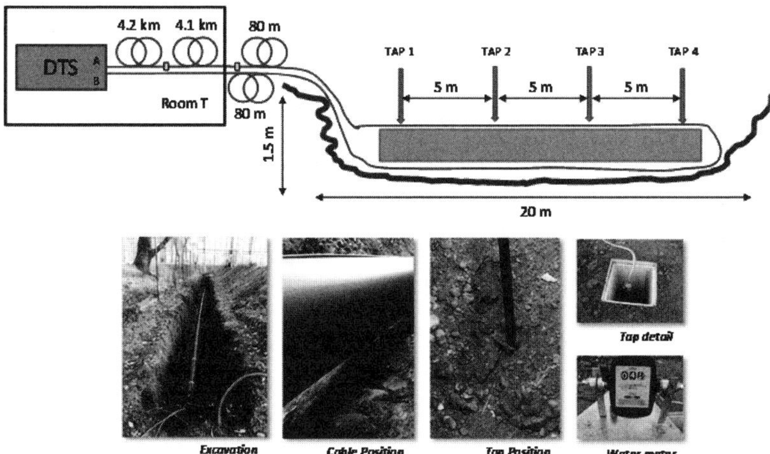


Fig. 7. Experimental set-up for RDTS in-field leakage detection assessment and details of the excavation, cable position and heated water leakage simulation.

intrusion detection. A multimode optical fiber cable (graded-index 50/125 multimode optical fiber) has been laid down along a buried pipe and connected to the RDTS system in loop configuration, as illustrated in Fig. 7, also reporting other details of the excavation and connection to heated water supply reproducing pipeline leakage events. Additional MM fiber spools have been added to simulate a sensing distance up to ~9 km. The leakage simulation has been implemented by inserting warm water at 30°C through tap 2 shown in Fig. 7, with a flow rate of 18 liters/min, with a total volume of 45 liters.

Fig. 8 reports the temperature distribution versus distance *before* and *after* the leakage, clearly pointing out the capability of the RDTS systems to detect even rather small temperature variation at about 1.5 meter underneath (measurement time 4'30'', spatial resolution 2 m, temperature resolution 0.1 C).

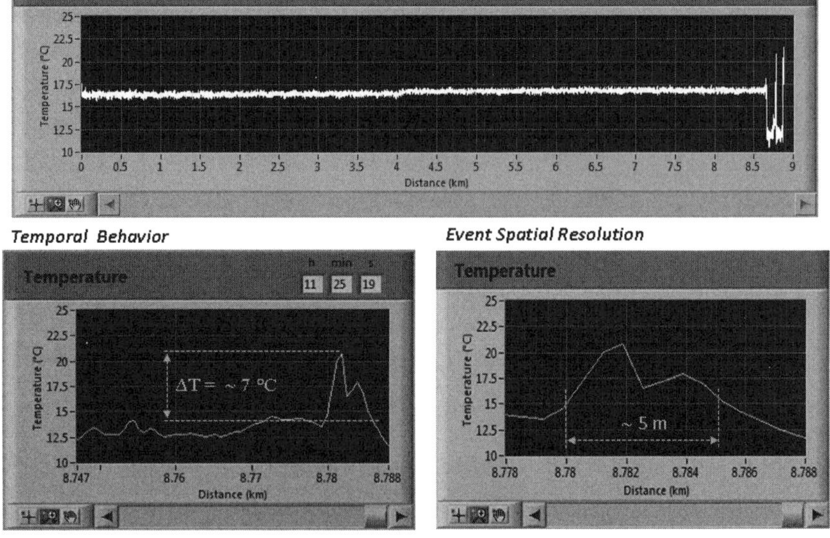


Fig. 8. Measured temperature distribution along the pipe before (graph at the top) and after (graphs at the bottom) the leakage event, clearly showing leakage detection at ~8.782 km.

Fig. 9 reports the temperature variation (versus time) induced by the leakage event at the leakage location, clearly demonstrating the effectiveness of the detection system in a timescale of few minutes.

It is interesting to note that the RDTS systems can also be effectively used for intrusion detection along buried pipelines; in fact, excavation

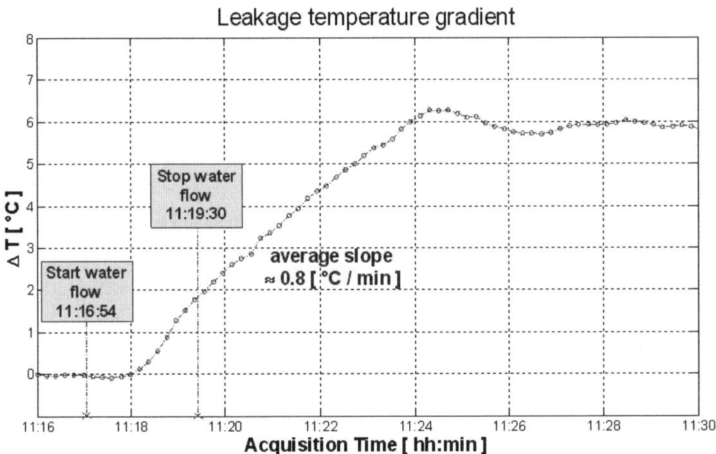


Fig. 9. Temperature variation induced at ~8.782 km by the leakage event.

activities carried out for tapping oil along a pipeline would expose the pipeline itself to temperature variations which can clearly indicate the execution of intrusion activities.

This feature has been assessed by digging around the buried pipe for a linear extension of about 5 meters. Fig. 10 reports the temporal and spatial behavior of the temperature distribution around the open section of the pipe, clearly showing the intrusion even in the case of a rather small temperature difference between the dug (exposed to air) and underneath regions.

From what we discussed, the possibilities offered by RDTS technology for in-field leakage and intrusion detection in buried pipelines are evident, and, although really attainable performance can be sometimes difficult to simulate and specific field-trials are needed such as the abovementioned one, this explains why this technique is more increasingly deployed in industrial applications dealing with pipeline monitoring, especially in the oil-gas industry.

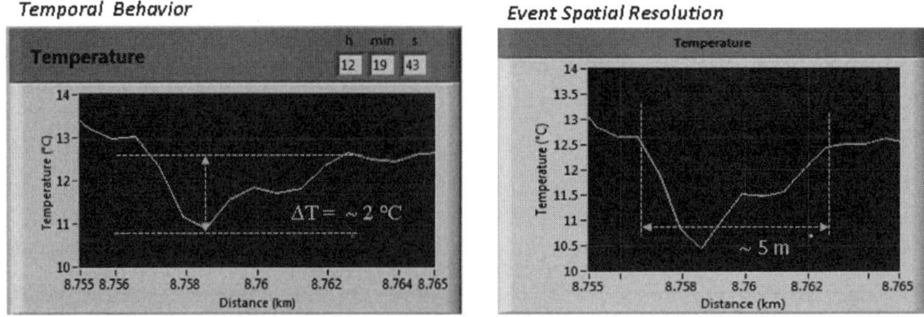



Fig. 10. Measured temporal and spatial behaviour of the temperature distribution around the open section of the pipe after excavation.

## 4. Advanced Raman DTS Solutions: Raman DTS using Single-Mode Fibers and Advanced Coding Techniques

In this section we briefly report on advanced developments which can greatly enhance the related performance of RDTS systems, making them very attractive for industrial applications in strategic sectors. In particular we will describe advanced RDTS systems based on standard single-mode fibers and optical pulse coding, which can be competitive with respect to Brillouin Optical Time-Domain Analysis (BOTDA) sensors in many long-range applications requiring temperature monitoring only, such as leakage detection in oil and gas pipelines and power cable monitoring.

### 4.1. *Optical Pulse Coding for RDTS Systems*

As discussed in Section 2, conventional RDTS systems are mostly based on Optical Time-Domain Reflectometry (OTDR), where the distributed temperature profile along an optical fiber is measured by injecting an optical pulse into the sensing fiber and detecting the backscattered Stokes (or Rayleigh) and anti-Stokes signals. However, since the intensity of the spontaneous Raman scattering in RDTS is very low (of the order of 60-70 dB below the peak power used for fiber interrogation with meter-scale resolution values), low optical power levels typically impact the receiver, leading to measurements with low Signal-To-Noise Ratio (SNR), and consequently to poor sensing performance.

The use of optical pulse coding techniques[18] has been proposed some years ago to improve the sensing capabilities of RDTS systems operating over multimode fibers[19] and over single-mode fibers[18-20] using externally modulated semiconductor lasers. Optical pulse coding method is based on the use of sequences of pulses that follow specific patterns according to particular codes. The optimum codes known for distributed optical fiber sensing are sequences based on Simplex codes[21] and complementary-correlation Golay codes[22]. Those are linear codes in which the measured coded traces correspond to the linear combination of single-pulse traces, each of them delayed according to the bit pattern defined by the code itself. In order to retrieve the single-pulse fiber response, a linear decoding process has to be followed, which depends on the specific code used in the system. The single-pulse trace that is obtained after decoding is expected to have a significantly better SNR in comparison to the standard measurement method, with an SNR increment that is defined as the *coding gain* and depends on the number of bits used in the pulse sequence. A better SNR clearly leads to RDTS measurements with better temperature resolutions and longer sensing distances in comparison to conventional single-pulse systems.

Usually RTDS systems employ multimode fibers as a sensing medium, allowing for measurements with an enhanced SNR thanks to the higher backscattered power and higher nonlinear threshold that characterize multimode fibers compared to single-mode fibers. Even though optical pulse coding further enhances the SNR of RDTS measurements when multimode fiber are used, the method does not solve the limitation imposed by the modal dispersion, which degrades the sensor performance in terms of spatial resolution at long sensing ranges.

On the other hand, when optical pulse coding is use in RDTS operating over single-mode fibers, the low backscattering levels as well as the relatively low power (a few hundred mW) of semiconductor lasers partially cancel out the benefits brought by coding in terms of SNR enhancement. In particular, for RDTS systems operating over standard single-mode fibers with optical pulse coding, the maximum usable peak power level before onset of nonlinearities is approximately 3-4 W, which is far above the maximum power of semiconductor lasers.

Recently a new coding scheme was proposed for RDTS systems exploiting high-power pulsed lasers (such as Q-switched and rare-earth doped fiber lasers) and the low chromatic dispersion of single-mode fibers[23,24]. Such a method is based on Simplex cyclic codes, and allows for long-range distributed temperature measurements with high spatial/temperature resolutions.

## 4.2. *Cyclic Pulse Coding for RDTS Systems over Single-Mode Fibers*

As previously mentioned, several techniques based on optical pulse coding have been proposed in the last years to enhance the SNR of RDTS systems[20], as well as of Brillouin-based distributed optical fiber sensors[25]. In order to achieve 1 meter spatial resolution, distributed optical fiber sensors based on standard coding methods require sequences of 10 ns intensity-modulated pulses, which is equivalent to a modulation frequency of 100 MHz. Unfortunately this modulation frequency is incompatible with the laser technology used for high-performance RDTS systems, such as rare-earth doped fiber and Q-switched lasers. These type of lasers are actually characterized by high peak power (several tens of Watt), but low repetition rates (a few hundred kHz) and very low duty cycles (typically 0.1%), making impossible the implementation of conventional pulse coding methods when such lasers are used. To overcome this limitation, recently a new coding technique based on cyclic Simplex codes has been theoretically proposed and experimentally validated in [23] and [24], respectively. The method consists in launching optical pulses into the fiber at a low repetition rate, so that the whole sensing fiber can be 'filled' with a large number of intensity pulses.

The main advantage of cyclic codes for RDTS applications with respect to other coding schemes is their compatibility with low-repetition rate lasers, exhibiting very low duty cycle values. They are based on a quasi-periodic intensity modulation pattern of the laser light, requiring only an additional external modulator to be used as a fast chopper when direct pulsed laser modulation cannot be implemented. Similarly to standard Simplex codes, the coding gain, i.e. the SNR enhancement, provided by cyclic Simplex codes is equal to $(L+1)/(2\sqrt{L})$ where $L$ is the code length[23].

It is important to point out that this coding method can also allow RDTS systems to improve the spatial resolution down to sub-meter scale using conventional time-domain techniques.

### 4.3. *Long-Range Coded RDTS with Meter Scale Spatial Resolution*

A scheme with the implementation of an RDTS using cyclic Simplex coding is shown in Fig. 11. The light source is a high-power rare-earth doped fiber laser (50 W maximum peak power) operating at 1550 nm, with 10 ns pulse width (allowing for an attainable spatial resolution of 1 m) and a maximum repetition rate of ~250-300 kHz.

A Variable Optical Attenuator (VOA) is used to adjust the peak power injected into the sensing fiber in order to avoid nonlinear effects.

An Acousto-Optic Modulator (AOM), which is triggered by an FPGA-based generator, is used to generate the cyclic Simplex codeword to be launched into the fiber, stopping laser pulses ('0' bits) or letting them through ('1' bits).

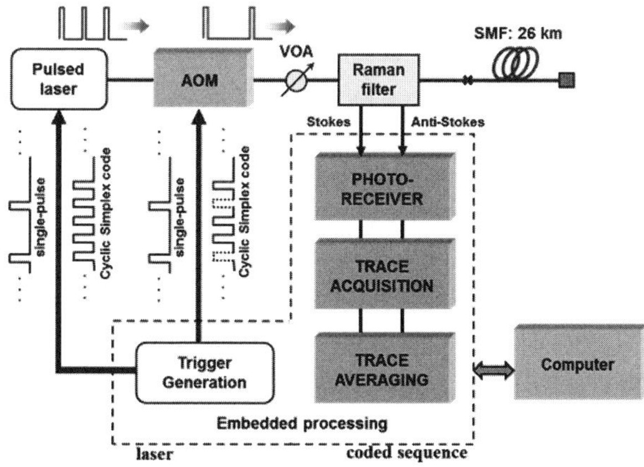


Fig. 11. RDTS system employing cyclic Simplex coding and a 26 km-long single-mode fiber.

108    G. Bolognini et al.

The cyclic pulse sequence is continuously sent into a 26 km standard SMF by using a suitable optical filter, which also allows the separation of the backscattered Stokes and anti-Stokes components into two parallel branches at the receiver. Each branch is composed of a high-sensitivity avalanche photodiode (APD), trans-impedance and high-gain amplifier stages, and an FPGA-controlled analog-to-digital converter (ADC) which is connected to a computer for data processing.

Measurements with coding are compared to the ones obtained by the conventional technique, using the same peak power (35 dBm) and the same acquisition time (about 30 seconds, equivalent to the acquisition of 100k time-averaged traces), so that the SNR enhancement provided by the cyclic codes can be estimated.

Considering that the used optical fiber is 26 km-long, a pulse repetition rate of ~230 kHz has been set in the laser, allowing 71 bits to propagate simultaneously along the fiber. Fig. 12 shows the acquired coded anti-Stokes and Stokes traces, where it is possible to observe the repetition period of the bits every 435 m, being in agreement with the laser repetition rate.

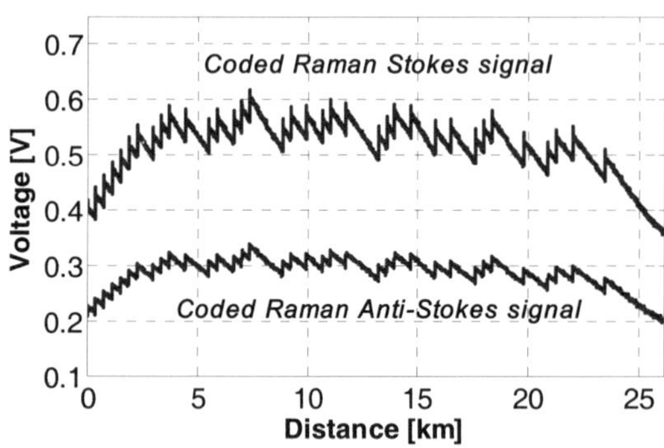



Fig.12. Coded Stokes and anti-Stokes traces, with 71 bit Simplex cyclic codes.

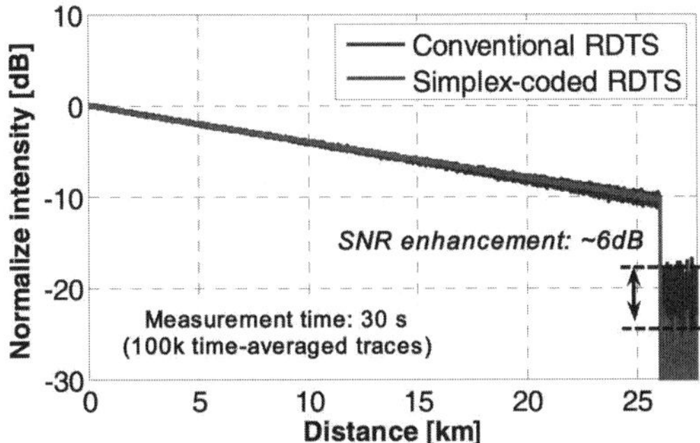


Fig. 13. Anti-Stokes traces obtained with both standard single-pulse technique and cyclic Simplex coding (trace after decoding).

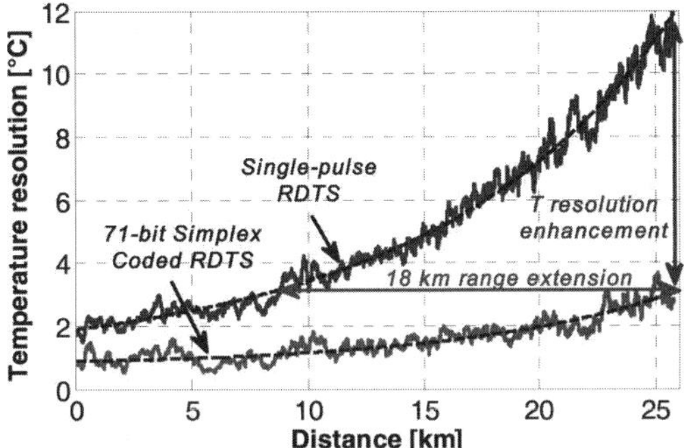


Fig. 14. Temperature resolution as a function of the distance, for both single-pulse and cyclic-Simplex coded RDTS.

The single-pulse fiber response for the anti-Stokes and Stokes components can be then simply recovered using a linear decoding process[23]. Comparing the anti-Stokes trace measured with the standard single-pulse scheme with the one obtained after decoding the Simplex-coded traces, the SNR enhancement provided by the code is evident, as reported in Fig. 13, resulting in an experimental coding gain of 6.0 dB, which is in good agreement with the theoretically expected SNR enhancement (6.3 dB).

Note that in the conventional case, the use of single-mode fibers leads to a relatively low electrical SNR, limiting considerably the best attainable temperature resolution of the RDTS (in this case equal to 12°C at 26 km distance), as reported in Fig. 14. However, the use of the cyclic optical code improves the resolution down to 3°C at the same distance. It is worth noticing that the conventional single-pulse RDTS scheme offers a temperature resolution of 3°C only at ~8 km distance; therefore, it can be concluded that the proposed technique offers a sensing range enhancement of about 18 km (when 71 bit are used), which agrees with the coding gain and the two-way fiber losses (~0.2 dB/km at 1550 nm).

It is important to point out that the use of single-mode fibers and optical cyclic coding allows for temperature resolution values of the same order as what is reported in the literature using single-pulse RDTS operating over multimode fibers[26], but with an improved spatial resolution. This is a very important feature, since the cyclic coding technique enables already-installed single-mode fibers to be used for long-range RDTS, without the need of installing new cables containing multimode fibers, and also providing better spatial resolution.

To estimate the real spatial resolution achieved in this experiment, the last few meters of fiber (near 26 km distance) have been placed inside a temperature-controlled chamber which has been set to 55°C, while the rest of the fiber is kept at 25°C (room temperature). Fig. 15 shows the temperature variation of 30°C, where a spatial resolution of ~1.0 m can be observed (calculated as the distance corresponding to a 10%-90% of the temperature change).

Comparing the amount of pulse dispersion for 10 ns pulses after propagation along single-mode and multimode fibers of similar length (26 km in this case), in single-mode fibers the pulse experiences a minimum temporal spread due to chromatic dispersion only

Fig. 15. Temperature profile vs distance, near ~26 km, with cyclic-Simplex coded RDTS.

(~15 ps/nm·km around 1550 nm), resulting in an expected spread smaller than 0.04 ns at the far fiber-end (laser line-width ~0.1 nm). On the other hand, when using multimode fibers the higher modal dispersion (up to several ns/km for graded-index fibers[27]) is expected to induce a significantly higher pulse broadening (~26 ns at the fiber end), exhibiting also a great variability depending on mode coupling conditions into the fiber. Considering that the backscattered light is also affected by similar spread, doubling the impact of dispersion, the spatial resolution of RDTS over multimode fibers is degraded to values up to about 5 m at 26 km, significantly worse than with single-mode fibers, which allow for an estimated resolution smaller than 1.08 m (as confirmed by the spatial resolution of ~1.0 m reported in Fig. 15).

## 5.  Conclusions

From what has been discussed above, it emerges that, thanks to their unique features, Raman-based distributed temperature sensors constitute a technology with an extremely large variety of application domains, and, indeed, one of the most successful photonic-enabled technologies from the point of view of industrial application. Many issues, that in early DTS laboratory experiments were not important, are becoming critical for industrial manufacturers and their customers, such as correct fiber cabling design, long-term reliability issues (e.g. due to wavelength-dependent losses), stability of the interrogation unit (and in particular of the opto-electronic components) to varying conditions in rough environments, power consumption, and so forth. The technical challenges that are posed by the growing range of DTS applications to scientists and engineers in the next few years include the ability to reach longer distances, the possibility to work with single-mode fiber with short measurement times and of achieving smaller resolution than the meter-scale. The integration of the sensor reading unit is another point which is currently sought-after, both by academia and industry, since it potentially allows for low-cost mass-scale sensor production and very large volumes of deployed DTS for instance in road tunnels, homes and buildings.

## Acknowledgments

We would like to thank Alessandro Signorini and Tiziano Nannipieri for their support in the field trials and Raman DTS development. We also thank Stefano Faralli and the courtesy of Fibersens S.r.l. for providing the Raman DTS system and part of the experimental data. We finally acknowledge Enzo Marzilli, Eugenio Fedeli and Rete Ferroviaria Italiana (RFI) for technical discussion and for their support.

## References


1. B. Culshaw and A. Kersey, *J. Lightwave Technol.,* vol. **26**, no. 9, pp. 1064-1078 (2008).
2. A. H. Hartog and A. P. Leach, *Electron. Lett.,* vol. **21**, no. 23, pp. 1061–1062 (1985).
3. J. P. Dakin and D. J. Pratt, *Electron. Lett.*, vol. **21**, no. 13, pp. 569-570 (1985).
4. K. Kikuchi, T. Naito, and T. Okoshi, *IEEE J. Quantum Electron.*, vol. **24**, no. 10, pp. 1973–1975 (1988).
5. G. Bolognini, J. Park, M. A. Soto, N. Park and F. Di Pasquale, *Meas. Sci. Technol.,* vol. **18**, pp. 3211-3218 (2007).
6. A. F. Fernandez, P. Rodeghiero, B. Brichard, F. Berghmans, A. H. Hartog, P. Hughes, K. Williams and A. P. Leach, *IEEE Trans. Nucl. Sci.*, vol. **52**, no. 6, pp. 2689-2691 (2005).
7. A. Kimura, E. Takada, K. Fujita, M. Nakazawa, H. Takahashi, and S. Ichige, *Meas. Sci. Technol.*, vol. **12**, no. 7, pp. 966–973 (2001).
8. M. Jaaskelainen, "Temperature Monitoring of Geothermal Energy Wells," in *Proc. of SPIE* vol. **7653**, p. 765303 (2010).
9. M. A. Farahani, T. Gogolla, *J. Lightwave Technol.*, vol. **17**, no. 8, pp. 1379-1391 (1999).
10. J. Park, G. Bolognini, D. Lee, P. Kim, P. Cho, F. Di Pasquale, and N. Park, *IEEE Photon. Technol. Lett..*, vol. **18**, pp. 1879-1881 (2006).
11. P. R. Stoddart, P. J. Cadusch, J. B. Pearce, D. Vukovic, C. R. Nagarajah, and D. J. Booth, *Meas. Sci. Technol.*, vol. **16**, no. 6, pp. 1299–1304 (2005).
12. S. W. Tyler, J. S. Selker, M. B. Hausner, C. E. Hatch, T. Torgersen, C. E. Thodal, and S. G. Schladow, *Water Resour. Res.*, vol. **45**, W00D23 (2009).
13. K. Suh and C. Lee, *Opt. Lett.*, vol. **33**, no. 16, pp. 1845–1847, Aug. (2008).
14. M. A. Soto, A. Signorini, T. Nannipieri, S. Faralli, G. Bolognini, F. Di Pasquale, *J. Lightwave Technol.*, vol. **30**, no. 8, pp. 1215-1222, Apr. (2012).
15. D. J. Lockwood and L. Pavesi, *Silicon Photonics II: Components and Integration, vol. 119 of Topics* in Applied Physics Springer (2010).

16. A. Signorini, M. A. Soto, S. Faralli, G. Bolognini, F. Di Pasquale, E. Fedeli, E. Marzilli, R. Tarelli, "Progetto e sviluppo di sensori distribuiti di temperatura in fibra ottica per rivelazione di incendi in gallerie ferroviarie", *1° Convegno Nazionale, Sicurezza ed Esercizio Ferroviario*, Roma, March 20, 2009 .

17. M. Nickles, B. Vogel, F. Briffod, S. Grosswig, F. Sauser, S. Luebbecke, A. Bais, T. Pfeiffer, *11ᵗʰ SPIE Annual International Symposium on Smart Structures and Materials,* March 14-18, 2004, San Diego, California, USA, pp. 18-25.

18. M. A. Soto, T. Nannipieri, A. Signorini, G. Bolognini, F. Di Pasquale, A. Lazzeri, F. Baronti, R. Roncella, *IEEE Sensors Conference*, Limerick, October 2011.

19. M. A. Soto, P.K. Sahu, S. Faralli, G. Bolognini, F. Di Pasquale, B. Nebendahl and C. Rueck, *Electron. Lett.*, Vol. **43** No. 16 (2007).

20. J. Park, G. Bolognini, D. Lee, P. Kim, P. Cho, F. Di Pasquale, and N. Park, *IEEE Photon. Tech. Lett.*, **18**, 1879-1881 (2006).

21. M. D. Jones, *IEEE Photon. Technol. Lett.*, 15, 822–824 (1993).

22. M. Nazarathy et al, *J. Lightwave Technol.*, vol. **7**, No. 1, pp. 24-38 (1989).

23. F. Baronti, A. Lazzeri, R. Roncella, R. Saletti, A. Signorini, M. A. Soto, G. Bolognini and F. Di Pasquale, *Electron. Lett.* **46**, 1221-1223 (2010).

24. M. A. Soto, T. Nannipieri, A. Signorini, A. Lazzeri, F. Baronti, R. Roncella, G. Bolognini, F. Di Pasquale, *Optics Lett.*, vol. **36**, Issue 13, pp. 2557-2559 (2011) .

25. M. A. Soto, P. K. Sahu, G. Bolognini, F. Di Pasquale, *IEEE Sens. J.* 8, 225-226 (2008).

26. A. Signorini, S. Faralli, M. A. Soto, G. Sacchi, F. Baronti, R. Barsacchi, A. Lazzeri, R. Roncella, G. Bolognini, F. Di Pasquale, "40 km Long-Range Raman-Based Distributed Temperature Sensor with Meter-Scale Spatial Resolution", *Optical Fiber Communication Conference*, Technical Digest OFC 2010, San Diego, USA, paper OWL2 (2010).

27. M. Webster, L. Raddatz, I. H. White, and D. G. Cunningham, *J. Lightwave Technol.*, vol. **17**, no. 9, pp. 1532-1541 (1999).

# PHOTONICS FOR DETECTION OF CHEMICALS, DRUGS AND EXPLOSIVES


Alessandro Garibbo[1] and Antonio Palucci[2*]

*[1]Selex Elsag S.p.A. – via Pieragostini, 80 – 16151 Genova - Italy*
*[2]ENEA – Frascati -Italy*
**Email: antonio.palucci@enea.it*


Detection of chemicals, drugs and explosives is a relevant task for countering terrorism and organized crime by preventing their effects on airports, land borders, seaports, through the whole supply chain or in mass transit, as in metro or railway stations.

Photonics and spectroscopic techniques are commonly used to retrieve information on atomic and molecular energy levels, molecular geometries, chemical bonds, interactions of molecules, and related processes. For their capacity to investigate at structure of the matter level these technologies are deemed an excellent instrument for safety and security protection, inspection and investigation.

Even if threat is changing continuously and photonic systems may not work properly in certain circumstances, it makes sense to continue to invest in photonics, since photonics remains one of the best means to investigate molecular structures of compounds; for this reason, photonic sensors are nowadays considered as the first detection technology for hazardous materials as explosive compounds.

## 1. Introduction

Photonics and spectroscopic techniques are commonly used to retrieve information on atomic and molecular energy levels, molecular geometries, chemical bonds, interactions of molecules, and related processes. Therefore, spectra are used to identify components of a sample (qualitative analysis) and/or to measure the amount of a target material in a sample (quantitative analysis).

For their capacity to investigate at structure of the matter level, discriminating among different substances and molecular arrangements, these technologies are deemed an excellent instrument for safety and security protection, inspection and investigation.

Photonic sensors can be fruitfully applied to detect high energetic materials (i.e. explosives) as well as chemicals biological agents, drugs and their precursors, because they also offer the most appreciable advantage of high sensitivity without sample preparation.

In some cases, photonics sensors may also be used when target substances are present in traces, dispersed in vapors or even concealed in luggage or under clothes.

The chapter is structured as follows: paragraph 2 presents addressable threats and the need for early detection of chemicals, drugs and explosives; paragraph 3 discusses about detection types and modes; paragraph 4 deals with some chemical and physical properties of chemicals, drugs and explosives making them detectable through photonics; paragraph 5 illustrates some of the most popular and promising detection techniques based on photonics; paragraph 6 presents a synthesis of research programs currently active in this field and paragraph 7 contains the conclusions.

This chapter has been written based only on open sources, i.e. public domain, unclassified, non-sensitive, commonly available knowledge sources.


## 2.  Addressable Threats

Because of their capacity of investigating matter at the structure level, photonics and related technologies can help addressing some of the threats related to biological and chemical warfare agents, explosives and drugs.

The detection of these compounds is a relevant task for countering terrorism and organized crime by preventing their effects in airports, land borders, seaports, through the whole supply chain or in mass transit, as in metro or railway stations. To this respect, photonic sensors can contribute significantly to tackle the possible intrusion of a threat in controlled areas, due to their capability to monitor the intrinsic features of the target substances.

## 2.1. *Biological and Chemical Warfare Agents*

Biological Warfare Agents (BWAs) include any organism or toxin found in nature or modified (i.e. "weaponized") that can be used to incapacitate, kill, or otherwise impede an adversary. Biological weapons are characterized by low visibility, high potency, substantial accessibility, and relatively easy delivery.

Any of a large number of bio agents might be employed, ranging from bacteria such as Bacillus anthracis, Brucella, Yersinia pestis, and Franciscella tularensis to viruses such as Variola and Ebola. Also of concern are a number of bio toxins, including Staphylococcal enterotoxin B and Botulinum. Genetic engineering could vastly increase the number of potential bio-agents.

Chemical Warfare Agents (CWAs) and Toxic Industrial Compounds (TICs) are poisonous vapors, aerosols, liquids and solids that have toxic effects on people, animals or plants. A lethal CWA is designed to kill, injure or incapacitate the enemy, or to deny entering a particular area.

Dissemination and dispersion methods for both BWAs and CWAs are multiple. BWAs and CWAs can be released by bombs or sprayed from aircraft, boats and vehicles.

Some CWAs may come in binary preparation[1], in general packed into ammunitions in which two precursors become mixed to produce the agent just before its use. Because the precursors are generally significantly less hazardous than the CWAs themselves, this method makes handling and transporting the munitions significantly simpler and safer. Nerve Gases (NGs) - actually aerosols or vapors - are highly toxic, chemically stable, and easy to disperse compounds. They produce rapid physiological effects both when absorbed through the skin or through the respiratory tract[2]. NGs are also fairly easy to synthesize, even in binary preparations, and the raw materials (precursors) required for their manufacture (pesticides, insecticides, flame retardants, sterilants for food etc.) are inexpensive and readily – and legally – available in large quantities.

The release of TICs from a chemical facility may be accidental, as happened on December 3 1984, when more than 40 tons of methyl isocyanate gas (a pesticide precursor) leaked from a pesticide plant in Bhopal[3], India, immediately killing at least 3,800 people and causing significant morbidity and premature death for many thousands more, or deliberate, as a consequence of an act of terrorism. This last is one of the most dreadful scenarios in emergency preparedness studies.

BWAs and CWAs are considered "Weapons of Mass-Destruction" (WMD). The 1925 Geneva Protocol prohibits the use of chemical and biological weapons in war[4]. The Biological and Toxin Weapons Convention[5] (BTWC), signed in 1972 and entered into force in 1975, and the Weapons Convention[6] (CWC), signed in 1993, ban the development, stockpiling, production, transfer and use of respectively BWAs and CWAs. However, WMD have been used in war (during the Iran-Iraq war of 1984/88, Iraq made use of CWAs against Iran[7]) and by terrorists.

For this reason, BWAs and CWAs are relevant threats also for Homeland Security (HS).

In July 1993, a liquid suspension of Bacillus anthracis was aerosolized from the roof of an eight-story building in Kameido, Tokyo, Japan, by members of the religious group Aum Shinrikyo (AS). The Kameido incident is the first documented instance of bioterrorism with an aerosol containing Bacillus anthracis[8]. In June 1994, AS members released sarin gas (a NG) from several sites in the Kaichi Heights area in Matsumoto, Nagano, Japan. Eight people were killed and over 200 harmed. On March 20 1995, sarin was introduced into Tokyo's subway system by AS members. The coordinated attack on five trains killed 13 commuters, seriously injured 54 and affected 980 more. Some estimates claim as many as 5,000 people were injured.

In some cases, BWAs dispersion was made through the ordinary postal service. In 2001, beginning on September 18, letters containing Bacillus anthracis spores were mailed to several news media offices and two Democratic U.S. Senators, killing five people and infecting 17 others.

## 2.2. *Explosives*

According to US National Counterterrorism Center 2011 NCTC Report on Terrorism[9], bombing attacks accounts for 40% of terroristic attacks perpetrated worldwide during 2011 (4,150 bombing attacks out of 10,283 attacks) and for 53% of associated deaths (6,724 deaths for bombing out of 12,533 deaths).

Recently, it became quite popular to use various chemical formulations to produce Home-Made Explosives (HMEs), i.e. non-commercial and non-military explosives. People performing such manufacturing are driven by a variety of reasons, ranging from carefree attitude to juvenile delinquency and international terrorism[10]. Many recipes - generally making use of common household chemicals - do exist on publicly available sources, as commercial literature or the web.

By the way, most of these sources are unreliable and treacherous.

Although there are not standard explosives associated with Improvised Explosive Devices (IEDs), HMEs are frequently used.

There are many different scenarios in which conventional bombs or IEDs are used. The main areas are transport means and transport ways, including air transport, railway, sea and river transport, roads, important buildings, hotels, theaters, sport halls and other crowded places[11].

IEDs are frequently used by insurgents in unconventional warfare (e.g. Iraq, Afghanistan); mainly as remote-controlled roadside bombs, often arranged in daisy-chain, in vehicle-borne IEDs and in suicide vests.

IEDs are actually one of the weapons of choice for an opponent who seeks an asymmetric advantage to avoid fighting against conventional strengths. The adversary exploits his use of IEDs mainly for demonstrative purposes. Consequently, IEDs employed by an adversary as a tactical weapon can have strategic effect[12].

In some cases, attacks with explosive devices were performed through the ordinary postal service. On March, 31, 2011, in Leghorn, Italy, a parcel bomb was delivered to the barracks of the Folgore parachute brigade. The brigade's chief of staff, a lieutenant colonel, opened the parcel that immediately exploded. The officer lost three fingers and his sight in one eye. Responsibility for the attack has been claimed by the Informal Anarchist Federation (FAI). On December, 9,

2011, in Rome, another parcel bomb was delivered to the headquarters of Equitalia, the Italian tax collection agency. The bomb exploded in the hands of the director general, who was wounded in the hand and eye.

## 2.3. *Drugs*

Drug trafficking is a global illicit trade involving the cultivation, manufacture, distribution and sale of substances which are subject to drug prohibition laws[13].

The three major international drug control treaties, the Single Convention on Narcotic Drugs of 1961 (as amended in 1972), the Convention on Psychotropic Substances of 1971, and the United Nations Convention against Illicit Traffic in Narcotic Drugs and Psychotropic Substances of 1988, are mutually supportive and complementary.

The 1988 United Nations Convention against Illicit Traffic in Narcotic Drugs and Psychotropic Substances (UN_Conv) extends the control regime to precursors.

In recent years it has become increasingly evident that terrorism and drug trafficking are intertwined. The terms "narco-terrorism" and "narco-terrorists" have started being used to describe this interface between terrorist organizations and narcotics smugglers. The UN_Conv recognizes the links between illicit drug traffic and other organized criminal activities which undermine the stability, security and legitimacy of sovereign states.

Customs services have found that drugs are most often smuggled in suitcases and bags with false bottoms. Drugs are also transported inside various household items, vehicles or freight containers and are often hidden in clothes and mailed inside letters, packages and parcels.

## 3. Detection Types and Modes

Depending on the types of substances to detect, their accessibility (a legal check-point control is something dramatically different from a collection of intelligence elements performed during a covert operation), operating constraints (e.g. the risk of a contamination or of an explosion), available technologies and other circumstances, two main detection types and three main detection modes may be chosen. Types and modes may be combined together in multiple arrangements.

### 3.1. *Bulk Detection (type)*

Bulk detection looks for the explosive (or for the drug) itself in the explosive device (or in the drug package) and requires large amounts of explosive (or drug) to properly work. Bulk detection is generally performed through X-Rays, beams of neutrons and Tera Hertz at a distance from the target ranging from some millimeters (point detection) to some tenth of meters (stand-off detection).

### 3.2. *Trace Detection (type)*

Trace detection looks for small quantities needs trace amounts of substances (drugs, chemicals, explosives, bacteria, etc.) in gas phase (target substance vapor pressure is a crucial issue) or in the form of particles. Trace detection is generally performed through photonic sensors (including LIDARs), olfactory sensors and canines at a distance from the target ranging from some millimeters (point detection) to some kilometers (stand-off detection).

### 3.3. *Point Detection (mode)*

Point detection occurs when the detection instrument is located very close (generally at a distance ranging from some millimeter to some centimeters) to the target substance (even in particulate or in vapor phase) or to the object to investigate.

### 3.4. *Stand-Off Detection (mode)*

Stand-off detection occurs when the detector is located at a safety distance from the target substance (even in particulate or in vapor phase) or to the object to investigate. The concept, and measure, of safety distance is heavily dependent on the specific application and detection scenario.

## 3.5. *Distributed Detection (mode)*

Multiple trace detection devices (in principle, trace detectors) may be distributed or spread over an area to be monitored and controlled (for instance, a metro station) and collectively managed by a common communication platform (sensor network), either wired or wireless.

As an alternative to a wireless sensor network, Viola et al.[14] proposed an architecture based on a single hollow optical fiber to detect nerve gases in an indoor environment, such as a metro station. The research was made in the framework of the "SENSEFIB" Corporate Project of Finmeccanica S.p.A.. The technical demonstrator is based on a single fiber, equipped with multiple sampling holes and somehow deployed inside a metro station. An IR laser source and an IR detector along with a sample pump (allowing air continuously flowing into the hollow fiber) are located at the fiber extremities. The line sensor is constituted by a widely tunable External Cavity Quantum Cascade Laser (EC-QCL), coupled to IR thermoelectrically cooled Mercury Cadmium Telluride (MCT) fast detectors by means of an Infrared Hollow Core Fibers (HCF). The detector measures the radiation emitted by the IR source after having travelled through an absorption cell, where it has been partially absorbed by the target compound. In this scheme, the quantum cascade laser and the HCF are the IR source and the absorption cell, respectively.

Absorption is wavelength dependent and compound specific. Traces of target compounds are thus revealed from the recognition of their spectral fingerprint.

## 4. Chemistry and Physics of Sensing

Photonics or spectroscopic techniques – also including the ones based on Tera Hertz - are commonly used to retrieve information on atomic and molecular energy levels, molecular geometries, chemical bonds, interactions of molecules, and related processes. Therefore, spectra are used to identify components of a sample (qualitative analysis) and/or to measure the amount of a target material in a sample (quantitative analysis).

Respect to other kind of sensors for BWAs, CWAs, explosive and drugs, photonic ones exhibit some attractive advantages such as immunity to electromagnetic interference and capability to support stand-off operations.

Photonic sensors can be of several different designs (to utilize different components of the optical signal) such as intensity based, interferometric, polarization, spectroscopic, pulse shape or arrival time based. The most appealing sensors adopt on board laser technologies due to the intrinsic features to be high selectivity in the excitation mode, high power to be delivered to the target and capability to operate in stand-off mode. Conversely, in-situ spectroscopic detectors can be very compact, being matched to fiber optic technology offering the capability to develop miniaturized photonic sensors that can answer to the request to be miniaturized, portable, user-friendly and as robust as to be operated in different harsh environments.

In any case, interrogation of different types of chemical hazardous materials requires ascertaining the unknown chemical substances' identity followed by the institution of appropriate countermeasures against the potential harm/toxicity of the substance. New emerging challenges from organized crime and terrorist groups underline the vital need for sensitive and selective (i.e. yielding virtually no false-positive nor false-negative results) rapid identification of substances while adapting a non-contact approach to the hazard.

These challenges do often require identifying concealed hazards without opening a suspected packaging in order to minimize harmful exposures.


### 4.1. *Biological and Chemical Warfare Agents*

CWAs – and a relevant number of TICs - are organic molecules containing a few non-metallic atoms, typically P, S, Cl, F, As. This implies the relevance of spectroscopy techniques for detection.

The deadliest form of a chemical or biological attack is aerosolized agents dispersed into the atmosphere. An aerosol attack would work optimally as a fine mist of 1-5 µm-sized particles, since particles in this size range find optimum inhalation and retention[15]. As a matter of fact, the same size range, considered as a range for wavelengths, belongs to infrared (IR) spectrum. This implies the relevance of LIDARS for direct detection[16]. On the other hand, when ultraviolet (UV) radiation is used as illumination source, the radiation may induce fluorescence from aerosolized material within the light beam path. This induced fluorescence may indicate that a cloud is biological in its nature.

Several point and stand-off detection technologies, covering a broad region of the electromagnetic spectrum are applicable for detection of BWAs and/or CWAs. These technologies include spectrally resolved Ultraviolet Laser Induced Fluorescence (UV-LIF) at several different excitation wavelengths, Infrared Depolarization, Long-Wave Infrared (LWIR) Differential Scattering (DISC)[17] and Remote Raman Spectroscopy[18].

Given the presence of background aerosols, all such techniques are subject to false alarms, and given various limitations in atmospheric propagation (attenuation and scattering), these are invariably range-limited. An additional limitation on optical sensors is their inability to discriminate live versus dead microbes, thus opening the potential for false alarms.

Detection in general is difficult owing to natural clutter and anticipated low concentrations of subject material. Typical detection architectures include a nonspecific early-warning trigger, a rapid identifier, and a confirming step, often in a laboratory. Modern technologies include the use of elastic scatter and ultraviolet laser-induced fluorescence for triggering and standoff detection. Standoff techniques are currently limited to LIDARS that interrogate clouds of particles.

### 4.2. *Explosives*

Explosive substances are rather simple molecules, virtually all of them containing H, C, O and N. Organic peroxide explosives, as the TATP, are the only explosives not containing N. Other explosives may contain, in addition: Ba, B, Fe, Al, F, Si, P[19]. This implies, in principle, the relevance of spectroscopy techniques for detection.

The majority of high-explosive formulations use inorganic or organic nitrate or nitro functional groups as the oxidant. The correlation between nitrogen and oxygen is roughly linear. The nitrogen content of a wide variety of nitrogen-containing explosives is ranging between 20% and 43%, while the oxygen content ranges between 37% and 53%. This suggest that dual analysis of the oxygen and nitrogen content might provide a more reliable indication of high explosives than techniques based on nitrogen content alone, even though characterization of HMEs is very difficult a task, due to the broad variety of mixed substances and the often uncontrolled stoichiometry of their mixtures due to homemade, likely coarse and inaccurate, often incomplete fabrication process. As a matter of fact, this kind of dual analysis does not work for nitrogen-free explosives, as the TATP[20].

Molecular spectroscopic techniques can be used to uniquely identify molecules in the vapor phase, but the low vapor pressure of many explosives (see Table 1 for some meaningful examples) limits their use. Moreover, explosives are typically deployed in powdered polycrystalline form and dispersed in a binder material that further lowers vapor pressure. This requires very high sensitive detection methodologies and the capability of discriminating explosives related fingerprints from the ones due to impurities, additives and binders. On the other hand, valuable forensic and intelligence information may be gained by the identification of possible precursor impurities in explosive samples.

In an effort to deter terrorism, The Montreal Convention of 1991[21] (M_Conv) mandated that complying nations ensure that all plastic explosives manufactured within their borders include chemical markers, or "taggants". The M_Conv Technical Annex specifies explosives that must be marked as those "formulated with one or more high explosives which in their pure form have a vapor pressure less than $10^{-4}$ Pa (equivalent to $10^{-6}$ mbar) at a temperature of $25°$ C." A substance with low vapor pressure is one that has low volatility, or likelihood to exist in vapor form. The M_Conv states that detection taggants "are intended to be used to enhance the detectability of explosives by vapor detection means". The addition of these volatile compounds to explosives by manufacturers is meant to increase the ability to detect explosives with intrinsic low vapor pressure.

Table 1. Chemical characteristics of some explosives[22,23,24].

| Sample | Molecular weight ($g*mol^{-1}$) | Vapor concentration (mbar; @ 20° C) | Formula | Structure |
|---|---|---|---|---|
| EGDN | 152.06 | $4.8\ 10^{-2}$ | $C_2H_4O_4N_2$ |  |
| NG | 152.06 | $3.16\ 10^{-4}$ | $C_3H_5N_3O_9$ | 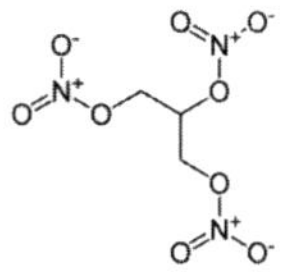 |
| RDX | 222.10 | $0.9\ 10^{-9}$ | $C_3H_6N_6O_6$ | 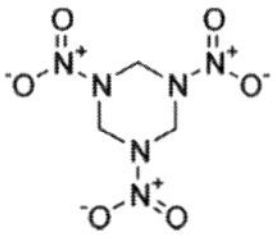 |
| TNT | 227.14 | $4.4\ 10^{-6}$ | $C_7H_5N_2O_8$ | 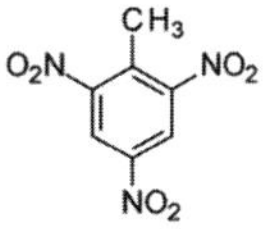 |
| PETN | 316.137 | $2.1\ 10^{-9}$ | $C_5H_8N_4O_{12}$ | 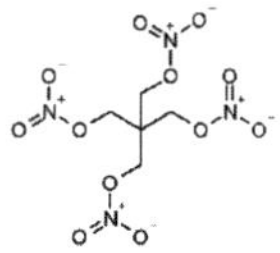 |
| HMX | 296.156 | $1.5\ 10^{-14}$ | $C_4H_8N_8O_8$ | 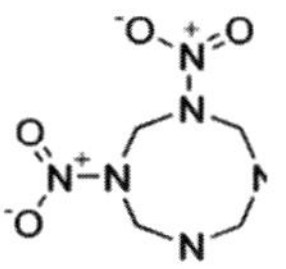 |
| TETRYL | 287.015 | $1.5\ 10^{-9}$ | $C_7H_5N_5O_8$ | 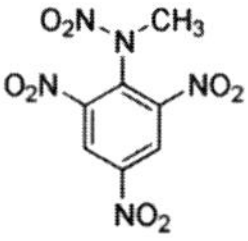 |
| TATP | 222.24 | $6.6\ 10^{-2}$ | $C_9H_{18}O_6$ | 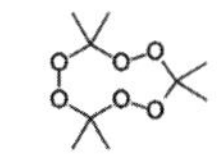 |
| Ammonium Nitrate | 80.04 | $1.3\ 10^{-8}$ | $NH_4NO_3$ | 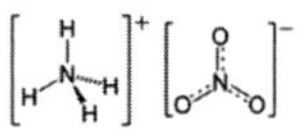 |

Testing for both the taggants and the actual explosive through vapor detection raises the probability of detecting the taggant or the explosive. However, there are explosives not containing taggants, including HMEs for IEDs and explosives manufactured by non-compliant countries or terrorist groups. Photonics technologies as Cavity Ring Down Spectroscopy (CRDS), Laser PhotoAcoustic Spectroscopy (LPAS) or Surface Enhanced Raman Spectroscopy (SERS) allow detecting at trace level those substances.

## 4.3. *Drugs*

Whether plant-based, such as heroin, cocaine or cannabis, or synthetic, such as the various amphetamines, illicit drugs are normally complex mixtures which rarely contain the drug alone[25]. As a consequence of the crude clandestine laboratory conditions under which they are produced, their chemical composition shows large variability. As well as containing the drug itself, samples may contain one or more of the three different types of key components: natural components (e.g. opium), by-products related to manufacturing (e.g. solvents) and cutting agents (e.g. benzocaine, a pharmaceutical product commonly used by dentists, used for cutting cocaine).

Although all samples of the same drug prepared in the same way could be expected to contain the same impurities (excluding cutting agents which may be added at any stage in the distribution chain), their relative concentrations may show large variations. Frequently, drugs are shipped mixed with large quantities other chemicals (e.g. tocopherol), intended to confound chemical detection by adding spectral noise.

Illicit drugs (e.g. morphine, cocaine, heroin, metamphetamine) are, by themselves, rather simple molecules containing H, C, O, N and Cl (only in chlorinated compounds); benzenic and/or heterocyclic rings are often present. Some of these molecules are polycyclic. This would suggest using spectroscopy-related techniques for detection, given the peculiar, well-known spectroscopic features of these molecules. However, the high level of spectral noise added by the presence of various by-products and different cutting agents makes spectroscopy a more viable technique for investigative and forensic purposes than for check-point inspection.

In fact, customs and check-point inspections are still performed using dogs or visually, or using imaging tools, especially those based on millimetric or Tera Hertz waves[26,27].

## 5. Sniffing Techniques: Local and Stand-off

Some of the most popular and promising detection techniques based on photonics are summarized in the following.

### 5.1. *Laser PhotoAcoustic Spectroscopy*

Photoacoustic (PA) effect, occurring even in solid compounds, has been known since 1880, when A. G. Bell[28] observed the audible sound produced by chopped sunlight impinging an optically absorbing material. A new impulse to this technique arrived after the sixties with the application of tunable laser devices as the $CO_2$ laser in 1968, due to the high beam quality profile, narrow spectral bandwidth and high power emission of laser with respect to the conventional light sources for spectroscopy. Because of the high power emitted (order of watts) and its tunability on strong fundamental vibrational transition lines, it turned out to be ideal source to push the sensitivity of PA detection in the ppb (parts per billion) range or even below.

A classic experimental set-up in LPAS technique is shown in Fig. 1, where the laser source is frequency stabilized in the single emission IR mode along the 9-11 μm wavelength range. The signal collected by a high sensitive microphone is synchronized by a lock-in amplifier.

Giubileo et al.[29] reported the detection of photo-acoustic spectroscopic characterization and identification of some classical explosives in solid phase (2,4-DNT; 2,6-DNT; HMX; TATP; PETN), at CO2 laser wavelengths (9.2 – 10.8 μm). As it can be observed in Fig. 2 - the three samples showing complex absorption spectra - it is quite difficult extracting spectral signatures to be unique for each sample.

In order to perform an unambiguous recognition of trace explosive compounds, a chemometric approach was suggested by the authors and applied to experimental results as a whole. As a matter of fact, in Fig. 3

the first two Principal Components are reported, thus allowing discriminating among the ten samples investigated.

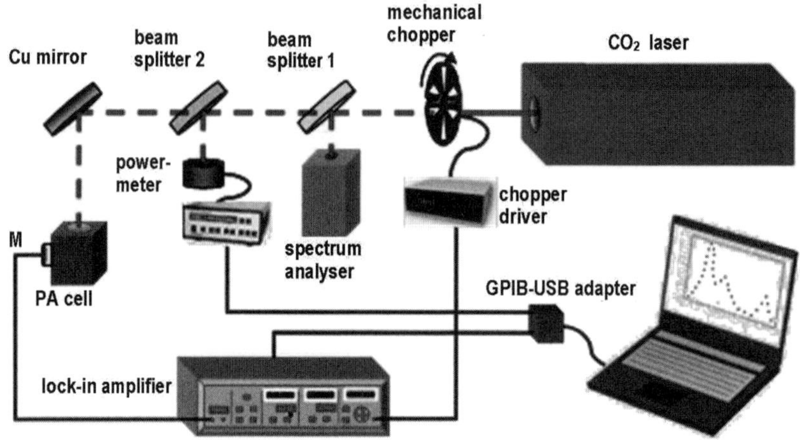


Fig. 1. Typical LPAS experimental set-up.

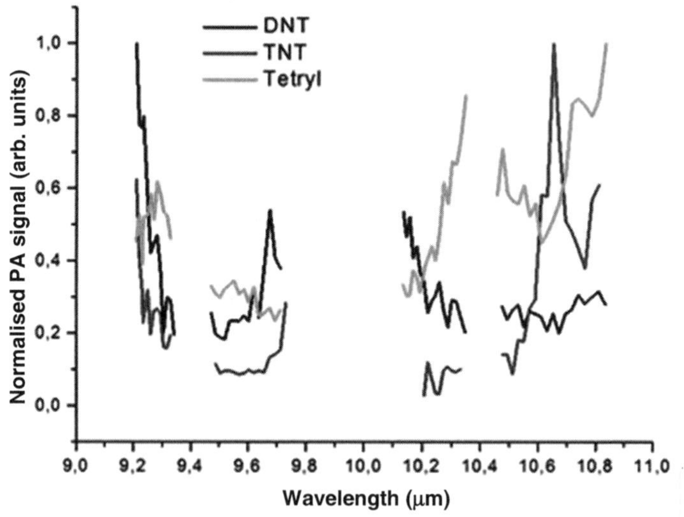


Fig. 2. Normalized LPAS spectra measured on DNT, TNT and Tetryl samples.

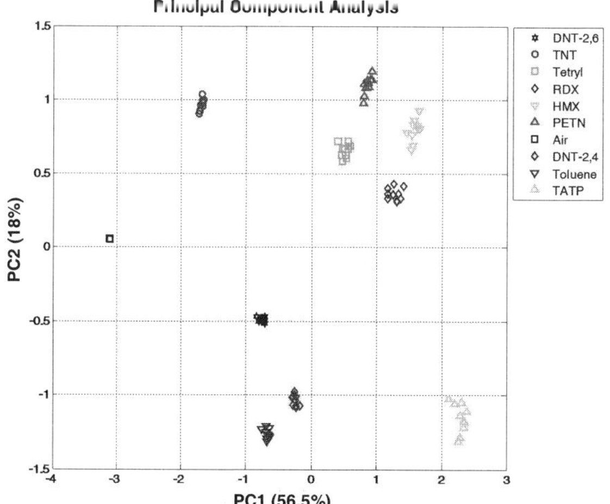

Fig. 3. 2D scatter plot of the photoacoustic measurements.

## 5.2. *Laser Induced Breakdown Spectroscopy*

Plasma generated by intense laser pulse leads to the atomization and ionization of sample material, which temperature can exceed 10.000 K comprises in its fast decay the spectral emissions from the excited species that in turns can be fruitfully used for identification and possibly quantification of the sample composition. LIBS technique is ultra-rapid, does not require sample preparation and therefore is applicable for in-situ analysis, also at high repetition rate (KHz range). This allows analyzing even rapidly moving objects, as is the case for objects transported by a belt conveyor.

A LIBS apparatus is typically formed by a pulsed solid state laser tightly focused on the target. The emitted radiation collected both on-axis and off-axis is spectrally dispersed by a high resolution (monochromatic) device, as shown in Fig. 4. Ad-hoc optical ray-tracing allows to optical match these few optoelectronics components and to optimize the detection capabilities. The few blocks of components allow for the development of compact, miniaturized and ruggedized apparata for field operation.

High energetic materials contain mainly C, H, and O, while N is also present in almost all the high explosives. Specifically, explosives are rich in N and O, and poor in H and C with respect to other organic substances. LIBS spectra from energetic materials normally contain atomic lines from these four elements and molecular bands of CN and $C_2$, since molecular emission can be attributed to both the native C=C and C-N bonds.

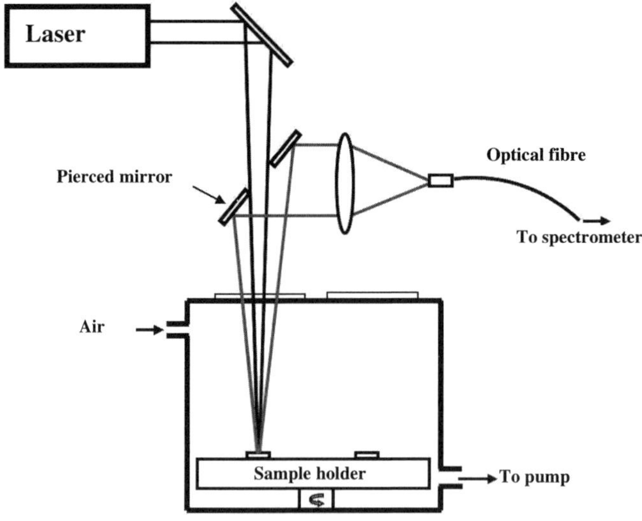


Fig. 4. Schematic layout of a LIBS apparatus in coaxial configuration (transmission and collecting optical paths are on the same axis).

As an example, a low resolution spectrum of TNT sample is shown in Fig. 5 along with the indication of main explosive molecular bands and the presence of inorganic interferences from the substrate where the sample was deposited[30].

One of the main drawbacks in the rapid LIBS detection of energetic materials, when performed in air surrounding, is the interference of air components on the LIBS spectra.

Therefore, the approach to use common spectral ratios of oxygen and nitrogen relative to carbon and hydrogen or other spectral algorithms often fails, thus to require a procedure for recognition of explosives based on properly constructed algorithms or on chemometrics[31].

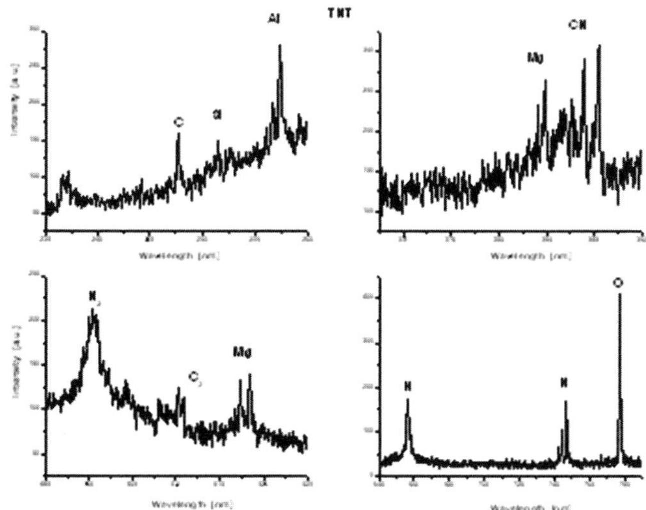

Fig. 5. Characteristic spectral parts of TNT traces on aluminum support.

LIBS technique has already been exploited for explosive detection form in-situ to far field applications as reported by Gottfried et al.[32]. Laserna and coauthors demonstrated the capability to detect explosive compounds up to 100 m distance[33]. Conversely, the main concern remains in the implementation of LIBS technique in civil scenario due to the severe restrictions for eye safe lasers.

Nonetheless, this technique remains appealing for the possibility to improve optical performances with the application of new and compact laser sources with short pulse emission (femto-second regime).

## 5.3. *Laser Raman Spectroscopy*

The spectral fingerprints are unique features of each molecule, due to its internal atomic bonds and molecular structure (atomic bonds, bonds strengths, masses and stereoscopic arrangement).

Assignation of vibrational modes to precise emission frequencies is therefore a peculiar belonging of distinguished molecules. In case of molecules exhibiting dipole moment's changes during vibrational motions, Raman technique do apply.

With Raman spectroscopy, a pulsed laser beam is focused on the sample of interest and the reflected light is collected by the instrument. While most of the light is reflected at the same wavelength as the original laser (called Rayleigh scatter), a small portion of the light is scattered at different wavelengths in a very specific manner (called Raman scatter), based on the molecular composition of the chemical (either solid or liquid). For just a small portion of the light is reflected inelastically (i.e. Raman scattered), one of the drawbacks of this technique is the low cross section of the process, which is in the order of $10^{-7}$ to $10^{-8}$ cm$^2$/sr.

The second drawback is the frequency domain of this spectrum being very close to the laser excitation, thus to require very efficient cut-off filters allowing eliminating all the "tail" of the laser but allowing as much as possible of the Raman emission spectrum to pass.

The third drawback of this technique is a consequence of the first one, thus the necessity to operate with not so low a level of laser energy emission. In turn, the main advantages are the possibility of real time recognition of the investigated target and the possibility to operate in stand-off mode[34], even through semi-transparent and translucent sealed containers. Even if different instruments based on Raman technique are already available on the market (Ahura Scientific, B&W Tek, DeltaNu, Enwave Optronics, InPhotonics, Ocean Optics, PerkinElmer, Raman Systems, Renishaw, Senspex, Smiths Detection, etc.) for local or contact detection, very few are presently the realizations in the remote sensing application (LIDAR Raman) and all of them in prototypal state at research institutions laboratories.

Raman technique also applies for contact and for stand-off applications, with different performances ranging from bulk to trace level detection capabilities.

A typical LIDAR Raman apparatus for stand-off detection is depicted in Fig. 6, while in Fig. 7 is shown the spectrum of Raman emission from a sample of ammonium nitrate, together with the assignation of the characteristic emission bands.

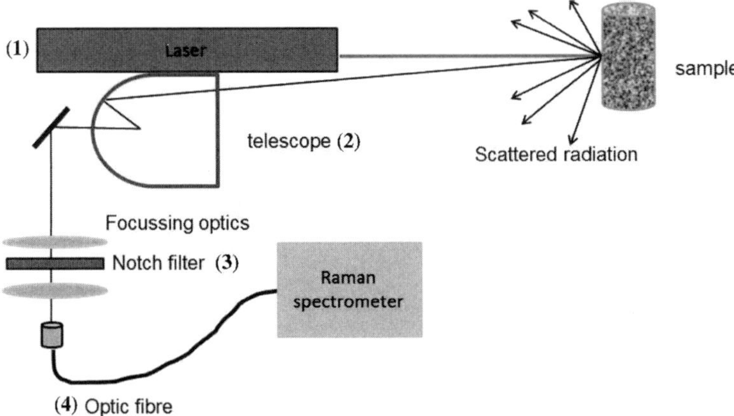


Fig. 6. LIDAR Raman apparatus: (1) solid state laser transmitter; (2) receiving telescope; (3) cut-off (notch) filter; (4) fiber optic.

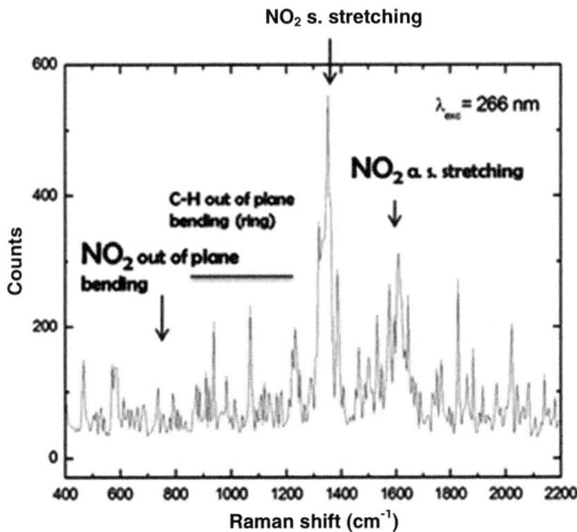


Fig. 7. Ammonium nitrate spectrum collected with a Lidar Raman apparatus.

Different realizations are reported in the literature[35,36,37,38,39].

Due to the very low cross section of Raman emission, in general, but more critically for explosive compounds, the total efficiency of the stand-off system becomes a strong concern in order to design the apparatus. In particular, not only the cut-off filter plays a strong role in signal collection, but all the optical matching among the spot size to the target and the collecting elements (telescope, fiber optics and monochromator's slit) have to be considered implementing a ray tracing optical design.

The main characteristics of stand-off research apparata are compared in terms of maximum distance reached, laser wavelength for excitation and respective energy per pulse emitted.

Table 2. Comparison among stand-off LIDAR apparata.

| Explosive (or precursor/surrogate) | Stand-off distance [m] | Laser wavelength [nm] | Energy [mJ] | Repetition rate [Hz] | Year |
|---|---|---|---|---|---|
| $NH_4NO_3$-$KNO_3$-$KClO_4$ | 50-100 | 532 | 600-6000 | 20 | 2011 |
| NM-TAPT-MEKP-MTD | 20-55 | 532 | 280-450 | 10 | 2008 |
| TAPT-UN-PETN-RDX-TNT | 30 | 248 – 266 – 355 | 20 | 13 | 2008 |
| DEM-Mes | 1.5-2.2 | 228 | 20 | 25 | 2003 |

The laser power necessary to Raman scatter to be effective becomes a strong limitation for apparata to be implemented for use in non-military environments, particularly if we are interested in scanning or pointing to persons. The severe European regulation on eye safe dramatically limits the dose of impinging laser light to the persons IEC 60825 (2001)[40]. This concern applies also in case of similar techniques as LIBS where intense laser dose are released in a very small but concentrated area.

Nonetheless, it's important to underline the potentiality of this technique that relies also in the merging possibility with other photonics techniques in the same apparatus as a LIBS or a LIF, in order to retrieve complementary information from the investigated target.

The in-situ or contact version of Raman technique finds in the market different devices for bulk detection, thus limiting the possibility of detection to large quantities of materials. Conversely, the new trend in this technique is the development of plasmonic devices (metal nanostructures) that enhance Raman emission, thus raising detection limits. The Surface-Enhanced Raman Scattering (SERS) effect, described more than 30 years ago[11], allows scaling the detection limit down to very low levels. It has been observed that the Raman scattering signal is strongly enhanced when the molecule is dispersed over a metal substrate with a nano-scale corrugated surface. Strong electromagnetic fields are generated when the incident light (typically CW near IR laser of few mW power emission), striking on metal nanostructures, excites the collective motion of conduction electrons (localized surface plasmons). When Raman scattering is subject to this intensified field, magnitude of induced dipole increases, thus enlarging the cross-section for inelastic scattering. The surface enhancement effect is maximized when the localized surface plasmon resonance frequency is the same of the excitation laser.

In Fig. 8 is shown a commercial metallic substrate for SERS application. The SERS substrates are gold-coated Si wafers with an array of hollowed inward (i.e. concave) pyramids.

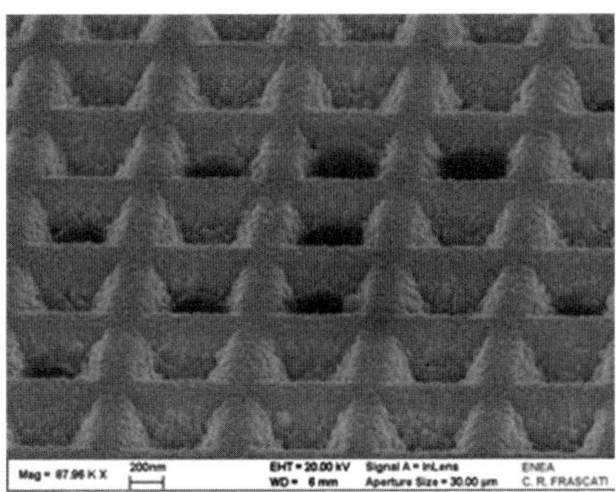


Fig. 8. Commercial metallic substrate for SERS application [Mesophotonics, Klarite, D3 Technologies Ltd].

The regularity of the structure guarantees the uniformity of the Raman signal enhancement without "hot spot" phenomena, while the pyramid walls offer a large space for analyte adsorption.

The potential of SERS spectroscopy for trace analysis has been explored actively in the recent years[42] and several methods have been developed to produce substrates with ordered and uniform nanostructures from different materials.

                         *A. Garibbo and A. Palucci*

As an example, in Fig. 9 are reported different SERS spectra belonging to TNT solution in acetonitrile at different acquisition times, where the explosive molecule can be identified by the symmetric and the asymmetric $NO_2$ stretching vibrations, observed at 1340 to 1360 $cm^{-1}$ and at 1535 to 1620 $cm^{-1}$.

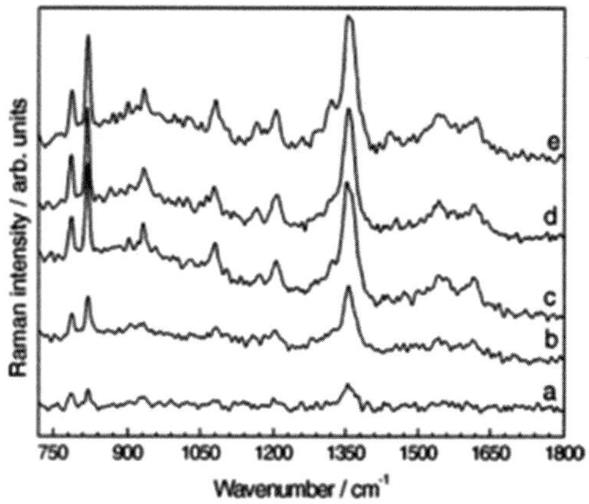


Fig. 9. Surface-enhanced Raman spectra of TNT. The TNT mass probed by laser is 200 pg. The acquisition times are 1 s (curve a), 3 s (curve b), 5 s (curve c), 7 s (curve d), 10 s (curve e).

An enhancement factor of $10^4$ of nano-scale textured substrate, with respect to the conventional Raman technique performance, can be roughly estimated by the ratio between the minimum quantity that can be detected on the latter (2 μg) and 200 pg. The surface enhancement is so high that it is still possible to identify the TNT by the presence of the $NO_2$ scissoring and stretching modes in the spectrum accumulated just in 1 s.

Similar results has been obtained also for NG and TATP detected down to 400 pg, thus allowing stating the strong potentiality of this technique to be suitable for detection of explosives at the trace level, as required in a practical detection instrument for routine analysis of security and forensic diagnostics.

## 5.4. *LIDAR/DIAL Technique*

The optical remote sensing technique, universally known as LIDAR from the sixties of XX Century[43], still remains a valid tool for real time remote sensing of suspicious compounds without contact with the sample. This peculiarity became relevant in Security applications in respect to the detection of peculiar compounds used in preparation of IEDs inside or outside the well-known bomb factories.

A LIDAR (LIght Detection And Ranging) system is essentially a laser radar[44]. In particular, DIAL (DIfferential Absorption Lidar) is a well-established sensor for the detection of range resolved profiles of atmospheric trace gases, both of natural and anthropic origin[45]. This technique is based on the detection of the backscattered photons from laser pulses transmitted at two different wavelengths (ON and OFF) and displayed general advantages in comparison to in situ sensors, namely:

- Single-ended detection of the number density of the targeted compound in range resolved profiles (range resolution of about 1 m),
- Continuous detection in time, allowing the assessment of the temporal variation (time resolution of few s),
- Probe-less measurement, thus eliminating the possibility of modifying the sample,
- Flexibility of probing direction, since the latter does not depend on natural illumination sources.

These advantages allow localizing remotely a vapor source and monitoring its activity leading up to several kilometers away from the instrument. In addition, the DIAL technique is self-calibrated, once the absorption spectrum of the targeted compound has been characterized in laboratory. Acetone, nitrimethane and hydrogen peroxide are the main components employed in IED preparation; therefore the selection of the ON and OFF transmission wavelengths is the main concern in designing a LIDAR/DIAL apparatus, in particular to avoid the strong atmospheric water vapor absorption (Fig. 10).

Their comparison with the atmospheric transmittance shows that the best spectral bands for the simultaneous detection of the three precursors are found around 3.5 and 7.5 $\mu$m[46].

**Gas phase IR spectra**

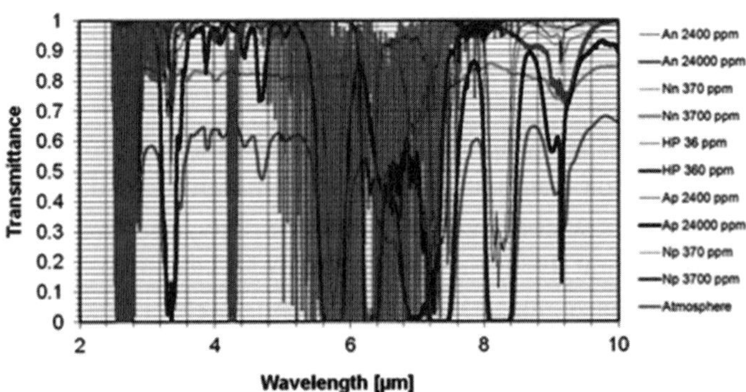


Fig. 10. Transmittance, along an optical path length of 1 m, of: acetone (A), nitromethane (N), hydrogen peroxide (HP) – in concentrations corresponding to 1 and 10% of the vapor pressure – and the atmosphere (U.S. Standard Atmosphere) between 2 and 10 µm.

The DIAL apparatus comprises a pulsed laser that can operate at the ON and OFF wavelengths. Having in mind the need for a tunability range in the mid IR range, the relatively high energy of the pulses and good beam quality, an Optical Parametric Oscillator (OPO) Laser, which can transmit 12.5 mJ at 3300 nm, has been selected and tested at laboratory stage (Fig. 11).

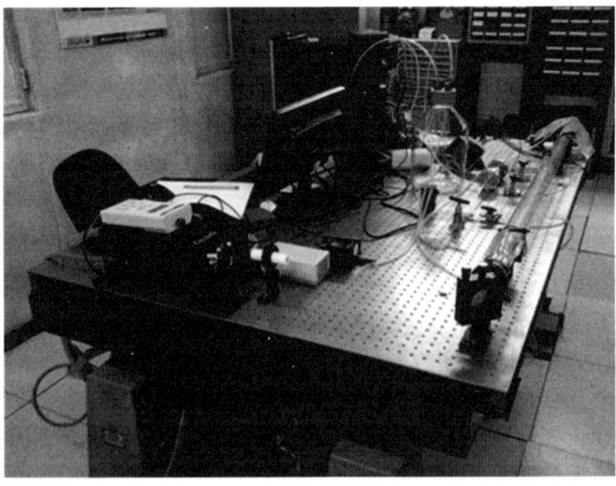

Fig. 11. Table top DIAL arrangement for laboratory tests on selected gases.

## 6. Research Trends

After the 9-11 events, the first action to counteract the terrorism in the USA was the creation, in 2003, of the Department of Homeland Security. Successively, in 2004, the EC realized the need to enhance citizen's security throughout EU countries launching the PASR (Preparatory Action to Security Research) resulted in the more robust and settled Security Program Call in the Seventh Framework Programme (FP7).

Among the four missions selected, the Security for Citizen was addressed to enhance prevention, detection and explosive detection, thus supporting the development of new and more effective instruments.

In Table 3 are reported the projects that consider the development of optical sensors for explosive and their precursors detection. In turn, this mirrors that optoelectronics devices are the state of the art for detection capabilities that needs to be specifically designed for the need and operational field where to be applied.

Table 3. List of European projects funded under the Security Call with the main goal to implement optical devices for explosive detection.

| Project | Photonic technique | Components |
| --- | --- | --- |
| ISOTREX FP6 | LIBS, CRDS, LPAS, Raman | Explosives |
| CUSTOM FP7 | LPAS | Drugs |
| LOTUS FP7 | Multipass Infra Red Absorption | Explosive precursors and compounds |
| BONAS FP7 | Lidar/Dial, SERS, QPAS | Explosive precursors and compounds |
| OPTIX FP7 | LIBS, Raman, IR | Explosives |
| EMPHASIS FP7 | Raman | Explosive precursors and compounds |
| FORLAB FP7 | LIBS, Raman, LIF | Explosives |
| DIRAC PF7 | IR Absorption Spectroscopy | Drugs |

## 7. Conclusions

Photonics or spectroscopic techniques are commonly used to retrieve information on atomic and molecular energy levels, molecular geometries, chemical bonds, interactions of molecules, and related processes. Spectra are used to identify components of a sample (qualitative analysis) and/or to measure the amount of a target material in a sample (quantitative analysis).

Depending on nature of threat and on operating requirements photonics allows detection both in stand-off and point or "in-situ" modes.

But two caveats are in order. First, there is no single "silver bullet", i.e. a sensor, or a system, of whatever nature, able to detect, at any time and in any circumstances (including any humidity, wind and weather conditions), any kind of threat in any operational situation in any type of environment. Second, what works properly in a laboratory, or even in a significant test-bed, no matter about how much complex and "realistic" this latter would be, not necessarily works in the real world. This depends on multiple factors, both technical (including environmental) and human.

Even if threat is changing continuously and photonic systems may not work properly in certain circumstances, it makes sense continuing investing in photonics, since photonics remains one of the best means to investigate molecular structures of compounds; for this reason photonic sensors and Tera Hertz equipment are nowadays considered as the first detection technology for hazardous materials as explosive compounds.

## Acknowledgments

The authors acknowledge the financial support of the EC for the activities carried on in the framework of the project ISOTREX (Integrated System for On-line Trace EXplosives detection in solid and vapor states) PASR-2006 Contract SEC6-PR-203600 and the project BONAS (BOmb factory detection by Networks of Advanced Sensors SEC-2010.1.3-3 contract N°: 261685).

The authors also wish to warmly acknowledge:

- The whole staff of the ENEA UTAPRAD-DIM Laboratory for their valuable support in the experimental activities,
- The whole Mindsh@re community of Finmeccanica, along with the SERIT platform members, as a valuable source of knowledge,
- Professor Antonello Cutolo, for his restless activity and personal committment in fostering research and innovation.

# References

1. The Militarily Critical Technologies List Part II: Weapons of Mass Destruction Technologies (ADA 330102), Chemical Weapons Technology - U.S. Department of Defense, Office of the Under Secretary of Defense for Acquisition and Technology, February 1998.
2. S. Franke, Manual of Military Chemistry Volume I-Chemistry of Chemical Warfare Agents, ACSI-J-3890, Chemie der Kampfstoffe, East Berlin, UNCLASSIFIED Technical Manual (AD849866), April 1968.
3. E. Broughton, The Bhopal disaster and its aftermath: a review, Environmental Health: A Global Access Science Source 2005, 4:6 doi:10.1186/1476-069X-4-6, http://www.ehjournal.net/content/4/1/6.
4. http://www.un.org/disarmament/WMD/Bio/pdf/Status_Protocol.pdf.
5. http://www.opbw.org/convention/documents/btwctext.pdf.
6. http://www.opcw.org/chemical-weapons-convention/.
7. http://www.un.org/Depts/unmovic/new/documents/technical_documents/s-2006-701-munitions.pdf.
8. H. Takashi, P. Keim, A.F. Kaufmann, C. Keys, K.L. Smith, K. Taniguchi, et al. "Bacillus anthracis bioterrorism incident", Kameido, Tokyo, 1993. *Emerg. Infect. Dis.*, Jan. 2004, available from: URL: http://wwwnc.cdc.gov/eid/article/10/1/03-0238_article.htm.
9. http://www.nctc.gov/docs/2011_NCTC_Annual_Report_Final.pdf.
10. M. H. Lefevre, "Chemical and Physical Characterization of Non-regular Explosives", NATO International Symposium on Countering Improvised Explosive Devices (IED), Madrid, October 2004.
11. P. Mostak and M. Stancl, "Methods of Mitigation of IED Blast Effect, NATO International Symposium on Countering Improvised Explosive Devices (IED)", Madrid, October 2004.
12. NATO STANAG 2295 Allied Joint Doctrine for Countering Improvised Explosive Devices AJP-3.15 (B).

13. United Nations Office on Drugs and Crime (UNODC): http://www.unodc.org/unodc/en/drug-trafficking/index.html.

14. R. Viola, N. Liberatore, D. Luciani, S. Mengali and L. Pierno, "Distributed nerve gases sensor based on IR absorption in hollow optical fiber", Optics and Photonics for Counterterrorism and Crime Fighting VI and Optical Materials in Defence Systems - Technology VII, edited by Colin Lewis, Douglas Burgess, Roberto Zamboni, François Kajzar, Emily M. Heckman, Proc. of SPIE Vol. 7838, 78380H· © SPIE· CCC code: 0277-786X/10/$18 · doi: 10.1117/12.865009 (2010).

15. C. J. Peters, R. Spertzel, and W. Patrick, III, "Aerosol technology and biological weapons", in Biological Threats and Terrorism: Assessing the Science and Response Capabilities.Washington, D.C.: National Academy Press, pp. 66–77, (2002).

16. D. P. Greenwood, et al., "Optical Techniques for Detecting and Identifying Biological-Warfare Agents", Proceedings of the IEEE 97.6: 971-989© IEEE, (2009).

17. NATO Report RTO-TR-SET-098 Laser Based Stand-Off detection of Biological Agents.

18. E. C. Cull, M. E. Gehm, B. D. Guenther and D. J. Brady, "Standoff Raman spectroscopy system for remote chemical detection, Chemical and Biological Sensors for Industrial and Environmental Security, edited by Arthur J. Sedlacek, Steven D. Christesen, Roger J. Combs, Tuan Vo-Dinh, Proc. of SPIE Vol. 5994.

19. B. M. Dobratz, Properties of chemical explosives and explosive simulants, US DOE Technical Report UCRL--51319; UCRL—51319 (REV.1)

20. J. Yinon, Forensic and Environmental Detection of Explosives; John Wiley & Sons: Chichester, UK, (1999).

21. The Convention on the Marking of Plastic Explosives for the Purpose of Detection, signed in Montreal in 1991, available from the International Civil Aviation Organization (ICAO), Committee on Marking, Rendering Inert, and Licensing of Explosives et al., (1998).

22. P. A. Pella, Measurement of the vapor pressures of TNT, 2,4 DNT, 2,6 DNT and EGDN, J. Chem. Thermodynamics 9 301-305, (1977).

23. J. M. Rosen, C. Dickinson, Vapor Pressures and Heats of Sublimation of some High Melting Organic Explosives, J. of Chemical and Engineering Data 14 120 – 124, (1969).

24. B.C. Dionne, D.P. Rounbheler, E.K. Achter, J. R. Hobbs and D.H. Fine, Vapor Pressure of Explosives, J. Energetic Materials 4 447-472, (1986).

25. United Nations International Drug Control Programme - Drug Characterization/ Impurity Profiling Background and Concepts - Manual For Use By National Law Enforcement Authorities And Drug Testing Laboratories; http://www.unodc.org/pdf/publications/st-nar-32-rev1.pdf.

26. G. P. Gallerano, A. Doria, M. Germini, E. Giovenale, G. Messina, I.P. Spassovsky, J. *Infrared Milli TeraHz Waves,* 30, 1351–1361, (2009).

27. J. F. Federici, B. Schulkin1, F. Huang, D. Gary, R. Barat, F. Oliveira and D. Zimdars - THz imaging and sensing for security applications—explosives, weapons and drugs, Semicond. Sci. Technol. 20, S266–S280, (2005).

28. A. G. Bell, *American Journal of Sciences,* **20** 305, (1880).

29. G. Giubileo, F. Colao, and A. Puiu. Identification of Standard Explosive Traces by Infrared Laser Spectroscopy: PCA on LPAS Data. ISSN 1054660X, Laser Physics, Vol. 22, No. 6, pp. 1033–1037, (2012).

30. V. Lazic, A. Palucci, S. Jovicevic, C. Poggi and E. Buono, *Spectrochim. Acta,* Part B, **64**, 1028–1039, (2009).

31. V. V. Lazic, A. Palucci, S. Jovicevic and M. Carpanese, *Spectrochimica Acta* Part **66**, 644–655, (2011).

32. J. L. Gottfried, F. C. De Lucia Jr., C. A. Munson and A. W. Miziolek, *Anal. Bioanal. Chem.*, 395, 283–300, (2009).

33. P. Lucena, A. Doña, L. M. Tobaria, J. J. Laserna, *Spectrochim. Acta B: At. Spectrosc.*, **66**12–20, (2011).

34. S. D. Harvey, T. J. Peters and B. W. Wright, *Appl. Spectrosc.* 59:580-587, (2003)

35. P. Jander and R. Noll, *Applied Spectroscopy,* **63**, 5, 559-563, (2009).

36. J. Moros, J. A Lorenzo, P. Lucena, L. M. Tobaria and J. J. Laserna, *Spectroscopy Europe*, **22**, 18-22, (2010).

37. H. Östmark, M. Nordberg and T. E. Carlsson, *Applied Optics*, Vol. **50**, No. 28, 5592-5598, (2011).

38. T. H. Chyba, N. S. Higdon, W. T. Armstrong, C. T. Lobb, P. L. Ponsardin, D. A. Richter, B. T. Kelly, Q. Bui, R. Babnick, M. K. Boysworth, A. J. Sedlacek and S. D. Christesen, Field tests of the Laser Interrogation of Surface Agents (LISA) system for on-the-move standoff sensing of chemical agents. National Technical Information Service (NTIS), ADA481089, (2003).

39. A. K. Mishra, S. K. Sharma, T. E. Acosta and D. E. Bates, *Spectroscopy*, 1, (2011).

40. International Electrotechnical Commission (2001) International standard IEC 60825-1 Ed. 1.2 2001-8. International Electrotechnical Commission, Geneva

41. D. L. Jeanmaire and R. P. Van Duyne, *J. Electroanal. Chem.*, 84, 1, (1977).

42. A. Lucotti and G. Zerbi, *Sens.Actuators*, B2007, 121, 356.; S. Botti, L. Cantarini and A. Palucci. J. Raman Spectrosc., 41, 866–869, (2010).

43. G. Fiocco and L. D. Smullin, *Nature*, 199, 1275-1276, (1963).

44. A. Palucci, "Environmental and atmospheric monitoring by LIDAR systems" in: An Introduction to Optoelectronic Sensors, Series in Optics & Photonics - Vol. 7. G. C. Righini, A. Tajani and A. Cutolo Editors. World Scientific Publisher ISBN 978-981-283-412-6 (2009).

45. L. Fiorani, F. Colao and A. Palucci, *Optics Letters*, **34**, N° 6, 800 – 802 (2009).

46.  L. Fiorani, S. Babichenko, J. Bennes, R. Borelli, R. Chirico, A. Dolfi-Bouteyre, L. Hespel, T. Huet, V. Mitev, A. Palucci, M. Pistilli, A. Puiu and O. Rebane, Lidar Detection Of Explosive Precursors. Presented at the International Laser Radar Conference, 25-29 June 2012, Porto Heli, Greece.

# RESONANT HYDROPHONES BASED ON COATED FIBER BRAGG GRATINGS FOR UNDERWATER MONITORING


Giuseppe Quero, Alessio Crescitelli, Marco Consales, Marco Pisco*,
Antonello Cutolo, Vincenzo Galdi and Andrea Cusano

*Department of Engineering, University of Sannio, Corso Garibaldi 107, I-82100
Benevento, Italy
*Email: pisco@unisannio.it*

Agostino Iadicicco
*Department for Technologies, University of Naples "Parthenope," 80143
Napoli, Italy*


In this chapter, we report on recent numerical and experimental results obtained with Fiber Bragg Grating (FBG) hydrophones for underwater acoustic detection. The optical hydrophones under investigation consist of FBGs coated with ring-shaped polymers of different size. Via full-wave numerical simulations, we study the complex opto-acousto-mechanical interaction among an incident acoustic wave traveling in water, the optical fiber surrounded by the ring shaped coating, and the FBG inscribed inside the fiber. Our results fully characterize the opto-acoustic response of the optical hydrophone, and highlight the key role played by the coating in enhancing significantly its sensitivity, over the whole acoustic wave range of interest (up to 30 kHz) by comparison with a standard uncoated configuration. The coating material was selected and designed in order to provide mechanical amplification, via choice of its acoustic-mechanical properties. Our underwater acoustic measurements reveal the resonant behavior of these optical hydrophones, as well as its dependence on the coating size. By comparison with uncoated FBGs, responsivity enhancements were found, demonstrating the effectiveness of polymeric coating in tailoring the acoustic response of FBG based hydrophones.

## 1.  Introduction

Acoustic sensors are widely used in civilian and military applications as SONAR devices. The best established technology relies on Piezoelectric (PZT) Hydrophones which convert an acoustic wave into an output voltage directly related to the wave amplitude. Consequently, part of the electronics for the multiplexing and the related telemetry is immersed in water (and hence subjected to continuous harms and degradations) and is rather heavy and bulky because of the wiring. The scenario becomes even harsher when taking into account the water infiltrations which often occur during medium and long term explorations, with the consequent detrimental effects of performance degradation and malfunctioning.

For these reasons, since the first demonstration[1,2], fiber-optic acoustic sensors have been widely studied for several application areas, such as structural health, marine monitoring, as well as military and medical detections[3,4]. By comparison with conventional piezoelectric hydrophones, fiber-optic sensors have the advantages of providing high sensitivity, large dynamic range, convenience for multiplexing, flexibility of design, immunity to electromagnetic interference, absence of electrical parts, especially important for underwater applications[3,4].

Wild et al.[4], Culshaw et al.[5], Thursby et al.[6], and Kirkendall et al.[7], provide complete reviews of the state of the art of optical fiber acoustic and ultrasonic sensing, summarizing the massive majority of sensing methods and techniques currently available in the literature. Although fiber-optic hydrophones based on various principles of operation have been demonstrated[4], simpler approaches, based on the use of Fiber Bragg Gratings (FBGs) in standard optical fibers[8-11], have been proposed.

Besides the advantages, common to all fiber-optic sensors, FBGs allow easy configurations: can work in reflectivity mode, and offer excellent multiplexing capabilities. Acoustic monitoring for underwater applications is also an important area of interest, where FBG-based hydrophones have gained large popularity[12-19].

After the first demonstration by Hill et al.[13], several studies based on active FBG acoustic sensor configurations have been developed[12,14, 20-24].

Goodman et al.[20] proposed a Distributed Feedback (DFB) Fiber Laser (FL) hydrophone with a frequency responsivity of 108 dB re Hz/Pa.

Zhang et al.[21] proposed a DFB FL hydrophone based on double diaphragms, which exhibited a maximum responsivity of 7 nm/MPa, with a minimum detectable acoustic signal of 140 $\mu Pa(Hz)^{0.5}$ at 1 kHz. Ma et al.[22] proposed a DFB FL hydrophone with a novel mounting structure, which exhibited a responsivity up to 102.77dB reHz/Pa and a fluctuation of less than 1.5dB within 2.5 kHz. Even though DFB FL provides respectable sensitivities, and in-field demonstration of an FL hydrophone array has been also recently reported[23], sophisticated technologies are needed[25].

Other configurations, which provide substantial benefits in the sensor use and interrogation units, involve passive FBG schemes. Takahashi et al.[8, 15] proposed the use of FBGs as underwater acoustic sensors by monitoring the light intensity transmitted through the grating. The operating principle relies on the shift in the reflected/transmitted spectrum of the grating caused by the sound pressure. Improvements in the hydrophone response were obtained with temperature compensation[16-17], by using feedback control and multipoint sensing configurations[18,19]. Unfortunately, in the case of uncoated FBGs, the mechanical deformation of the fiber due to the incident pressure is limited by the high Young's modulus[26], leading to a poor responsivity at very low pressure levels.

Hocker[27-29] overcame this limitation, demonstrating that low-elastic-modulus ring-shaped coatings may produce substantial pressure responsivities. Starting from this work, FBGs coated with materials characterized by Young's modulus lower than that of the fiber glass were proposed as promising hydrostatic pressure sensors[26,30]. Although the theoretical/numerical analysis carried out by Hocker[27-29] was limited to the hydrostatic case and cannot rigorously be extended to the acoustic case, Cusano et al.[31,32], recently, have shown experimentally the evidence of acoustic responsivity optimization in the case of coated FBGs.

The static analysis[27-29] could be extended also to acoustic scenarios and we recently carried out a numerical study of FBGs coated with cylindrically-shaped polymeric materials, via numerical (finite-element) solution of the complex opto-acousto-mechanical problem in the acoustic wave range[33,34] (up to 30 kHz).

Based on these considerations, this chapter reports an accurate numerical analysis and an experimental characterization of coated FBG hydrophones. In particular, we designed a set of FBG hydrophones with ring-shaped coatings characterized by different size, in order to elucidate their impact on the acoustic response of the final device. These hydrophones were simulated, fabricated and experimentally characterized within the frequency range 4-35 kHz by means of a suitably equipped tank[35].

## 2. Numerical Analysis

In our simulations, in order to model the multi-physical (acoustic and mechanical) interactions between an acoustic wave propagating in water and the composite structure constituting the hydrophone, we use the multiphysics software package COMSOL Multiphysics®[36] based on the finite-element method (FEM). For the theoretical background used in this chapter, we remand to an our previous work[34].

The following sub-sections regard: the problem geometry (providing some details on the simulation parameters and computational tools utilized throughout), the representative results from harmonic and modal analysis and also the influence of the damping effect.

### *2.1. Problem Geometry and Methods*

The underwater acoustic sensor under investigation is schematically represented by a 3-D structure (Fig. 1) composed by an inner cylinder (the optical fiber), with height $h$ and radius $R_f$, and an outer annular cylinder (the ring shaped coating), with same height and coating radius $R_C$. From the acousto-mechanical viewpoint, no distinction is made between the core and the cladding of the fiber. The sphere of radius $R_W$ in Fig. 1 represents the water-filled acoustic domain surrounding the hydrophone. Basically, the full-wave 3-D investigation of the acoustic-mechanical response of the underwater acoustic sensor, under the effect of an incident acoustic plane wave, is carried out in COMSOL by solving, in the acoustic domain, the Helmholz equation:

$$\nabla^2 p + \left(\frac{\omega}{c}\right)^2 p = 0, \tag{1}$$

where $p$ denoting the frequency-domain pressure, $\omega$ denoting the angular frequency, and $c$ is the speed of the sound in the fluid.

Moreover, we use, in the mechanical domain, the principle of virtual work (equivalent to the equilibrium equation[36]) described by the Equation[37]:

$$\mu\nabla^2 u + (\lambda + \mu)\nabla(\nabla \cdot u) + \rho'_s \omega^2 u = -f, \tag{2}$$

where $\rho'_s$ is the material density, $\lambda$ and $\mu$ are the Lamè constants, which describe the elastic behavior of the medium and are related to the Young modulus and to the Poisson's ratio[34], finally $u$ and $f$ are the frequency-domain displacement vector and the applied dynamic load, respectively.

The resulting equations are coupled via the boundary conditions, described by the following Equations[37]:

1.  The normal displacement in the solid and fluid must be equal at the interface:

$$\rho_0 \omega^2 u \cdot \mathbf{n} = \frac{\partial p}{\partial n}. \tag{3a}$$

2.  The pressure must be in static equilibrium with the stress normal to the solid boundary:

$$\sigma_n = -p. \tag{3b}$$

3.  In an ideal fluid, no tangential stresses must occur at the boundary:

$$\sigma_{\perp} = 0. \tag{3c}$$

In Eqs. (3a-3c), $\mathbf{n}$ is the unit vector normal to the boundary surface, while $\sigma_n$ and $\sigma_{\perp}$ are the components of the stress vector normal and tangential to the boundary surface, respectively. Furthermore, within the mechanical domain, we assume perfect bonding between the optical fiber and the coating, which implies the additional boundary condition:

4.  The stresses and displacements at the fiber-coating interface must be continuous[38]:

                                    *M. Pisco et al.*

$$\sigma_c = \sigma_f, \quad U_c = U_f, \tag{3d}$$

where $U=U(t,x,y,z)$ is the displacement vector; $c$ and $f$ identify the coating and fiber, respectively.

Moreover, at the spherical surface of the water domain, a radiation condition is set, so as to mimic an infinite water domain. At the external boundary, an acoustic plane wave (of amplitude $p_0=1$ MPa) is set as source. The analysis is carried out within the frequency range 0.5-30 kHz.

The FEM-based 3-D analysis yields the strain distribution $\varepsilon$ along the $x$, $y$ and $z$ axes into the optical fiber where the FBG is inscribed. The strain components evaluated at the FBG location $(x, y, z)=(0, 0, 0)$ are then used in the optical model to calculate the Bragg wavelength shift according to the following Equation[39]:

$$\frac{\Delta\lambda}{\lambda_0} = \varepsilon_z - \frac{n_{eff}^2}{2}\left[p_{11}\varepsilon_x + p_{12}\left(\varepsilon_z + \varepsilon_y\right)\right], \tag{4}$$

where $\Delta\lambda$ is the Bragg wavelength shift, $\lambda_0$ is the central wavelength of the FBG, $p_{11}$ and $p_{12}$ are the elasto-optic parameters, $n_{eff}$ is the effective refractive index, and $\varepsilon_i$ $(i=x, y, z)$ are the Cartesian strain components evaluated at the FBG location.

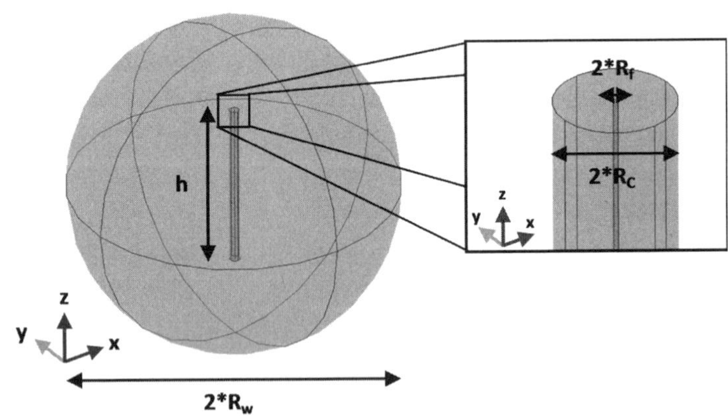


Fig. 1. 3-D geometry considered in the numerical simulations[34] (reproduced under the permission of OSA Publisher).

The observable "sensitivity" can be defined as:

$$S = \frac{\Delta\lambda}{\lambda_0 p_0} = \frac{1}{p_0}\left\{\varepsilon_z - \frac{n_{eff}^2}{2}\left[p_{11}\varepsilon_x + p_{12}\left(\varepsilon_z + \varepsilon_y\right)\right]\right\},\qquad(5)$$

where[26]: $p_{11}$=0.121, $p_{12}$=0.265 and $n_{eff}$=1.465.

Moreover, from the ratio between the calculated sensitivity and that pertaining to the bare (i.e., without coating) fiber in the hydrostatic case[26,27], a sensitivity gain may be defined as:

$$SensitivityGain = 20\log_{10}\left(\left|\frac{S}{S_{BARE}}\right|\right),\qquad(6)$$

where[26]: $S_{BARE}$=-2.76·$10^{-6}$ MPa$^{-1}$.

In order to gain a deeper insight into the sensor response and the underlying mechanisms, we also carried out a modal analysis, identifying the free vibration modes of the sensing structure. The analysis was still performed in COMSOL, by solving Eq. (2) in the absence of external loads[36].

For a comprehensive characterization of the sensor performance, we carried out a series of parametric studies, by varying the geometrical and physical parameters such as the coating height $h$, the coating radius $R_C$, the Young's module $E$, the Poisson's ratio $v$, the density $\rho'$, and the loss factor $\eta$.

In all the simulations below, we assume an optical fiber radius $R_f$=62.5 μm, and an identical length $h$ for the fiber and the coating. The remaining mechanical and acoustic parameters listed in Table 1 are kept constant.

Table 1. Mechanical and Acoustic Constants of the Analyzed Structure (reproduced under the permission of OSA Publisher) [34].

| Parameter | Value |
|---|---|
| Optical fiber Young's modulus ($E_f$) | 72 GPa |
| Optical fiber Poisson's ratio ($v_f$) | 0.17 |
| Optical fiber density ($\rho'_f$) | 2200 kg/m$^3$ |
| Water density ($\rho'_w$) | 997 kg/m$^3$ |
| Speed of sound in water ($c_w$) | 1480 m/s |

## 2.2. Resonant Behavior of the Coated Fiber Bragg Gratings: Harmonic and Modal Analysis

In order to analyze the opto-acoustical response of our proposed sensor, we begin considering a specific configuration with fixed coating size and material properties, and investigate the interaction with an acoustic wave from the phenomenological viewpoint. The configuration of interest features a ring-shaped coating with an outer radius $R_C$=1.25mm (i.e., 20 times larger than the fiber radius) and a height $h$=4 cm. The elastic properties of the coating, given in Table 2, are consistent with the nominal properties of a specific thermosetting polyurethane[32] (Electrolube UR5041).

The structure is excited by an acoustic plane wave normally-incident with respect to the cylinder axis. Numerical simulations are performed within the frequency range 500 Hz to 30 kHz (with steps of at least 1 kHz) of interest for typical SONAR applications. For all observables shown below, we also compute, as a reference, the zero-frequency values via the hydrostatic model[26,27].

Figure 2(a) shows the simulation results in terms of strain components along the $x$, $y$, and $z$ axes over the whole frequency range of interest. Four sharp peaks are clearly visible in the strain components at frequencies 5.6, 14.8, 21.8 and 28.7 kHz.

Table 2. Elastic Properties of the Ring-Shaped Coating (reproduced under the permission of OSA Publisher)[34].

| Coating Parameters | Value |
| --- | --- |
| Young's modulus ( $E$ ) | 78 MPa |
| Poisson's ratio ( $v$ ) | 0.3 |
| Density ( $\rho'$ ) | 1180 kg/m$^3$ |

For better visualization of the background response (away from resonances) that is hidden by the strength of the resonant peaks, we show in Fig. 2(b) a magnified detail with the strain scale saturated. Similarly, in Fig. 2(c), we show a magnified detail around 14.8 kHz, so as to illustrate with a finer sampling (down to 1 Hz) the line-shape of the resonance.

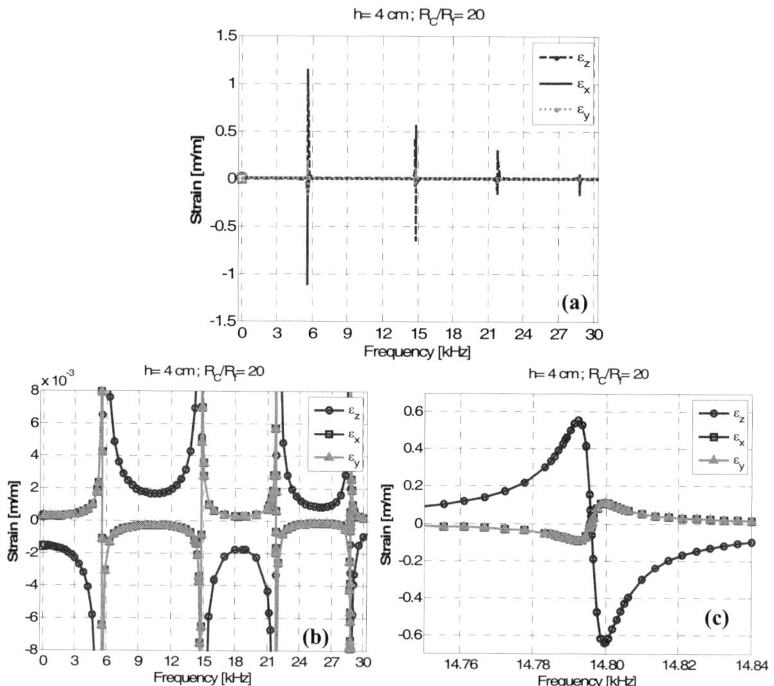

Fig. 2. (a) Spectra of the strain components; (b), (c) Magnified details along the strain and frequency axes, respectively[34] (reproduced under the permission of OSA Publisher).

All the observed resonances clearly resemble the Fano line-shape and may be accurately fitted with a Fano-type model[40]. Recalling that Fano resonances stem from the interference between a discrete state with a continuum of states[40], we can deduce that the resonances observed are attributable to the interaction between one of the vibration modes of the sensor and the impinging acoustic wave.

As an illustration of the opto-acoustic response of the sensor to the impinging plane wave, Fig. 3 shows the pressure distribution in water (on the sphere boundary) and the $z$-strain distribution on the cylinder surface, for different frequencies (10, 20 and 30 kHz) away from resonances. As evident from Figs. 3(a-c), the cylindrical sensor responds to the impinging plane wave through a mechanical deformation, according to its elastic properties. The resulting strain at the FBG location determines a Bragg wavelength shift according to Eq. (4).

The sensitivity of this basic transduction principle is strongly enhanced at the resonant frequencies, where a Fano-type strain response occurs at the FBG location.

The sensitivity of this basic transduction principle is strongly enhanced at the resonant frequencies, where a Fano-type strain response occurs at the FBG location. In Fig. 3(d-e), we show the pressure distribution in water and the $z$-strain distribution on the cylinder surface around the second resonance (14.792 and 14.800 kHz), with color scales suitably saturated so as to allow correct visualization of all distributions.

From the pressure distribution, it can be observed the outgoing pressure radiated by the excited resonance (the discrete state) which, interfering with the impinging plane wave (the continuum), generates the Fano-resonance phenomenon.

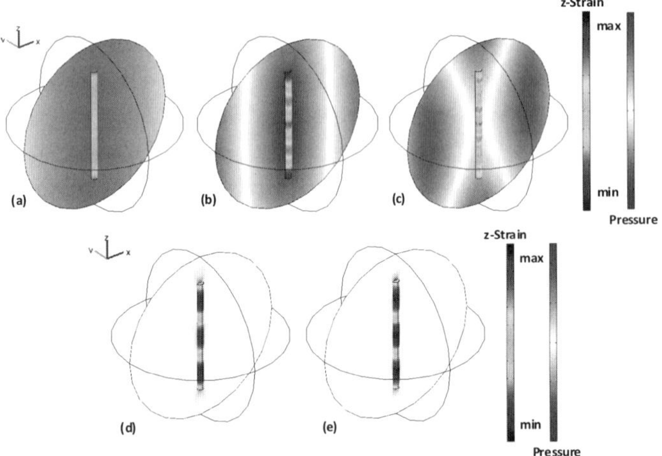


Fig. 3. Pressure distribution in water and $z$-strain distribution on the cylinder surface, for different frequencies of the acoustic wave, away from resonances: (a) 10 kHz, (b) 20 kHz, (c) 30 kHz, and around the second resonance: (d) 14.792 and (e) 14.800 kHz (reproduced under the permission of OSA Publisher)[34].

Figures 4(a-b) show the sensitivity (cf. Eq. (5)) and corresponding sensitivity gain (cf. Eq. (6)), respectively, as a function of frequency, computed on the basis of the $\varepsilon_x$, $\varepsilon_y$, and $\varepsilon_z$ strain distributions obtained from the simulations.

As expectable, at the resonant frequencies (5.6, 14.8, 21.8 and 28.7 kHz), the sensitivity also exhibits a resonant behavior, which retains the Fano-type line-shape (see the inset in Fig. 4(a)). The sensitivity gain in Fig. 4(b) is characterized by peaks of amplitude up to 110dB, superimposed onto a $\sim$ 50dB background which is slowly decreasing with frequency.

Note that, in Fig. 4, the sensitivity value at zero-frequency computed via the hydrostatic model[26, 27] is in very good agreement with our low-frequency predictions.

It is also worth noting that the high sensitivity values observed in Fig. 4 at the resonant frequencies strongly rely on the assumption of zero damping in the composite cylindrical structure, for which an excited resonance yields a lossless vibration. Moreover, the resonances are very narrow (e.g., the sensitivity gain in Fig. 4(b) exceeds 80dB over bandwidths narrower than 0.5 kHz).

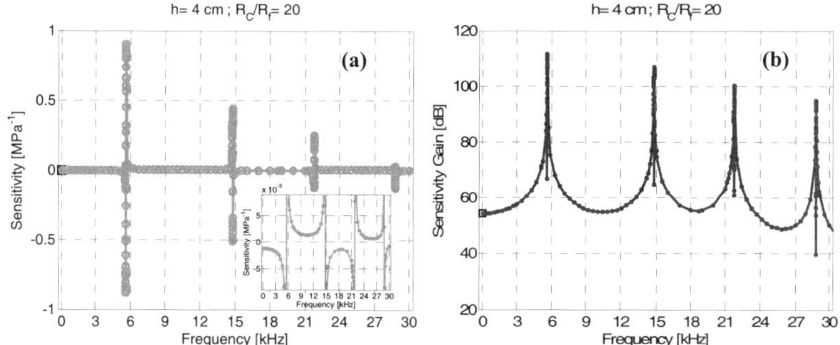

Fig. 4. (a) Sensitivity (cf. Eq. (5)) and (b) sensitivity gain (cf. Eq. (6)) spectra. The inset in (a) highlights the Fano-type line-shape of the sensitivity around the resonant frequencies[34] (reproduced under the permission of OSA Publisher).

To sum up, the numerical results above indicate that the ring-shaped coating is capable of significantly improving the sensitivity of the optical hydrophone, over the whole investigated frequency range, when compared with an FBG without coating. Furthermore, the hydrophone sensitivity spectrum also exhibits characteristic resonant peaks, which yield a further strong enhancement.

In order to identify the free vibration modes of the composite structure, and relate them with the sharp sensitivity resonances, we carried out a modal analysis of the optical hydrophone.

In particular, we computed the vibration modes of the composite cylindrical structure under investigation up to 30 kHz. Figure 5 shows the only non-zero ($z$) component of the longitudinal resonant modes, visualized in terms of deformed shape (retrievable from the displacement vector) and strain distribution (in color scale). More specifically, the composite cylindrical structure of the sensor supports four longitudinal vibration modes (at frequencies 5.7, 15, 21.9 and 27.9 kHz) characterized by even symmetry along the $z$-axis, and three longitudinal vibration modes (at frequencies 10.8, 18.7 and 25k Hz) with odd symmetry along the $z$-axis. It is interesting to note that the sensitivity peaks observed in Fig. 4 occur at frequencies associated to the even modes only.

The comparison between the results from modal and harmonic analysis therefore confirms that the resonant features characterizing the hydrophone sensitivity are attributable to the excitation (by the normally-incident pressure plane wave) of longitudinal vibration modes with even symmetry. From the phenomenological viewpoint, this observation somehow resembles the excitation of guided resonances in periodic or quasi-periodic photonic crystals[41,42]. In fact, more in general, we observed that only the longitudinal vibration modes of the composite structure that exhibit a symmetry matching that of the impinging pressure plane wave are excited. In the normal-incidence case, only modes with even symmetry satisfy this condition (since plane waves exhibit even symmetry in the plane orthogonal to the direction of propagation) but, for oblique incidence, the excitation of vibration modes with odd symmetry becomes possible.

As a concluding remark on this analysis, it should be noted that other vibration modes are supported by the composite structure (i.e., bending modes at frequencies 0.4, 1.0, 1.8, 2.9, 4.3, 5.7, 7.3, 9.0, 10.8, 12.7, 14.6, 16.5, 18.5, 20.5, 22.5, 24.5, 26.6 and 28.6 kHz), which, however, do not give rise to sensitivity enhancements.

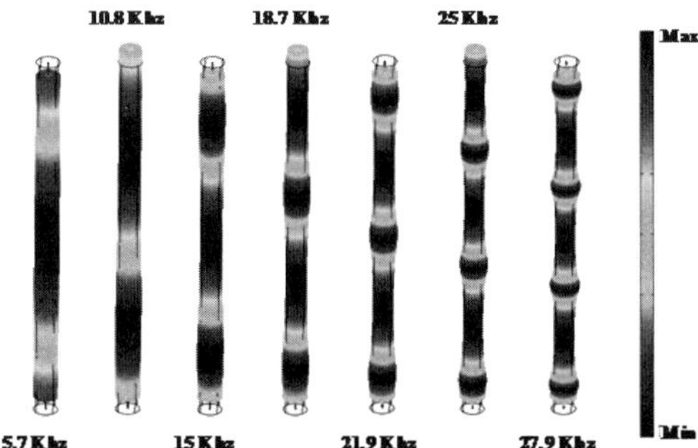


Fig. 5. Deformed geometry and z-strain distribution (in color scale) of the resonant modes[34] (reproduced under the permission of OSA Publisher).

Moreover, for different physical and geometrical parametric configurations, other modes (of same and different kinds) may occur in the frequency range considered (e.g., transversal modes may fall within the range if larger radii are assumed).

Nevertheless, an exhaustive study of the natural frequencies of free-standing (and composite) elastic cylinders[43] is beyond the scope of this chapter, which is instead focused on the effects induced by the excitation of such modes in the sensing performance.

## 2.3. *Effects of the Coating Properties on the Sensing Performance*

We now move on to exploring the influence of the geometrical parameters of the structure (height and diameter) on the hydrophone frequency response, by maintaining the coating physical parameters fixed (see Table 2).

Figure 6 shows the sensitivity gain spectrum pertaining to a configuration with same radii as before, but with coating height $h=1$, 3 and 4 cm. The response qualitatively resembles that in Fig. 4(b), with the number of resonant peaks changing with the cylinder height, and no significant changes in the background gain. In particular, the number of peaks tends to decrease when decreasing the height. From the modal analysis, the change in the number of sensitivity peaks is an effect of a frequency downshift occurring to the resonances when the cylinder height increases. Figure 7(a) illustrates such variation for the first three resonances: the resonant frequencies decrease with the coating height, and this decrease is faster for higher-frequency (e.g., third) peaks than for the lower-frequency (e.g., first) ones. In Fig. 7(b), with focus on the first resonance, we also show the effects of changing the coating radius ($R_C$=1.25, 2.5 and 5 mm). By increasing the coating height, a decrease of the resonant frequency is still observed, and the slowest decrease is obtained for the largest coating radius.

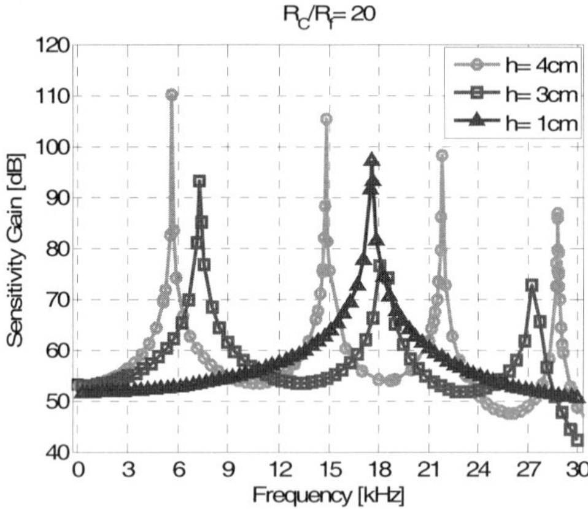


Fig. 6. Sensitivity gain spectra, for different coating heights[34] (reproduced under the permission of OSA Publisher).

Overall, regardless the coating radius, an increase of the height yields a decrease of the resonant frequencies, which eventually saturates.

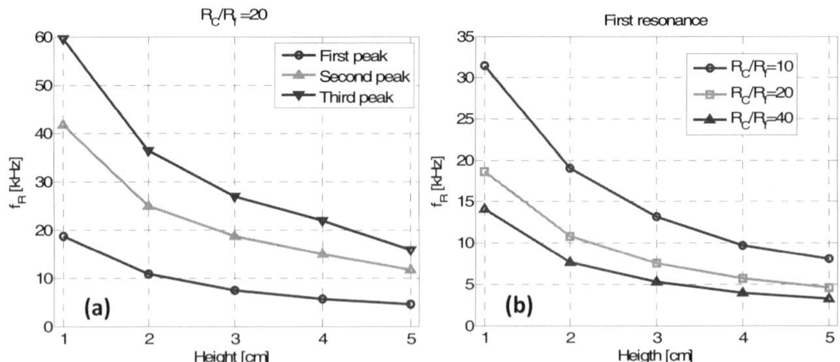


Fig. 7. (a) Resonant frequency as a function of coating height, for different peaks. (b) First resonant frequency as a function of coating height, for various coating radii[34] (reproduced under the permission of OSA Publisher).

Consequently, the height of the ring-shaped coating may be a useful parameter for tuning the gain sensitivity peaks without affecting meaningfully the background sensitivity.

Figure 8 shows the sensitivity gain spectrum pertaining to a configuration with a fixed height $h$=4 cm, and with coating radii $R_C$=0.625, 1.25, 1.875 and 2.5mm (i.e., ratios $R_C/R_f$=10, 20, 30 and 40, respectively), with the physical properties of the coating unchanged (Table 2).

Some similarities and differences with respect to the previous case (changing the coating height) are worth highlighting. Similarly as before, the general spectral features (background with resonant peaks) are preserved when changing the coating radius, and the number of resonant peaks increases with increasing the radius. Figure 9(a) shows the variation of the first three resonant peak frequencies with the coating radius. Also in this case, a decrease of the resonant frequencies is observed, with the first peak moving slower than the third. In Fig. 9(b), we show the variation of the first resonant peak with the coating radius, for different height values. As can be observed, the fastest decrease is obtained for the smallest height considered. Basically, increasing the coating radius (similar to the coating height) yields a decrease of the resonant peak frequencies.

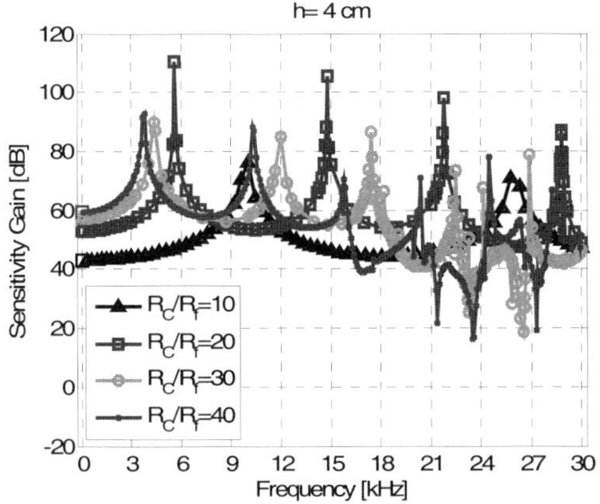


Fig. 8. Sensitivity gain spectra for different coating radii[34] (reproduced under the permission of OSA Publisher).

All the variations of the resonant frequencies with the geometrical features of the ring-shaped coating were found to be in substantial agreement with the free vibration modes variations of a finite-length elastic composite cylinder[43].

Moreover, increasing the coating radius has two further main effects on the sensor performance: *i)* the enhancement of the background sensitivity at low frequencies, and *ii)* the progressive reduction of the background sensitivity at high frequencies.

In connection with the former effect, observing the low-frequency (below 3 kHz) regions of the spectra in Fig. 8, it is evident that increasing the coating radius yields a significant enhancement of the sensitivity gain, up to a saturation value of about 60dB.

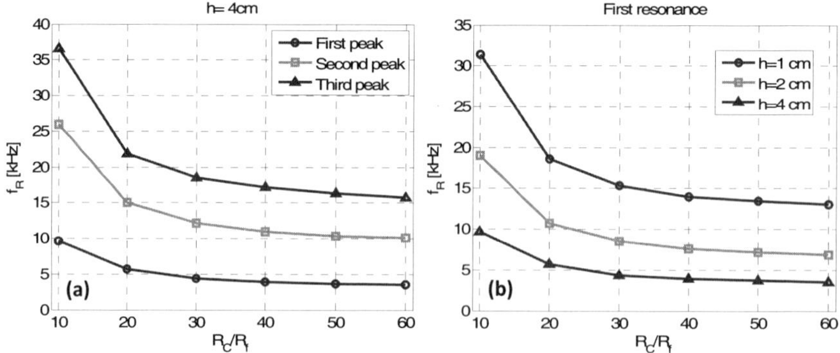


Fig 9. (a) Resonant frequency as a function of coating radius, for various peaks. (b) First resonant frequency as a function of coating radius, for various coating heights[34] (reproduced under the permission of OSA Publisher).

To better highlight this aspect, in Fig. 10(a) we show the minimum values of the sensitivity gain between the first and second peaks, for various coating radii (up to $R_C/R_f$=60). Basically, the sensitivity gain background increases with the coating radius until it reaches a constant value. This is in agreement with the hydrostatic model[26,27], and may be attributed to the role played by the ring-shaped coating in contrasting the stiffness of the optical fiber glass.

As for the latter effect, looking at the high-frequency (e.g., above 20 kHz, for $R_C/R_f$=40) regions of the spectra in Fig. 8, an overall deterioration of the sensitivity gain can be observed, in spite of the appearance of additional resonances. In this framework, it is worth mentioning the appearance of transversal resonances (e.g., at 26.5 and 27.7 kHz for $R_C/R_f$=30, and at 22.1 and 24.2 kHz for $R_C/R_f$=40) which, however, does not yield meaningful sensitivity enhancements. To highlight the deterioration of the sensitivity gain background, in Fig. 10(b) we show the spectra (extracted from Fig. 10) around the first local minima between, from which a low-pass behavior is clearly observed, with a cut-off frequency that decreases when the coating radius increases, and the previously shown low-frequency gain level increasing with the coating radius. Note that such low-pass behavior could not be predicted by the hydrostatic model[26,27], and is attributable to diffraction effects occurring when the acoustic wavelength approaches the cylinder diameter.

In order to explore the effects of the coating elastic properties, we compared, for a given coating size ($h$=4 cm and $R_C$ =1.25mm), the sensitivities obtained by changing its Young's modulus $E$, Poisson's ratio $v$, and density $\rho'$. The values used in the analysis are consistent with (but not strictly limited to) the nominal properties of certain specific polymeric materials (thermosetting polyurethanes)[32], which are particularly suited to underwater applications in view of their adhesion (on the fiber glass) and waterproof properties.

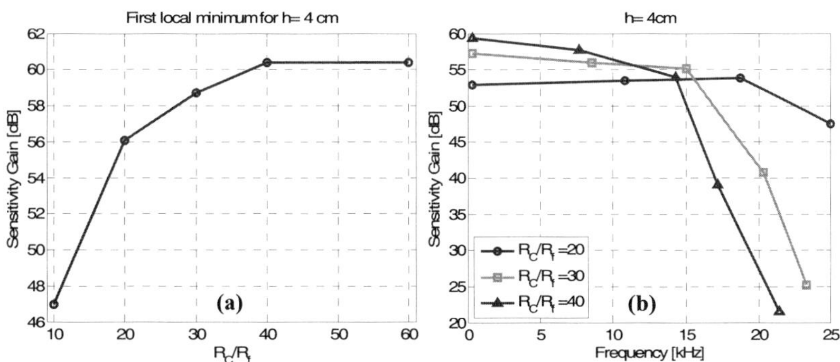


Fig. 10. (a) Sensitivity gain at first local minimum as a function of coating radius. (b) Sensitivity gain spectra around the first local minimum, for different coating radii[34] (reproduced under the permission of OSA Publisher).

We start by considering two basic materials: the former (identical to that considered so far, cf. Table 2) is characterized by $E$=78 MPa, $v$=0.3, and $\rho'$=1180 kg/m$^3$; the latter features the same Poisson's ratio, but $E$=970 MPa, and $\rho'$=1070 kg/m$^3$ (consistent with the thermosetting polyurethane Electrolube UR5528).

Moreover, in order to separate the effects of the Young's module variation from those due to the density variation, we also consider other two sets of physical parameters (not corresponding to specific existing materials) featuring $E$=500 MPa, $v$=0.3, $\rho'$=1070 kg/m$^3$, and $E$=500 MPa, $v$=0.3, $\rho'$=1180 kg/m$^3$, respectively. Figure 11(a) shows the sensitivity-gain spectra pertaining to these four materials, from which it can be clearly seen that increasing the Young's module yields a desensitization of the hydrophone, accompanied by a high-frequency shift of the peaks.

Much less significant are the effects of the density variations, limited to a slight sensitivity increase (for decreasing density) and frequency shift of the resonant peaks.

Finally, Fig. 11(b) illustrates the effects of the Poisson's ratio variations (from 0.3 to 0.4), for fixed $E$=78 MPa, and $\rho'$=1180 kg/m$^3$, which basically amount to a deterioration (for increasing values of $v$) of the sensitivity-gain background (away from resonances), and a slight frequency shift of the resonant peaks.

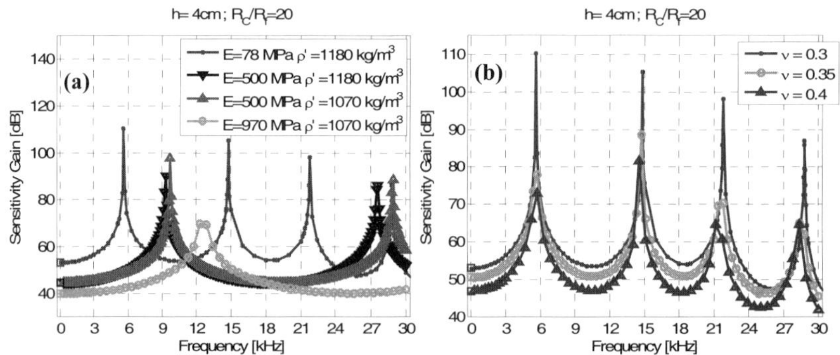


Fig.11. Sensitivity gain spectra, by varying (a) the coating Young's modulus $E$ and density $\rho'$ (for $v = 0.3$), and (b) the coating Poisson's ratio $v$ (for $E$=78 MPa and $\rho'$=1180 kg/m$^3$) [34] (reproduced under the permission of OSA Publisher).

Overall, the numerical results indicate that sensitivity enhancements may be obtained by using a ring-shaped coating characterized by low values of the Young's modulus, Poisson's ratio, and density. With reference to the Young's modulus and Poisson's ratio, this is in agreement with the hydrostatic analysis[26,27]. Basically, decreasing the Young's modulus and/or the Poisson's ratio reduces the bulk modulus $K$ of the ring-shaped coating (given by $k=E/[3(1-2v)]$, for homogeneous, isotropic, linear elastic materials) which, for a given acoustic pressure, yields in turn an increase of the coating compression, and hence an enhancement of the strain components acting on the FBG.

Further insight may be gained by considering the characteristic acoustic impedance[37]: $Z= \rho'c_c$ of the material composing the ring-shaped coating, where $c_c$ is the bulk speed of an elastic wave in a linearly-elastic solid. It can readily be noticed that the sensitivity optimization criteria emerged from the parametric analysis are consistent with a minimization of the coating impedance. It is also worth noting that such minimization implies an increase of the impedance mismatch with the surrounding water, which is detrimental in terms of increased signal reflection, but is largely compensated by the increase of the strain energy endowed to the FBG.

### 2.4. *Structural Damping and Directivity*

All the results discussed so far are obtained considering lossless materials. In what follows, we consider the effects of damping in both the glass fiber and the coating material. The structural damping is expressed in terms of a loss factor $\eta$ defined as the ratio between the imaginary and real parts of the complex Young's modulus[44].

For the optical fiber, such loss factor is assumed to be $\eta=0.002$. Figure 12 illustrates the effects of damping with reference to a ring-shaped coating featuring $R_C/R_f=20$, $h=4$ cm, $v=0.3$, $\rho'=1180$ kg/m$^3$, $R_e(E)=78$ MPa, and $\eta=0.1$. As expectable, it can be noted that the damping affects primarily the resonances (especially those at higher frequencies), yielding a strong sensitivity deterioration, while the baseline sensitivity gain turns out to be essentially unaffected. Since our model does not take into account the frequency-dependence of the coating elastic properties, the stronger deterioration effects observed at higher frequencies may be attributed to the corresponding longer acoustic paths in the coating. The effect of damping on the sensitivity must be carefully taken into account in the choice of the coating material, since a low Young's modulus (useful to increase the sensitivity, cf. Fig. 11(b)) is often competing with a low damping, because "soft" materials typically exhibit high damping.

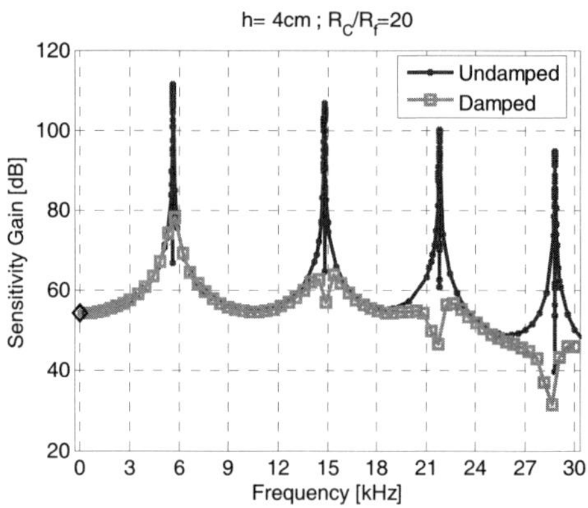


Fig. 12. Effects of the structural damping on the sensitivity gain spectrum, for a ring-shaped coating featuring ring-shaped coating featuring $R_C/R_f$=20, $h = 4$ cm, $v = 0.3$, $\rho'$=1180 kg/m$^3$, $\mathrm{Re}(E) = 78$ MPa, and $\eta$=0.1 (reproduced under the permission of OSA Publisher) [34].

We conclude our parametric studies by considering the effects of *oblique* incidence, with specific reference to a parameter configuration featuring $R_C/R_f$=20, $h$=3 cm, $v$=0.3, $\rho'$=1180 kg/m$^3$, $R_e(E)$=78 MPa, and no damping. In view of the cylindrical symmetry of the structure, we only consider variations of the angle $\theta$ between the incidence direction and the $x$-$y$ plane. Thus, $\theta = 0$ represents the normal-incidence case considered so far, while for $\theta = \pi/2$ the acoustic wave travels parallel to the cylinder axis.

Figure 13(a) compares the sensitivity gain spectra for $\theta = 0$, $\pi/4$ and $\pi/2$. While the lower-frequency part of the spectrum turns out to be very weakly affected, at higher frequencies (above 9 kHz), the sensitivity gain visibly decreases for increasing incidence angles. This is better quantified in Fig. 13(b), which shows the sensitivity gain as a function of the incidence angle, at selected frequencies (3, 15 and 25 kHz) away from resonances. Moving from normal ($\theta = 0$) to longitudinal incidence ($\theta = \pi/2$), the sensitivity gain gradually deteriorates, and the higher frequencies turn out to be more affected. This is consistent with the dependence on the angle of incidence of the Rayleigh scattering from an elastic cylinder[45], and may therefore be simply attributed to the reduced cross-sectional area exhibited by the cylinder with respect to the wavelength of the impinging plane wave. In connection with the resonances, the first and the second peaks do not experience significant frequency shifts. The main difference between normal and oblique incidence (highlighted in the inset of Fig. 13(a)) consists in the appearance of a new (weaker) resonant peak at 22.7 kHz. The modal analysis indicates that such peak is attributable to a longitudinal anti-symmetric mode (odd *z*-strain distribution along the *z*-axis), which can be excited only under oblique incidence.

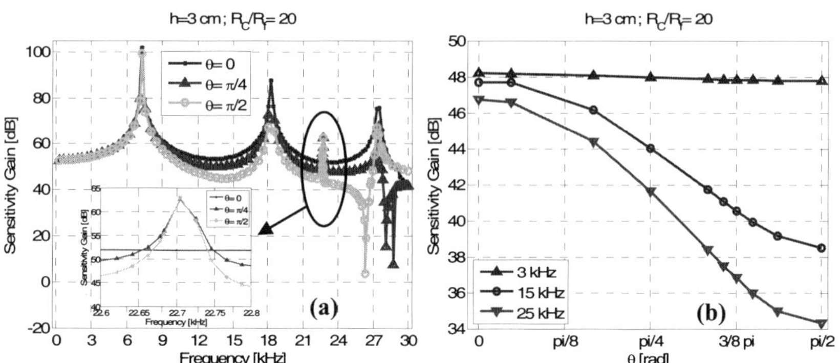


Fig. 13. (a) Sensitivity gain spectrum (magnified in the inset, around a resonant peak) pertaining to a ring-shaped coating featuring $R_C/R_f$=20, $h = 3$ cm, $v = 0.3$, $\rho' =1180$ kg/m$^3$, $E = 78$ MPa, and no damping, for various sound-wave incidence directions. (b) Sensitivity gain angular response at selected frequencies[34] (reproduced under the permission of OSA Publisher).

## 3.  Design and Fabrication of the Sensor

Following our previous numerical results, we designed a suitable set of FBG hydrophones, taking into account two main objectives:

1)   to experimentally demonstrate the resonant behavior in the sensitivity;

2)   to experimental analyze the influence of the cylindrical coating features on such resonant behavior;

The polymeric coating was properly selected to provide resonant features in the selected range (4-35 kHz). The selected coating material, characterized by relatively "low" Young's modulus, is Damival® E 13650. This material is a polyurethane resin, commonly used to overcoat piezoceramic acoustic transducers without degrading their performance, which exhibits a Young's modulus on the order of few hundred MPa, and acoustic impedance matching that of water.

In particular, here we focused the attention on two sensors with height fixed to 30 mm and coating diameter of 5 mm and 10 mm, respectively (here named 5-30 and 10-30 respectively).

FBG hydrophones taken into consideration in this chapter consist of standard FBGs written in single mode fiber coated by Damival material with cylindrical shape and different sizes. To this aim a modular holder was designed to achieve polymeric cylindrical coating on fiber grating with desired size (in terms of sensor diameter and height). Besides, the holder is designed to leave two conic regions (of the same material) on both ends of cylindrical region to reduce mechanical stress on the lead in and lead out fiber.

As described above, when a sound pressure is applied to a FBG hydrophone a shift in Bragg wavelength occurs depending on the actual strain acting on the sensor region. In order to detect sound pressure around an FBG, edge triggered narrowband interrogation was used. To this aim, a tunable laser was used to lock the operating wavelength on the linear edge of the grating response. Since spectrum of FBG shifts without changing its shape[8,31,32], the change in reflection at the laser wavelength is also expected to be proportional to pressure. The optoelectronic setup thus consists in a stable tunable laser (Yokogawa AQ2200-136), a 2x1 coupler capable to split transmitted and reflected signals whereas a photodiode provides an electrical output proportional to the optical power of the laser reflected from the grating edge. Finally the electric signal is amplified and stored at 400Ksample per second.

Field trials have been carried out in a professional tank at Whitehead Alenia Sistemi Subacquei's laboratory with size 11x5 m and depth 7 m. As schematically plotted in Fig. 14, the tank is equipped with two PZT acoustic sources capable to emit sound pressure in different frequency ranges with Gaussian behaviour and centred at 8 kHz and 28 kHz, respectively. The combination of the two acoustic sources permits a complete sensor characterization in the whole range 4-35 kHz. Besides the tank is equipped with PZT reference hydrophone exhibiting almost constant sensitivity of -209±0.9 dB (reV/μPa) in the investigated acoustic frequency range. Acoustic sources and hydrophones (FBG and PZT) are positioned at 3 m below the water level while the distance from the PZT sources is of approximately 2 m for the reference hydrophone and 3 m for the FBG hydrophone, respectively. The FBG hydrophones were vertically kept by a few gram weight and consequently the sound pressure was perpendicular to the fiber longitudinal axis.

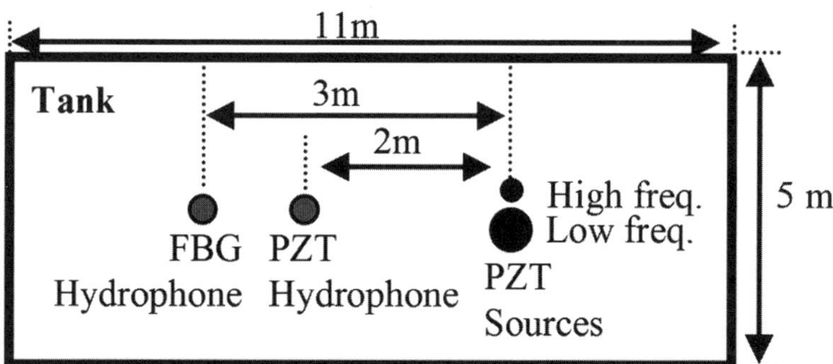


Fig. 14. Schematic view of instrumented tank[35] (reproduced under the permission of SPIE Publisher).

## 4. Experimental Results

We now move on to illustrating the experimental results[46] obtained using the FBG hydrophones 5-30 and 10-30, with the aim of validating the numerical predictions, described above. In particular, the attention is focused on the responses to an acoustic pressure pulse with frequency varying within the range 4-35 kHz, and amplitude ranging between 2 and 154 Pa.

### *4.1. Time Response*

Figure 15 shows the typical time response of sensor 5-30 to a 15 kHz acoustic pressure pulse with duration 0.5 ms. For comparison, the time response of the PZT reference hydrophone is also reported. The results clearly evidence the capability of both devices to detect not only the generated direct wave emitted by the acoustic source (i.e. the one that directly reaches the sensors without undergoing reflections), but also the weak contributions due to the tank sidewalls and water surface. The delay of the optical response (approximately 0.75 ms) is due to the longer distance from the acoustic sources at which the FBG hydrophone has been located with respect to the piezoelectric counterpart.

The different position of the two hydrophones is also the cause of the phase difference occurring between their responses and makes the optical fiber device to experience a lower intensity of the acoustic wave with respect to the piezoelectric one (32 Pa and 48 Pa, respectively). Similar results have also been observed for the sensor 10-30. In this case the optical response is less intense, thus evidencing that the coating diameter represents a key parameter ruling the sensing performance. This aspect, however, will be better investigated in the following.

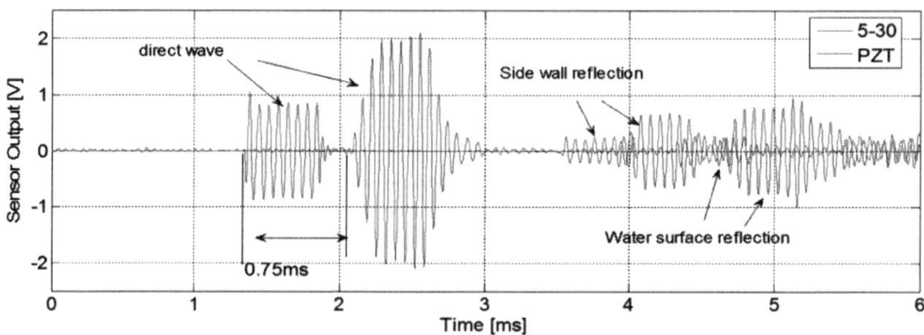


Fig. 15. Typical time responses of the sensors 5-30 to a 15 kHz acoustic pressure pulse with duration 0.5 ms[35] (reproduced under the permission of SPIE Publisher).

The linearity of the optical response towards acoustic signal was also investigated. To this aim the voltage applied to the PZT transducer was properly tuned in order to generate a series of acoustic pressure pulses with increasing intensities. The linearity analysis was carried out for all frequencies in the range 4-35 KHz. The pressure levels, however, was not the same in the whole range because of the resonant behavior of the exploited acoustic sources. In Fig. 16(a) the time responses of the sensors 5-30 to several acoustic pulses of increasing intensity at 15KHz and duration 0.5 ms has been shown. From time responses, amplitude of optical sensors response was thus retrieved by FFT analysis. Figure Fig. 16(b) reports the optical output intensity versus applied sound pressure at 15KHz for 5-30 and 10-30 sensors. The experimental data clearly reveal the good linearity of the sensors response, whose sensitivities (in terms of $\Delta V/Pa$) were estimated to be approx. 9.3 mV/Pa and 7.3 mV/Pa, respectively for the 5-30 and 10-30 sensors.

By considering the minimum detectable signal change that could be retrieved with the exploited set-up (approx. 18 mV), resolutions of approximately 2 Pa and 2.5 Pa can be obtained at 15 KHz for 5-30 and 10-30, respectively. However, it is worth noting that the presence of eventual environmental noise could significantly affect the minimum detectable signal change and, as consequence, the resolution values. In addition it has been verified that changing frequency of the emitted sound pressure does not affect the linear behavior of the optical response-acoustic intensity curves for both sensors.

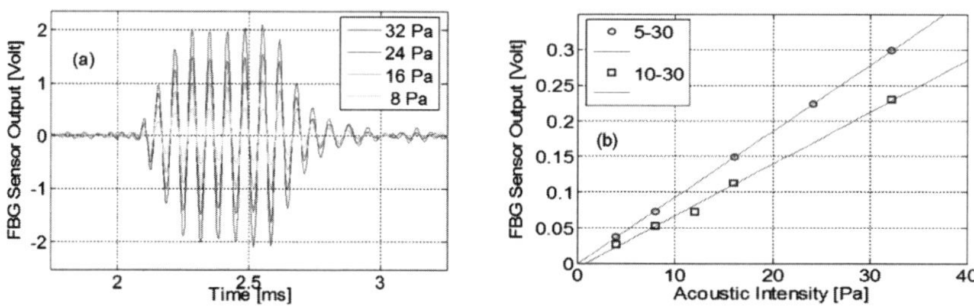


Fig. 16. (a) Typical temporal responses of the sensor 5–30 to a 15 kHz acoustic pressure pulse with increasing intensity; (b) optical response-acoustic intensity curves for the sensor 5–30 and 10–30[35] (reproduced under the permission of SPIE Publisher).

## 4.2. Frequency Domain Analysis

Next, we illustrate the spectral characterization of the fabricated hydrophones (5-30 and 10-30), obtained by measuring their optical signals in correspondence of sound pressure pulses with increasing frequency in the range 4-35 kHz (with step of 1 kHz), have been resumed in Figure 17. In particular, here the sensitivity (cf. Eq. (5)) of the optical sensors versus frequency is considered in term of $\Delta\lambda/(\lambda_0 \cdot P_0)$ (where $\Delta\lambda/\lambda_0$ is the relative Bragg wavelength shift and $P_0$ is the amplitude of sound pressure). For comparison with the reference PZT, also the sensitivities calculated as[33] dB (reV/μPa) have been shown in Fig. 17(b).

The results clearly evidence that, from a qualitative point of view, the experimental measurements strongly agree with the numerical results previously reported. As matter of fact, here several key factors are worth to be highlighted:

- first of all FBG hydrophone sensitivity exhibits a resonant behavior, with several peaks superimposed to background value slightly decreasing as acoustic frequency is increased. The obtained sensitivities are significantly enhanced with respect to the case of bare FBG hydrophones ($S_{BARE}= -2.76 \cdot 10^{-6}$ MPa$^{-1}$)[26] with an amplification of three order of magnitude for the first resonance and of two order of magnitude for background value. However, it is evident that the presence of damping causes the experimentally obtained sensitivity to be lower than the one predicted by numerical analysis. It is useful to stress the fact that, so far, this aspect could not be taken into account by means of the static model[29];

- in agreement with the simulations, resonances downshifted as a consequence of an increase of the cylindrical coating diameter from 5 mm to 10 mm.

- the background sensitivity of both FBG hydrophones exhibit a "low-pass" behavior with cut-off frequencies that decrease with the diameter. This is due to the acoustic diffraction whose effects are more relevant at lower frequencies when thicker coatings are exploited. Also this result is in good agreement with the numerical simulations and could not be predicted by the static model[29].

It is also worth noting from Fig. 17(b) that, independently from resonances, the FBG hydrophone resulted much more sensitive than the reference PZT one in the range 4-19 kHz and 4-21 kHz respectively for 5-30 and 10-30, whereas for higher frequencies the optical sensitivity strongly decreases probably due to the acoustic wave scattering. From a quantitative point of view it can be said that pressure resolutions of the order of a few Pascal can be obtained in correspondence of hydrophones resonances.

All the results here reported evidence the possibility to design and realize extremely sensitive optical fiber devices whose performance can be optimized and tailored by a proper selection of the coating features to suite the specific SONAR application.

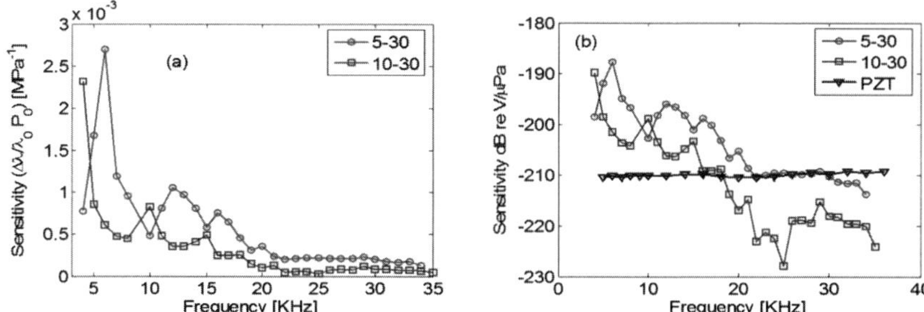

Fig. 17. (a) Sensitivity of the optical sensors versus acoustic frequency and (b) sensitivity of acoustic sensors compared with PZT reference hydrophone[35] (reproduced under the permission of SPIE Publisher).

## 5. Conclusions

In this chapter, we have reported the results of a characterization analysis of FBG hydrophones for underwater sound pressure detection, carried out within the frequency range 4-35 kHz. The optical hydrophones investigated consist of FBG sensors coated with cylindrically-shaped coating with different size.

Our numerical results fully characterize the opto-acoustic response of the optical hydrophone, and indicate that the coating may significantly enhance the sensitivity over the whole investigated frequency range, by comparison with an FBG without coating. The excitation of characteristic resonances (at frequencies related to the physical and geometrical parameters) of the cylindrical coating further improves the sensor performance. With the aid of modal analysis, we have associated the frequency peaks with longitudinal vibration modes supported by the coating. In particular, under normally-incident plane wave excitation, the structure exhibits sensitivity peaks associated to symmetric longitudinal modes, whereas anti-symmetric modes may also excited under oblique incidence.

Our experimental results confirmed the expected resonant behavior of such devices, and are in good agreement with the numerical predictions. The best performances have been obtained in the range 4-21 kHz, where acoustic scattering does not considerably affect the sensor response. Moreover, our results revealed the possibility to properly tailor the FBG hydrophones sensitivity via suitable choice of the coating geometry, thereby opening up new venues for the optimization and tailoring of acoustic performance of optical hydrophones for specific applications.

Finally, optical hydrophones based on coated FBG exhibited an excellent capability to detect acoustic waves in the investigated frequency range, with extremely high sensitivity.

## References


1. J. H. Cole, R. L. Johnson, and P. G. Bhuta, *J. Acoust. Soc. Am.* **62**, 1136 (1977).
2. J. A. Bucaro, H. D. Dardy , and E. F. Carome, *J. Acoust. Soc. Am.* **62**, 1302 (1977).
3. J. R. Lee and H. Tsuda, *Meas. Sci. Technol.* **17**, 645 (2006).
4. G. Wild and S. Hinckley, *IEEE Sensors J.* **8**, 1184 (2008).
5. B. Culshaw, G. J. Thursby, D. Betz, and B. L. Sorazu, *IEEE Sensors J.* **8**, 1360 (2008).
6. G. Thursby, B. Sorazu, D. Betz, W. Staszewski, and B. Culshaw, *Appl. Mech. Mat.* **1–2**, 191 (2004).
7. C. Kirkendall and A. Dandridge, *J. Phys. D: Appl. Phys.* **37**, R197 (2004).
8. N. Takahashi, A. Hirose, and S. Takahashi, *Opt. Rev.* **4**, 691 (1997).
9. D. C. Betz, G. Thursby, B. Culshaw, and W. Staszewski, *Smart Mater. Struct.* **12**, 122 (2003).
10. D. C. Betz, G. Thursby, B. Culshaw, and W. Staszewski, *Smart Mater. Struct.* **15**, 1305 (2006).
11. H. Tsuda, E. Sato, T. Nakajima, H. Nakamura, T. Arakawa, H. Shiono, M. Minato, H. Kurabayashi, and A. Sato, *Opt. Lett.* **34**, 2942 (2009).
12. P. Wierzba and P. Karioja, SPIE Conference Proceeding Vol. 5576 (International Society for Optical Engineering, 2004), pp. 348.
13. D. J. Hill and P. J. Nash, SPIE Conference Proceeding Vol. 4185 (International Society for Optical Engineering, 2000), pp. 33.

14. S. Foster, A. Tikhomirov, M. Milnes, J. van Velzen, and G. Hardy, SPIE Conference Proceeding Vol. 5855 (International Society for Optical Engineering, 2005), pp. 627.

15. N. Takahashi, K. Yoshimura, S. Takahashi, and K. Imamura, *Ultrasonics* **8**, 581 (2000).

16. S. Tanaka, H. Yokosuka, and N. Takahashi, SPIE Conference Proceeding Vol. 5855 (International Society for Optical Engineering, 2005), pp. 699.

17. S. Tanaka, H. Yokosuka, N. Takahashi, *Acoustical Science and Technology* **27**, 50 (2006).

18. N. Takahashi, K. Tetsumura, and S. Takahashi, SPIE Conference Proceeding Vol. 3740 (International Society for Optical Engineering, 1999), pp. 270.

19. H. Yokosuka, S. Tanaka, and N. Takahashi, *Acoust. Sci. Tech.* **26**, 456 (2005).

20. S. Goodman, A. Tikhomirov, and S. Foster, SPIE Conference Proceeding Vol. 7004 (International Society for Optical Engineering, 2008), pp. 700426-1.

21. W. Zhang, Y. Liu, F. Li, and H. Xiao, *J. Lightwave Technol.* **26**, 1349 (2008).

22. L. Ma, H. Yongming, L. Hong, and H. Zhengliang, *Photonics Technol. Lett.* **21**, 1280 (2009).

23. S. Goodman, S. Foster, J. van Velzen, and H. Mendis, SPIE Conference Proceeding Vol. 7503 (International Society for Optical Engineering, 2009), pp. 75034L1.

24. S. Foster, A. Tikhomirov, and J. van Velzen, *J. Lightwave Technol.* **29**, 1335 (2011).

25. G. A. Cranch, G. M. H. Flockhart, and C. K. Kirkendall, *IEEE Sensors J.* **8**, 1161 (2008).

26. D. J. Hill and G. A. Cranch, *Electron. Lett.* **35**, 1268 (1999).

27. G. B. Hocker, *Appl. Opt.* **18**, 3679 (1979).

28. G. B. Hocker, *Opt. Lett.* **4**, 320 (1979).

29. G. B. Hocker, *Appl. Opt.* **18**, 1445 (1979).

30. Y. Liu, Z. Guo, Y. Zhang, K. S. Chiang, and X. Dong, *Electron. Lett.* **36**, 564 (2000).

31. A. Cusano, S. D'Addio, A. Cutolo, S. Campopiano, M. Balbi, S. Balzarini, and M. Giordano, *Sens. Trans. J.* **82**, 1450 (2007).

32. S. Campopiano, A. Cutolo, A. Cusano, M. Giordano, G. Parente, G. Lanza, and A. Laudati, *Sensors* **9**, 4446 (2009).

33. M. Moccia, M. Pisco, A. Cutolo, V. Galdi, and A. Cusano, SPIE Conference Proceeding Vol. 7753 (International Society for Optical Engineering, 2011), pp. 775384.

34. M. Moccia, M. Pisco, A. Cutolo, V. Galdi, P. Bevilacqua, and A. Cusano, *Opt. Exp.* **19**, 18842 (2011).

35. M. Moccia, M. Consales, A. Iadicicco, M. Pisco, A. Cutolo, and A. Cusano, SPIE Conference Proceeding Vol. 7753 (International Society for Optical Engineering, 2011), pp. 775383.

36. COMSOL Multiphysics, User's Guide (COMSOL AB, 2008).
37. F. Ihlenburg, in *Finite Element Analysis of Acoustic Scattering*, Eds. J. E. Marsden and L. Sirovich (Springer, 1998), p. 6.
38. L. D. Landau and E.M. Lifshitz, in *Fluid Mechanics,* Eds. L. D. Landau and E.M. Lifshitz (Pergamon Press, 1987), p. 1.
39. C. J. S. de Matos, P. Torres, L. C. G. Valente, W. Margulis, and R. Stubbe, *J. Light. Tech.* **19**, 1206 (2001).
40. U. Fano, *Phys. Rev.* **124**, 1866 (1961).
41. K. Sakoda, *Phys. Rev.* **B 52**, 7982 (1995).
42. A. Ricciardi, I. Gallina, S. Campopiano, G. Castaldi, M. Pisco, V. Galdi, and A. Cusano, *Opt. Express* **17**, 6335 (2009).
43. G. M. L. Gladwell and D.K. Vijay, *J. Sound Vibration* **42**, 387 (1975).
44. C. F. Beards, in *Structural Vibration Analysis and Damping* (Butterworth Heinemann, 1996), p. 157.
45. A. Guran, J. Ripoche, and F. Ziegler, in Acoustic Interactions with Submerged Elastic Structures, Eds. A. Guran, J. Ripoche, and F. Ziegler (World Scientific, 1996), p.1.
46. M. Moccia, M. Consales, A. Iadicicco, M. Pisco, A. Cutolo, V. Galdi, and A. Cusano, "Resonant Hydrophones based on Coated Fiber Bragg Gratings", Journal of Lightwave Technology, vol. 30, pp. 2472-2481 (2012)

# LASER REMOTE SENSING FOR ENVIRONMENTAL APPLICATIONS


Antonella Boselli[1], Gianluca Pisani[2], N. Spinelli[2*], Xuan Wang[3]

*[1]CNISM and CNR-IMAA*
*c.da S.Loja-Tito-Scalo 85100 Potenza, Italy*

*[2]CNISM and Dipartimento di Scienze Fisiche,*
*Università di Napoli "Federico II",*
*via Cintia 21, 80126 Napoli, Italy*
**Email: spinelli@na.infn.it*

*[3]CNISM and CNR-SPIN*
*via Cintia 21, 80126 Napoli, Italy*


After the first applications of the laser remote sensing devices in the '60s, studies related to the atmosphere (long-range transport phenomenon, biomass burning, desert dust, and anthropogenic pollutants), volcanic aerosol, cloud investigations, bathymetric and topographic applications have greatly benefited from the use of technology mainly based on LIDAR device. Due to continuous advances in laser technology, LIDAR systems are increasingly adopted in all fields of remote sensing where range resolved capability is required. This chapter is devoted to the presentation of the state of the art of the LIDAR applications to the environmental monitoring.

## 1. Introduction

Environmental monitoring by laser remote sensing is one of the most relevant applications of photonics. Between the large number of different implementations of laser remote sensing devices, probably the most fertile application is the optical version of the RADAR, commonly named LIDAR (LIght Detection And Ranging).

     *A. Boselli et al.*

LIDAR technique is based on the same principle of the RADAR, consisting in the transmission of a short pulse of light though the target and in the analysis of the backscattered radiation (Fig. 1). The measure of the time of flight for light pulse to return back to the receiver allows determining the distance of the target, while the analysis of the intensity and of the spectral distribution of the detected radiation gives information on optical properties of the target and of the medium.

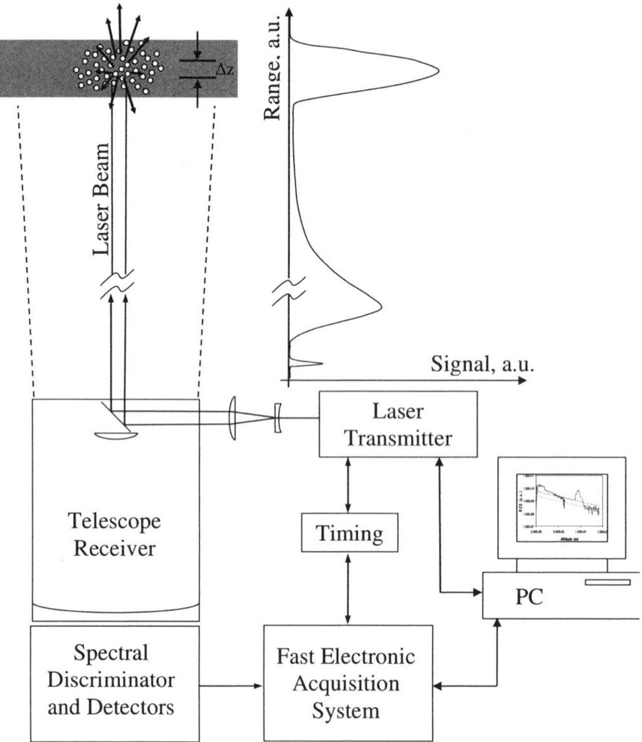


Fig. 1. Schematic drawing of the LIDAR setup. The altitude z of the target is obtained by measuring the time T elapsed between the transmission of the laser beam and the detection of the backscattered radiation through the relation z=cT/2.The spatial resolution $\Delta z$ of the LIDAR measurement depends on the laser pulse duration and the time resolution of the electronics.

The first LIDARs were used for atmospheric studies related to composition, structure, clouds, and aerosols. Initially based on ruby lasers, LIDARs represented one of the first applications of the laser.

Starting from the first implementation, the development of new experimental schemes, methodologies and applications of the LIDAR technique is grown at the same rate as the development of laser sources, sometimes also driving it. The improvement of lasers has been crucial for the LIDAR technology development and the characteristics of laser presently available allowed to free the huge potential of laser remote measurements and to open a large number of new perspectives.

In the past years the LIDAR technique has been applied in a lot of different fields through an uninterrupted development of experimental schemes by using different physical processes.

A rough classification of the experimental techniques presently implemented or under development and of the involved physical processes includes the following:

## 1.1 *Elastic Scattering*

LIDAR based on elastic backscatter is typically used for studies of aerosols and clouds. The backscattered radiation is detected at the same wavelength of the transmitted beam. The intensity of the received optical signal at a defined range is related to the backscattering efficiency of the target and to the beam extinction along the path. In these cases, extinction and backscattering coefficients of scatterers are the quantities of interest.

Elastic LIDARs prevalently make use of laser sources operating in the 300-1100nm range, often with multi-wavelength and angular scanning capabilities.

## 1.2 *Raman Scattering*

Raman LIDARs operate by making use of the inelastic scattering in order to identify the specific gas of interest. Devices combining elastic and inelastic scattering observation capabilities are used to retrieve atmospheric aerosol optical parameters. Moreover, the Raman inelastic scattering is at the basis of LIDAR measurements of water vapor and temperature vertical profiles in the atmosphere. Due to the $\lambda^{-4}$ dependence of the Raman scattering cross section, Raman LIDARs operate mainly in the UV region.

### 1.3  *Polarization*

In atmospheric studies, changes of the polarization status of the backscattered radiation with respect to the incident beam allow inferring information about geometrical properties and physical state of aerosol particles.

### 1.4  *Differential Absorption*

A differential absorption LIDAR (DIAL) transmits laser beams at two wavelengths: an on-line wavelength that is absorbed by the gas of interest and an off-line wavelength that is not absorbed. This methodology is used for range-resolved measurements of the concentration of gas in the atmosphere ($CO_2$, $O_3$, $SO_2$, NO, $NO_2$, Hg, $H_2O$). DIAL technique is applied in spectral regions where the absorption bands of molecule of interest lie. DIAL LIDARs operate typically in the UV region (290 – 450 nm) for $O_3$, $NO_x$ and $SO_2$, while 600 – 900 nm region is typically used for $H_2O$ and 9 - 11 $\mu$m for $CO_2$.

### 1.5  *Doppler Effect*

Doppler LIDAR is used to measure wind speed along the beam by measuring the frequency shift of the backscattered light. Doppler frequency shift is also at the basis of High Spectral Resolution LIDAR (HSRL). This last technique makes use of difference in the spectral bandwidth of scattered radiation from molecules and aerosol (see Fig. 2) in order to distinguish their relative contribution to the elastic signal HRSL LIDARs allow independent determination of the extinction and backscattering coefficient of particulate. Doppler wind LIDARs operate in the 1.5 - 2 $\mu$m spectral region and HRSL technique is applied mainly at 532 nm.

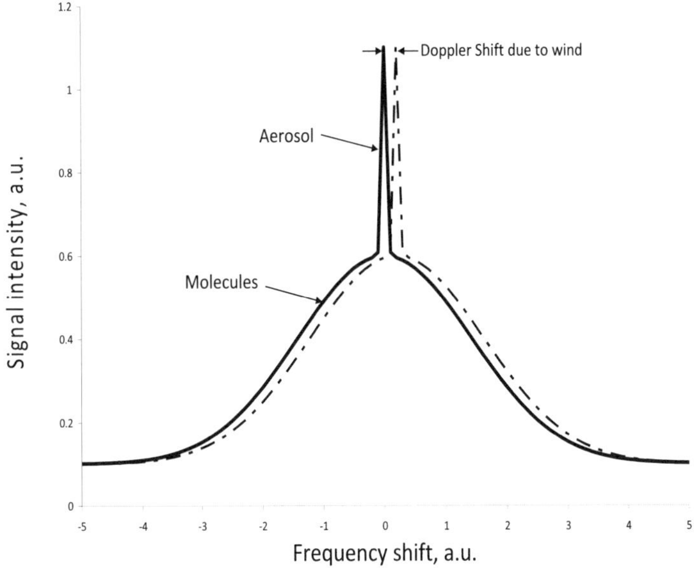



Fig. 2. Doppler shift of the spectral distribution of the elastically scattered radiation from molecules (Rayleigh component) and aerosol particles (Mie component) for zero (solid line) and non zero wind (dashed line) speed.

## 1.6 *Fluorescence*

Several implementations of the LIDAR are based on fluorescence from oil, pollutants, metal atoms, vegetation, biological aerosols and insects.

Typically, in these applications, UV wavelengths are employed.

## 1.7 *Reflection or Diffuse Reflection*

The most direct implementation of the LIDAR principle is in finding devices in which high resolution measurements of the target's distance is performed. In general 1.5 – 2 μm emitting lasers are employed in telemetry while simultaneous emission at 532 nm and 1064 nm are used in bathymetry.

Today's applications of LIDAR make often use of combinations of two or more of the above mentioned methods in ground based, airborne or spaceborne platforms.

These applications are related to climatological studies (atmospheric gas, atmospheric aerosol, including desert dust and volcanic ashes, studies on tropospheric and stratospheric clouds, wind measurements), as well as to agricultural and forest monitoring, to archaeological and architectonic heritage monitoring, to range finding and terrain mapping, to bathymetry and to several others non scientific applications as the traffic speed law enforcement for vehicle speed measurement.

This chapter is devoted to the presentation of the state of the art of major applications of the LIDAR to the environmental monitoring.


## 2. Atmospheric Aerosol, Volcanic Ash, Desert Dust

In the last decade, a large number of studies have been carried out aiming to understand direct and indirect effects of aerosol on the radiation budget and to reduce the uncertainty in climate forcing studies[2]. A great effort has been done in coordinating at international level systematic observations of aerosol optical parameters and their profiling[3,4,5]. Moreover, high interest has been devoted to develop and implement new and enhanced satellite sensors[6,7,8] in order to improve numerical models[9,10].

The LIDAR technique is actually recognized as a well established tool for aerosol research because it gives detailed information on the aerosol properties with high vertical and temporal resolutions. In this respect, long-term observational data result very useful in the analysis of temporal and spatial variations of aerosols and clouds and of their interactions in the atmosphere. These data are also very important to validate satellite observations and to evaluate global and regional aerosol models[11].

Different levels of knowledge of the aerosol properties can be achieved by arranging different configurations of the LIDAR system making it sensible to specific optical interactions.

In the case of LIDAR based on elastic or Raman scattering, a characteristic equation, the LIDAR equation, relates the received signal with both the scattering properties of the atmospheric species and the characteristics of the LIDAR instrument. In the following we report the LIDAR equation in the single scattering approximation:

$$P(\lambda,\lambda_L,z) = P_o \frac{c\tau}{2} \frac{A}{z^2} \xi(\lambda)\beta(\lambda,\lambda_L,z)\exp\left\{-\int_0^z [\alpha(\lambda,\varsigma) + \alpha(\lambda_L,\varsigma)]d\varsigma\right\} \quad (1)$$

where:

z is the spatial coordinate which is related to the time T elapsed from the light emission trough the relation: $z = cT/2$;

$\lambda,\lambda_L$ are the detected and the laser wavelengths, respectively, and $\lambda = \lambda_L$ if the elastic scattering is considered;

$P, P_0$ are the detected optical power and the output power of the laser, respectively;

c is the light speed;

$\tau$ is the temporal pulse-length;

$A/z^2$ is the solid angle of scattering covered by the telescope;

$\xi(\lambda)$ is the overall efficiency of the system at the wavelength $\lambda$;

$\beta(\lambda, \lambda_L, z)$ is the backscatter coefficient. It represents the fraction of the incident energy that is backscattered for the unity of the solid angle and for unity of atmospheric depth $(m^{-1}sr^{-1})$ by both aerosols and molecules; it and can be written as: $\beta(\lambda,z) = \beta_{mol}(\lambda,z) + \beta_{aer}(\lambda,z)$;

$\alpha(z, \lambda)$ is the extinction coefficient $(m^{-1})$ related to the total attenuation of the laser beam through the atmosphere. The extinction is due to both scattering and absorption processes by aerosols and molecules and can be written as the sum of four components:

$$\alpha(\lambda,z) = \alpha_{mol}^{sca}(\lambda,z) + \alpha_{mol}^{abs}(\lambda,z) + \alpha_{aerl}^{sca}(\lambda,z) + \alpha_{aer}^{abs}(\lambda,z) \quad (2)$$

Different forms of the basic LIDAR equation are applied to any specific LIDAR technique in order to take into account the physical processes involved and to derive different information by means of suitable retrieval procedures.

A LIDAR in the elastic backscatter configuration typically uses laser sources in the 300 - 1100 nm spectral range. This simple configuration of

                    *A. Boselli et al.*

LIDAR technology allows defining the height, evolution and structure of aerosol layers in the atmosphere[12].

The aerosol content in the atmosphere is related to the aerosol backscatter coefficient $\beta_{aer}(\lambda,z)$, which profile can be obtained from elastic LIDAR measurements by following the Klett-Fernald retrieval methods[13,14]. This method requires a hypothesis on the extinction-to-backscattering ratio (LR), a parameter depending on the aerosol microphysical properties, such as chemical composition, refractive index, shape, and particles size distribution[15].

The vertical structure of the lower atmosphere is marked by the Planetary Boundary Layer (PBL) representing the upper limit for aerosol dispersion[16]. Local circulations phenomena promote the development of aerosol layered structures within the PBL, related to complex patterns of pollutant transport[17,18].

To determine the PBL height from LIDAR, several methods have been proposed[19-22]. These studies pointed out the need to follow the time evolution of the LIDAR profiles and to perform a cross-comparative analysis of the results in complex situations with elevated aerosol layers in the atmosphere[17,23].

The possibility to follow the temporal evolution of aerosol structures by means of long records of vertically resolved LIDAR observations is also important to assess transport events.

Advanced multi-wavelength elastic/Raman LIDAR systems can provide a more relevant contribution to monitor aerosol optical and microphysical properties[24]. Using the methodology proposed by Ansmann et al.[25], the simultaneous measurement of the elastic backscatter and the Raman inelastic backscattered signals due to $N_2$ or $O_2$ molecules allows determining the profiles of both extinction and backscatter coefficients independently. LIDAR measurements of aerosol extinction coefficient at two transmitted wavelengths allow also estimating the spectral dependence of backscattering and extinction coefficients so giving information on aerosol particle size.

These more complex LIDAR measurements allow discriminating aerosol typology; they have been used together with aerosol transport models output to gain information on mixing processes in the atmosphere. Elastic/Raman LIDAR systems have been also used to study the evolution of potentially hazardous aerosols as dust aerosol or aerosol produced by volcanic activities. In this respect, several field campaigns have been carried out to characterize particles transported over long distances and to quantify their radiative effects.

In particular, vertically resolved LIDAR observations of dust were simultaneously performed close to the source of desert dust and in remote areas. By using advanced LIDAR systems with the main aim to study optical and microphysical properties of dust aerosol, field campaigns provided also a good opportunity to analyze the impact of dust on cirrus cloud formation and to study the heterogeneous ice formation[26,27]. Systematic LIDAR observations of Saharan dust outbreak over European region were performed from the EARLINET LIDAR network[28], in order to investigate dust modification processes during the transport on a continental scale.

Furthermore, in recent years it is grown the interest in the study of volcanic ash plumes dispersion.

Since the first observations of airborne volcanic ash layers in the 1990s measurements, LIDARs have improved our knowledge on volcanic ash aerosols released during explosive eruptions. In particular, the polarization LIDAR technique has been used to distinguish shape and thermodynamic phase of atmospheric aerosol[29,30,31]. The so called "linear depolarization ratio" $\delta$ can be used to characterize volcanic aerosols and to establish criteria for the discrimination from other aerosol types. This parameter is defined as the ratio of the backscatter coefficients corresponding to parallel (P) and cross-polarized (S) component of the detected signal with respect to the polarization plane of the emitted laser beam: $\delta_{\mathrm{V}} = \dfrac{\beta_{\mathrm{S}}(z)}{\beta_{\mathrm{P}}(z)}$ ).

Moreover, under suitable conditions, LIDAR measurements may estimate the column height of the volcanic plume. This parameter is one of the key elements needed to reliably forecast plume dispersal using volcanic ash transport and dispersal models.

Recently, studies of geometrical and optical properties of volcanic plumes as well as evaluations of spatial maps of the ash mass concentration have been performed[31,32]. These results can help to improve numerical dispersion models that need a first guess estimate of the ash emissions as input variable[33].

Airborne and ground-based LIDAR, sun/sky photometer, and surface in-situ measurements of aerosols and trace gases allowed to characterize volcanic ash plumes over Europe during the eruption of the Eyjafjallajökull volcano on April/May 2010[34,35]. This strong volcanic ash event provided also a good opportunity to investigate the impact of volcanic eruptions on cirrus cloud formation in the atmosphere by using LIDAR observation combined with model results[36].

A special effort during this recent eruption was made by the EARLINET network. Multi-wavelength Raman LIDAR observations performed in most part of the Europe allowed studying the impact of volcanic ash on a continental scale. The performed measurements allowed to characterize in near real-time geometrical, optical and microphysical properties of the volcanic plume and to provide high-quality data set for evaluating satellite data and aerosol dispersion models for this kind of volcanic events[37,38].


## 3. Clouds

Aerosol and clouds have a fundamental role in climate change. The incomplete knowledge of clouds is an important source of uncertainty in climate change prediction[2]. Clouds act as greenhouse gas by reflecting part of the solar radiation and absorbing thermal radiation from Earth. Clouds occupy about 50% of the sky in the average[39]; in particular, cirrus clouds are responsible for an average coverage of about 20%[40,41].

Due to their altitude and composition, cirrus clouds are difficult to study by airborne or satellite microwave radars, but they can be characterized by LIDAR systems based on different optical techniques. LIDAR based on elastic backscatter has the ability to determine spatial structure of clouds and their boundaries.

The depolarization LIDAR technique[42], thanks to its ability to provide information on particle shape, is a powerful tool to discriminate between liquid water and ice clouds. Cloud extinction coefficient profiles and their optical depth are provided by Raman or High Spectral Resolution

LIDAR (HSRL), while Differential Absorption LIDARs (DIAL) are able to determine the vertical profiles of water vapor mixing ratio.

The availability of LIDAR apparatuses simultaneously operating with several of the above mentioned capability has greatly enhanced the potentiality for a better knowledge of clouds properties. Alternatively, multiple remote sensors combine LIDAR with millimeter-wave radar or radiometer measurements (LIRAD) methods to derive quantitative data on visible and infrared optical depths[43-46].

Cloud properties at a global scale are presently derived from satellite borne LIDARs as Cloud-Aerosol LIDAR and Infrared Pathfinder Satellite Observation (CALIPSO)[47].

Depolarization and backscatter intensity measurements allowed to quantify seasonal and spatial variations of clouds, the associated radiative impacts, water and ice cloud discrimination[41], to estimate liquid water content information and also to study the spectral dependence of the backscattering signals from cirrus between 532 nm and 1064 nm[48].

## 4. DIAL LIDAR for Atmospheric Gas

An application of the LIDAR technique being able to determine the distribution and concentration of gaseous species in the atmosphere is the DIfferential Absorption LIDAR (DIAL), which is based on the molecular absorption process.

In the DIAL technique laser pulses at two different wavelengths, $\lambda_{on}$ and $\lambda_{off}$, are transmitted into the atmosphere over the same path simultaneously, or closely in time ($\sim$1 sec), so that the atmospheric volume under observation can be considered unchanged. The wavelength, $\lambda_{on}$, is chosen in the region where the interesting gas has a strong absorption. The wavelength, $\lambda_{off}$, is not absorbed by the species being measured and it is close to $\lambda_{off}$ so as to consider that the scattering and extinction by other molecules and particles at both wavelengths is the same.

The comparison of the LIDAR signals at the two wavelengths allows obtaining the direct determination of the number density profile of the gas under observation.

The LIDAR signals backscattered at two wavelengths are described by the following equations:

$$P(\lambda_{on},z) = P_0(\lambda_{on}) \frac{A_0}{z^2} \frac{c\tau}{2} \beta(\lambda_{on},z) \xi(\lambda_{on}) \exp\left\{ -2\int_0^z \left[ \alpha_{\lambda_{on}}(z') + n(z')\sigma_{\lambda on} \right] dz' \right\} \quad (3)$$

$$P(\lambda_{off},z) = P_0(\lambda_{off}) \frac{A_0}{z^2} \frac{c\tau}{2} \beta(\lambda_{off},z) \xi(\lambda_{off}) \exp\left\{ -2\int_0^z \left[ \alpha_{\lambda_{off}}(z') + n(z')\sigma_{\lambda off} \right] dz' \right\} \quad (4)$$

where:
- $n(z)$ is the density of the species of interest;
- $\sigma_\lambda$ is the absorption cross section of the gas of interest;
- $\beta(\lambda)$ is the backscattering coefficient;
- $\alpha_\lambda$ is the extinction coefficient.

If the two wavelengths are close together and nearly simultaneously transmitted in the atmosphere, from the ratio of the two LIDAR returns we can obtain:

$$n(z) = \frac{1}{2\Delta z} \frac{1}{\sigma_{\lambda_{on}} - \sigma_{\lambda_{off}}} \ln\left( \frac{P_{off}(z+\Delta z)P_{on}(z)}{P_{on}(z+\Delta z)P_{off}(z)} \right) \quad (5)$$

where $\Delta z$ is the vertical resolution.

The DIAL technique features high sensitivity (~1 ppb). In principle, it can be applied to monitor any atmospheric specie provided that the suitable laser sources and detectors are available.

In the past, DIAL application was limited to UV spectral range, because of the lack of tunable laser sources over a large infrared spectral range (up to several μm). Tunable laser sources (e.g. Ruby laser, Ti:Sapphire laser, Nd:YAG and Excimer pumped dye lasers and Alexandrite lasers) were employed relied on the coincidence of the laser wavelengths with pollutants absorption lines[49]. Nevertheless, the absorption line of the gas under study must be carefully chosen in considering: i) the line strength and minimizing; ii) the interference from other atmospheric pollutants; iii) the sensitivity on atmospheric temperature and pressure changes. These conditions are fundamental requirements for DIAL applications[50]. The development of continuously tunable IR sources for DIAL application in the atmosphere was a great challenge, because of the large number of pollutants displaying absorption lines in such a spectral region. The research in the field of innovative laser sources allowed the realization of laser systems based on non linear crystals that can generate IR radiation by frequency mixing and optical parametric amplification. The development of advanced light sources such as pulsed Optical Parametric Oscillators (OPO) and Amplifiers (OPA) and the availability of a new generation of photomultipliers and photodiodes operating in near IR range opened new perspectives for LIDAR measurements, so encouraging the development of DIAL systems in a spectral region important for monitoring many pollutant species[51].

Using ground-based and airborne systems as well as systems that are suitable for long-term deployment in space, the DIAL technology has been successfully applied to the study of molecules as $H_2O$, $SO_2$, $NO_2$, $CO_2$, $NO$, $O_3$. Moreover, the capability of the DIAL technology in revealing forest fire and minimizing false alarm events with the additional advantage of higher degree of eye safety provided by lasers operating in the infrared has been also demonstrated[52].

An innovative DIAL system using UV laser source provided regular observations of ozone vertical distributions in the troposphere under both daytime and nighttime conditions[53]. Moreover, DIAL measurements carried out in the infrared region allowed to track plumes coming from sources such as petrochemical plants and landfill sites demonstrating the suitability of DIAL for monitoring Volatile Organic Compound (VOC) emissions[54]. Finally, new multi-wavelength DIAL technologies using tunable dye laser pumped by Nd:YAG laser allowed accurate measurements of $SO_2$, $O_3$ and aerosol for a volcanic emission event[55].


## 5. Wind

Wind velocity measurements have many applications. The knowledge of wind field is crucial to find the best location for aeolian power stations and for their optimization and management. The wind influence by hills, forestry, complex terrain and also by other turbines must be characterized to select proper turbines and to maximize the power production.

An accurate knowledge of tropospheric wind data is needed for several climate questions and weather forecasts. Wind measurements are relevant in the study of the atmosphere dynamics, for the validation of atmospheric models and for predicting the future state of the atmosphere[56,57,58].

The interest for wind field measurements is also related to:
- Air turbulence, wind shear and gust fronts characterization
- Air pollution monitoring
- Wind tunnels characterization
- Aircraft true airspeed, aircraft wake vortices detection

Several techniques for wind speed measurement by LIDAR have been implemented[59] that can be classified as direct motion, coherent Doppler, direct Doppler and geostrophic wind detection techniques.

Direct detection of the motion is possible by using the definition of velocity (derivative of displacement). For direct motion detection, wind tracers are needed to track the motion. Tracers can be aerosols, clouds, plumes or any inhomogeneous distribution of tracers[60,61].

A scanning LIDAR can be used to take two images $I_1(x,y)$ and $I_2(x,y)$ of tracers at two different times $t_1$ and $t_2$; displacements $\sigma_x$ and $\sigma_x$ in x and y coordinate can be determined by maximizing the correlation coefficient:

$$c\left(\sigma_x,\sigma_y\right)=\int\int I_1\left(x,y\right)I_2\left(x-\sigma_x,y-\sigma_y\right)dxdy. \qquad (6)$$

Then the component of the average wind speed in the xy plane can be derived as $\sigma_x/(t_2-t_1)$ and $\sigma_y/(t_2-t_1)$, respectively.

The direct detection of the motion by LIDARs can provide wind measurements with high resolution in lower atmosphere, industrial plant or wind tunnels.

Geostrophic wind is the horizontal wind velocity for which the Coriolis force exactly balances the pressure gradient force (horizontal pressure force). Geostrophic wind can be retrieved by measuring temperature and pressure, from which gradients can be determined[62]. It can be obtained if the horizontal distribution of density and temperature or pressure is available and these data, in principle, can be obtained from LIDARs based on elastic backscattering.

Doppler shift is the apparent frequency change of radiation due to the relative motion of scattering or absorbing particle with respect to the source or radiation. In Fig.2 the Doppler shift of the spectral distribution of the elastically scattered radiation from molecules (Raleigh component) and aerosol particles (Mie component) for a given wind velocity is shown. In general, aerosol scattering signal is used for Doppler LIDAR wind measurements, due to its narrow bandwidth and strong signals.

Two main implementations of Doppler wind LIDARs are based on heterodyne and direct detection methods, respectively. In the heterodyne detection (coherent detection) the return signal is optically mixed with a local oscillator. As a result, a signal at the difference frequency is generated which is the Doppler shift due to the scattering of moving particles.

In direct detection Doppler LIDAR no local oscillator is used. Instead, optical filters or spectrum analyzers are used to resolve the spectral components of the received signals, so converting the Doppler frequency shift to a change in optical intensity of the detected signal[63,64,65,66].

The direct detection method requires a narrowband LIDAR transmitter with stable frequency and very sophisticated receiving devices.

Although the direct detection Doppler LIDAR could be applied to aerosol or molecular scattering or both, in practice the aerosol scattering is used due to the higher intensity of the signal. For that reason this technique is applied mainly in the lower atmosphere where aerosol scattering dominates.

Resonance fluorescence Doppler LIDAR is a direct detection method using the resonance fluorescence from atoms, e.g., Na, K, Fe, in the mesosphere and lower thermosphere (70-120 km of altitude). In these LIDARs the atomic absorption lines act as natural frequency analyzers.

In any implementation of the Doppler wind LIDAR, the directly measured speed is the velocity component along the line of sight of the radiation beam, therefore the estimation of the vector wind velocity requires at least three independent measurements along different directions. To do that, in principle, three or more LIDAR systems are required. In practice two LIDAR beam scanning techniques can be used to determine the vector wind velocity, both based on the assumption of the horizontal homogeneity of the wind field over the sensed volume during the time of measurements, they are:

- the Velocity-Azimuth-Display technique, consisting in a conical scan of the target volume[67];

- the Doppler-Beam-Swinging (DBS) techniques consisting in pointing the LIDAR beam along three or four suitably chosen different directions within the target volume[59].

Any optical or spectroscopic methods that can distinguish frequency shift or difference could be applied to direct Doppler LIDAR. Therefore, new methods are still being proposed.

The recent development of fibre lasers has enabled the development of wind LIDAR devices with an increased capacity, degree of accuracy and reliability. Moreover wind LIDAR is a field still going through a rapid growth especially for airborne and spaceborne applications[56,68].


## 6. Airborne Laser Bathymetry

Hydrographical survey can be usually done by multi-beam and side-scan sonar technologies, imaging spectrometers or hyperspectral sensors and Airborne Laser Bathymetry (ALB). Surface-borne multi-beam sonar systems are more efficient in deeper waters. Hyperspectral image and ALB are more suitable in offshore areas. Imaging spectrometers are passive sensors that measure reflected sunlight from water, objects in the water and the bottom of water. Multispectral sensors collect images of a scene and provide access to several tens to hundreds of very narrow spectral channels nearly simultaneously. It can determine navigation hazards, bottom type and also water optical conditions but less information about the water deep. Due to the fact that airborne LIDAR can be fully operational with light aircraft, ALB can be considered an operational tool, at least for hydrographical survey in littoral region, since the operational time and costs are fair.

An ALB uses LIDAR technology to measure water depths. In principle, a laser transmitter on board of an aircraft fires a laser pulse towards the water surface where a fraction of its energy is back reflected to the receiver while the remaining energy propagates trough the water column down to the sea bottom and is reflected itself. The time delay between the surface and bottom returns is the direct measure of the water depth (Fig. 3).

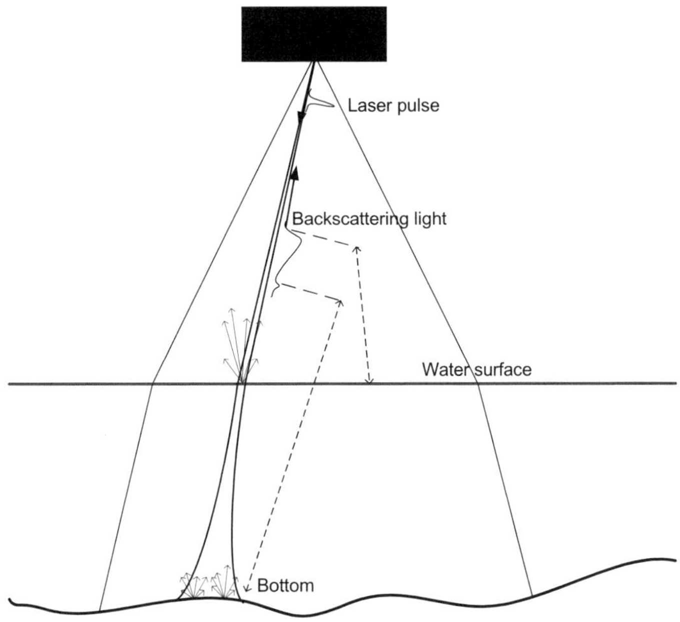


Fig. 3. Layout of the LIDAR bathymetry operating principle.

The maximum water depth reached by ALB is a function of the bottom return strength, which is limited by the laser energy losses due torefraction at the water surface and sea bottom, scattering and absorption as the pulse travels through the water column. The shape of the seabed and the water clarity affect substantially the sounding depth as well. Typically, LIDAR bathymetry is suitable for measuring depths equal to 2-3 times the site's Secchi visible depth[69,70].

In general, green and infrared (IR) pulsed beams are used in ALB[71]. For the clear open sea water, the maximum transparency wavelength is around 400 nm (blue). While for the typical coastal water green light has the least attenuation[72] and it is chosen for bottom detection. Infrared light is more appropriate for detection of the sea surface position because it penetrates very little in water. The IR beam can be collimated and scanned together with the green beam or it may be broader and pointed in nadir direction. In practice, the infrared surface lidar echo is mainly due the interface reflection component. The IR volume backscatter is considerably weaker because of 3 orders of magnitude higher attenuation in the water at this wavelength with respect to green beam, therefore the lidar echo source is so near to the interface that the corresponding timing error is generally negligible.

The Raman backscattered signal from green beam excited in the water at the interface may also be used to determining the surface position[73].

A large beam spreading occurs in the water column because of multiple scattering phenomena. The scattering causes the beam to expand into a cone whose interior angle increase with depth.

The receiver system consists of a telescope, optical filters, light detectors, amplifiers and analog-to-digital converters.

Infrared (1064 nm), green (532 nm) and green induced red water Raman (645 nm) are detected by three different receiver channels. Because there is a difference of more than six orders of magnitude in signal amplitude between strong water surface returns and much weaker bottom returns, each of the two elastic channels (IR 1064 nm and green 532 nm) uses two detectors: a APD for shallow water and a PMT for deep-water.

The altitude of aircraft is chosen by trade-off the swath and the coverage rate for a specific laser system. Typical ALB measurements are performed from 200-500 meter of altitude. An optical scanner provides coverage of a broad swath under the aircraft track trough different modes: circular or zigzag patterns are commonly used. To avoid large pulse timing errors from both surface and bottom returns the maximum scanner nadir angle is typically 15-20 degrees corresponding to surveyed swaths widths almost equal to one half of the aircraft altitude.

Considering a typical value of the laser beam divergence and hence the corresponding spot size, the horizontal resolution is the order of 2 m; this resolution is obtained with laser pulse repetition rates lower than

1000 pulses per second. The coverage density can be significantly increased with higher repetition rates (5-10 kHz).

Gross coverage rate can be of the order of several thousand $m^2$/s.

Pulse widths <10 ns are needed to provide a vertical resolution suitable for shallow depths sounding. Such a pulse, with an appropriate digitizer and leading-edge detector, can provide sub-decimeter measurement precision[74].

The maximum significant depth depends critically on water clearness[75]. Other limitations are related to system parameters, such as laser pulse energy, spectral bandwidth and field of view of the receiving system, but also to eye-safety considerations and environmental conditions. In order to achieve eye-safe operation the beams are expanded to a diameter of the order of meters at the water surface.

Depth penetration slightly increases with the pulse energy, but the combination of eye-safe energy density requirements and geometrical effects limit the usable pulse energy to values of the order of 3 mJ.

As a simplified rule, one can expect successful operations to depths 2-3 times the Secchi depth.

Generally, airborne LIDARs are not able to detect small objects under the water, but the detection probability for small objects (< 1 m) can be increased by increasing the survey density.

A channel corresponding to the green-excited water Raman backscatter wavelength in the red portion of the spectrum can constitute a great advantage. In fact, the Raman inelastic scattering process arising from a vibrational mode of the O:H bond in water molecules, allows to obtain a relative weak return that with a high-gain, low-noise channel, is suitable for altitude up to about 400 m and can be used for surface location. Wind speed and sea-surface wave slopes don't affect this signal, but Raman surface returns from very shallow, clear water should be discriminated from bottom reflected red energy. The broad-band, green-excited fluorescence arising from dissolved organics, various pigments, and the chlorophyll tail will be also detected in the red channel. This return represents a useful extra signal if the source is in the water, while it can constitute a problem if coming from the sea bottom.

Airborne LIDAR bathymetry represents an accurate and powerful tool to perform and support services in both coastal waters and land, also in areas where sonar cannot survey. However, ALB application is limited by water clarity and cannot replace sonar use; instead it should be used together with sonar in region where both methodologies can overlap.

Particular care must be devoted to ALB system design, operation procedures and data analysis in order to build a device useful for sea bottom detection. Moreover, an accurate LIDAR for bathymetry requires a careful design of both hardware capability and software development and the establishment of procedures for the automatic analysis of the data.

## 7. Other Air/Space-Borne Applications

LIDAR apparatus plays a key role in the scientists and technicians toolboxes and can be used on both air and space based platforms. At the turn of the 1970s, the first apparatus flew on board of aircrafts, in order to obtain a better spatial coverage (from regional up to global scale), and, afterwards, the first space based atmospheric LIDAR were launched on board of Spacelab in 1994.

The main drawback of air/space-borne LIDARs is the living costs operation, which turns often into little chance of flight. Further disadvantage for flying LIDAR is complexities of the system itself that often should be unmanned and have space and power constraints.

Problems with vibrations, G-force, temperature and power supply stability must be also considered when a LIDAR system is designed for air/space-borne operation. Last but not least, there are the eyes safe considerations; actually threshold level must be considered for each flying level. To match these demands several steps can be taken like increasing the beam divergence, reducing the laser power, choosing the right eye safe wavelength or varying the altitude of the aircraft.

### 7.1. *3D mapping*

A LIDAR system on board airplanes can fire the laser beam directly toward the Earth's surface utilizing scanning technology so that LIDAR can probe the ground swept left and right with respect to the line of flight. Baltsavias[76] and Akermann[77] showed that a great degree of automation in retrieving of raw X, Y, and Z coordinates data than conventional photogrammetry can be implemented. The resulting data collected in such a pattern can be used to create a high-resolution 3D model of both the atmosphere and ground (Digital Terrain Model – DTM) surface. Due to extremely short laser pulse and precise sensor platform orientation and position, LIDAR provides highly accurate models, with accuracy of the order of the mm and cm. To ensure that collected data represent the actual ground position, it is necessary to relate the LIDAR system to a reference system at ground, hence differential GPS needs with at least two stations, the first one at ground and a second one co-located with the LIDAR sensor in the aircraft. The accuracy of GPS location of the aircraft will be reflected in the quality of final data. Hence, LIDAR data do not require aerial triangulation and ortho-rectification. The majority of the commercial systems can collect between 2,000 and almost 100,000 records per second. The spatial resolution of the digital model depends on data point density that is function of the pulse repetition frequency, the scanning angle, the flight altitude and the speed of the aircraft. Most of the commercial systems measures time of flight of the emitted pulses and the echoes return, but some of them can also measure the intensity of the reflected light carrying information on the surfaces properties where the light has been scattered. To perform this kind of measurements calibrated sensor must be utilized, since the absolute value of detected signal depends on the gain of the system.

## 7.2. *Temperature Measurement*

Recently, LIDAR application to monitor flight-critical air parameters (such as temperature, density, and pressure) in all weather conditions demands has been proposed by Fraczek[76]. Through a four channels detecting system (one channel for elastic backscatter radiation, two rotational Raman echoes[79] of the atmospheric molecules, e.g. $N_2$, and the fourth one for Raman echoes from water vapor molecules) clouds and aerosol, air temperature and density of the moist air can be retrieved.

### 7.3. *Forest Area Monitoring*

In recent years, one of the main aims at measuring the biophysical/physiological features of the forest ecosystem is the evaluation of their health in order to evaluate the sustainability. Remote sensing techniques have demonstrated a wide applicability to measure the physical/chemical properties of the investigated target at ground level. In particular, forest physical and structural features can be easily estimated and mapped[80,81,82]. Airborne laser remote sensing represents the state of art of technology to estimate forest biophysical parameters at different scales (up to single tree) since, unlike the multispectral and hyperspectral sensor, it provides information on the vertical structure of vegetation.

LIDAR data have the great advantage that there is no shadowing in the measurement, so species classification can be retrieved from range corrected LIDAR intensity[83,84,85] and important forest parameters such as canopy heights, stand volume and the vertical structure of the forest canopy can be retrieved[86,87]. High resolution airborne LIDAR was also used to monitor the health state of the urban trees in stressful environment[88]. LIDAR and multi/hyperspectral data provide complementary information (height resolved and spectral resolved, respectively)[89,90,91]. Hence these technologies represent a promising and challenging approach in the study of forests.

### 7.4. *Fluorescence LIDAR*

An application of LIDAR remote sensing is the fluorescence LIDAR that is a non-invasive technique based on Laser-Induced Fluorescence (LIF technique) to the outdoor environment. In particular, chlorophyll fluorescence emission spectra have peaks at around 685 and 740 nm and they can be used to monitor the health state of vegetation[92,93]. For instance, D'Ambrosio et al. measured the chlorophyll content and the fluorescence induction kinetics at 690 nm and 735 nm in leaves of some broadleaf tree species during the autumnal chlorophyll breakdown[94]. A joint research project by the Technical University of Budapest and the University of Karlsruhe for a non-destructive stress detection of plants, proposed in 1996 the Karlsruhe CCD-OMA fluorescence that was based on a 10mW He-Ne laser, a spectrograph and a CCD-line detector[95]. In 2000, Saito[96] proposed a system that was based on a frequency-doubled, Q-switched Nd:YAG laser operating at 532nm and 10ns pulse width. Result of chlorophyll concentration from LIF LIDAR were compared with results from a High Performance Liquid Chromatography (HPLC) indicating that the LIF imaging LIDAR could be used for macroscale measurements, such as the distribution maps of the average chlorophyll mass or the overall vegetation activity.

Several other applications of the LIF LIDAR have been proposed.

Petroleum oils absorb ultraviolet light and the process of fluorescence emission occurs primarily in the visible region of the electromagnetic spectrum at different spectral range with respect to other natural substances such as chlorophyll. This consideration allows using LIDAR LIF to distinguish oil from water or biological substances in the case of an oil spill emergency[97]. Moreover, the knowledge of the nature and concentrations of pollen, fungal spores, viruses, bacteria, and animal or plant matter in near real-time has become of interest because of its effects on human health and climate changing. Traditional methods of characterization of those species involve impact onto a suitable substrate that are then analyzed trough optical (microscopy, SEM) or chemical methods[98].

## 7.5. *Atmospheric Measurement*

Airborne LIDARs have been used since the early eighties. One of the first airborne LIDAR apparatus was loaded on board the DC-8 platform in DIAL configuration. The system was used to measure ozone and aerosol distributions[99]. Some other examples of airborne system for atmospheric measurements can be found elsewhere[100,101,102].

High Spectral Resolution LIDAR is also used in airborne systems to configuration the aerosol characteristics can be retrieved without assumptions on the aerosol extinction-to-backscatter ratio[103,104], offering both high vertical and temporal resolution.

Also Doppler LIDAR can be used in airborne configuration. Wekker reported on detailed spatial structure of the atmospheric flows and associated aerosol features within a segment of the Salinas Valley in November 2007[105].


## 7.6. *Volcanic Plume*

Due to its extreme flexibility, the airborne LIDAR is the right instrument to perform measurements over extended regional areas in case of volcanic plume emissions. One of the first volcanic ashes plume observed by airborne system was that one emitted from the last eruption of the Mt. La Soufriere, St. Vincent, on 17 April 1979[106]. The in situ LIDAR measurements performed on January and June 1990 on the plume emitted by Mount Redoubt, Alaska, was described by Hobbs[107] that used a dual wavelength LIDAR on board a Convair C131A aircraft.

However, in the last century the "most investigated by LIDAR" big eruption was that one from Mt. Pinatubo on June 1991. Winker and Osborn reported on two distinct layers ranging from 17 and 26 km sounded by an airborne depolarization LIDAR at 532 nm and flying between Barbados Island and the Florida peninsula some weeks after the eruption[108]. However, the most spectacular demonstration of the benefit that airborne LIDAR could give to atmospheric sciences became in 2010 when the Eyjafjallajökull volcano erupted on March the 20th in Iceland for almost three months. During these days, several ground based and airborne LIDARs were operated all over the Europe to monitor the airspace safety and to increase the knowledge on volcanic ash properties[109,110,111].


## 7.7. *LIDAR in Space*

First application of LIDAR in space was performed by NASA on board of the Apollo 15, 16 and 17 missios[112,113], when the spacecraft altitude above the lunar surface was measured with a laser altimeter.

The first space borne LIDAR for atmospheric measurements was that one of the LIDAR In-Space Technology Experiment (LITE). A Nd:YAG laser system with doubled and tripled frequency generators was used with a 1 m diameter telescope[114]. LITE was designed by NASA/Langley Research Center for flying on the Space Shuttle Discovery on mission

STS-64 in September 1994 and it was the first attempt to validate the technologies required for operational spaceborneLIDARs[115] in order to observe clouds, tropospheric and stratospheric aerosols, to collect information on their dynamic range and variability with much greater resolution than available at that time from orbiting sensors[116,117,118].

The first space-based LIDAR that used a diode pumped Nd:YAG laser was the Mars Orbiter Laser Altimeter (MOLA) developed at the National Aeronautic and Space Administration (NASA) Goddard Space Flight Center (GSFC) for the Mars Observer, a mission that unfortunately did not reach the Mars orbit in 1993. A copy of the same instrument was then utilized in the next mission, the Mars Global Surveyor (MGS), launched on November 7, 1996[119].

Another LIDAR equipped mission for Earth planet research was the ICESat/GLAS (Ice, Cloud and land Elevation Satellite/Geoscience Laser Altimeter System) experiment[120], which has been launched in January 2003. GLAS was a laser altimeter designed to measure ice-sheet topography and associated temporal changes, as well as atmospheric and surface properties[121,122]. It had on board three Nd:YAG lasers operating at fundamental (for surface altimetry and dense cloud heights) and doubled frequency (for the vertical distribution of clouds and aerosols).

The Cloud Aerosol LIDAR and Infrared Pathfinder Satellite Observations (CALIPSO) mission was launched in April 2006. The main goal of this mission is to determine base and the top of the clouds and aerosol layers and their overlap, identify the composition of clouds and the presence of subvisible (or "invisible" clouds), and to estimate the abundance and sources of aerosols[123,124,125]. The Cloud-Aerosol LIDAR with Orthogonal Polarization (CALIOP) on board of the spacecraft utilizes a Nd:YAG laser operating at 532 and 1064 nm. The receiver telescope is 1 meter in diameter and spectral selection system splits the collected elastic signal in three channels: one corresponding to the 1064nm backscatter intensity and two channels corresponding to parallel and orthogonally polarized components of the 532nm backscattered signal. Being CALIOP, as much as LITE and GLAS, a backscatter LIDAR, it cannot provide direct measurement of measurements of aerosol optical properties profiles. For this reason some effort has been done to provide retrieval algorithms for LIDAR equation inversion[126,127,128].

Finally, the first ESA LIDAR in space mission will be a high spectral resolution Doppler wind LIDAR for sampling of the 3-dimensional wind field[129]. The primary aim of Aeolus is to provide global observations of vertical wind profiles from the troposphere and lower stratosphere. The payload of the ADM-Aeolus mission will be the Doppler wind LIDAR ALADIN (Atmospheric LAser Doppler INstrument), which makes use of a pulsed UV laser (355 nm, circularly polarized) and a 1.5 m telescope.

## References

1. G. Fiocco, L. D. Smulli'n, *Nature*, **199**, 1275 (1963).
2. D. Randall, R. Wood, S. Bony, R. Colman, T. Fichefet, J. Fyfe, V. Kattsov, A. Pitman, J. Shukla, J. Srinivasan, R. Stouffer, A. Sumi and K. Taylor, In *Climate Change 2007: The Physical Science Basis. Contribution of Working Group I to the Fourth Assessment Report of the Intergovernmental Panel on Climate Change*. Ed. S. Solomon, D. Qin, M. Manning, Z. Chen, M. Marquis, K. Averyt, M. Tignor, and H. Miller,  Cambridge University Press, Cambridge, United Kingdom and New York, NY, USA (2007).
3. J. Bösenberg, A. Ansmann, J. M. Baldasano, D. Balis, C. Böckmann, B.  Calpini, A. Chaikovsky, P. Flamant, A. Agard, V. Mitev, A. Papayannis, J. Pelon, D. Resendes, J. Schneider, N. Spinelli, T. Trickl, G. Vaughan, G. Visconti, M. Wiegner, [A. Dabas, C. Loth, J. Pelon, (eds.)], in *Advanced in Laser Remote Sensing*, **155**, (2001).
4. B. N. Holben, T. F. Eck, I. Slutsker, D.  Tanré, J. P. Buis, A. Setzer, E. Vermote, J. A. Reagan, Y. J. Kaufman, T. Nakajima, F.  Lavenu, I. Jankowiak and A. Smirnov, *Remote Sens. Environ.* **66**, 1, (1998).
5. GAW, Report 178, Hamburg, Germany (2007).
6. F. Barnaba and G. P. Gobbi, *Atmos. Chem. Phys*, 4, no. 9-10, 2367 (2004).
7. K. Schepanski, I. Tegen, B. Laurent, B. Heinold and A. Macke, *Geophysical Research Letters*, **34**, 18, L18803 (2007).
8. D. M. Winker, M. A. Vaughan, A. Omar, Y. Hu, K. A. Powell, Z. Liu, W. H. Hunt and S. A. Young, *J. Atmos. Ocean. Technol.* **26**, 11, 2310 (2009).
9. WMO Secretariat, *Atmospheric Research and Environment Branch* (2011).
10. S. Kinne, M. Schulz, C. Textor, S. Guibert, Y. Balkanski, S. E. Bauer, T. Berntsen, T. F. Berglen, O. Boucher, M. Chin, W. Collins, F. Dentener, T. Diehl, R. Easter, J. Feichter, D. Fillmore, S. Ghan, P. Ginoux, S. Gong, A. Grini, J. Hendricks, M. Herzog, L. Horowitz, I. Isaksen, T. Iversen, A. Kirkevåg, S. Kloster, D. Koch, J. E. Kristjansson, M. Krol, A. Lauer, J. F. Lamarque, G. Lesins, X. Liu, U.  Lohmann,

V. Montanaro, G. Myhre, J. Penner, G. Pitari, S. Reddy, O. Seland, P. Stier, T. Takemura and X. Tie, *Atmos. Chem. Phys.*, **6**, 1815 (2006).

11. D. Wang, D. Wu, D. Liu and J. Zhou, *J. Opt.Soc. Korea.*, **14**, 444 (2010).

12. V. Santacesaria, F. Marenco, D. Balis, A. Papayannis and C. Zerefos, *Nuovo Cimento*, **C 21**, 6, 585 (1998).

13. J. D. Klett, *Appl. Opt.* **20**, 2, 211 (1981).

14. F. G. Fernald, *Appl. Opt.* **23**, 652 (1984)

15. J. Ackermann and J. Atmos, *Ocean Technol.*, **15**, 1043 (1998).

16. R. B. Stull, Kluwer Academic Publishers – Dordrecht/Boston/London (1998).

17. A. Boselli, M. Armenante, L. D'Avino, M. D'Isidoro, G. Pisani, N. Spinelli and X. Wang, *Boundary-Layer Meteorol.*, **132**, 151 (2009).

18. I. G. McKendry and J. Lundgren, *Prog. Phys. Geogr.*, **24**,(3), 329 (2000).

19. C. Flamant, J. Pelon, P. H. Flamant and P. Durand, *Boundary-Layer Meteorol.*,**83**, 247 (1997).

20. K. L. Hayden, K. G. Anlauf, R. M. Hoff, J. W. Strapp, J. W. Bottenheim, H. A. Wiebe, F. A. Froude, J. B. Martin, D. G. Steyn and I. G. McKendry, *Atmos. Environ.*, **31**, 2089 (1997).

21. L. Menut, C. Flamant, J. Pelon and P. H. Flamant, *Appl. Opt.*, **38**, 6, 945 (1999).

22. C. Senff, J. Bösenberg, G. Peters and T. Schaberl, *Contrib. Atmos. Phys.*, **69**, 161 (1996).

23. M. Sicard, C. Pérez, F. Rocadenbosch, J. M. Baldasano and D. García-Vizcaino, *Boundary-Layer Meteorol.*, **119**, 135 (2006).

24. E. Giannakaki1, D. S. Balis, V. Amiridis and C. Zerefos, *Atmos. Meas. Tech.*, **3**, 569 (2010).

25. A. Ansmann, M. Riebesell, U. Wandinger, C. Weitkamp, E. Voss, W. Lahmann and W. Michaelis , *Appl Phys B*, **55**, 18 (1992).

26. A. Ansmann, M. Tesche, D. Althausen, D. Müller, P. Seifert, V. Freudenthaler, B. Heese, M. Wiegner, G. Pisani, P. Knippertz and O. DubovikJ. Geophys. Res., **113**, D04210, (2008)

27. M. Tesche, A. Ansmann, D. Muller, D. Althausen, I. Mattis, B. Heese, V. Freudenthaler, M. Wiegner, M. Esselborn, G. Pisani and P. Knippertz, *Tellus,* **61B**, 144 (2009).

28. A. Papayannis, et al., *J. Geophys. Res.* **113**, D10204 (2008).

29. D.M. Winker and M. T. Osborn, *Geophys. Res. Lett.*, **19** (2), 171 (1992).

30. K. Sassen, J. Zhu, P. Webley, K. Dean and P. Cobb, *Geophys. Res. Lett.,* **34** (L08803), 4 (2007).

31. G. Pisani, A. Boselli, M. Coltelli, G. Leto, G. Pica, S. Scollo, N. Spinelli and X. Wang, *Atmos. Environ.*, **62,** 34, (2012).

32. S. Scollo, A. Boselli, M. Coltelli, G. Leto, G. Pisani, N. Spinelli and X. Wang *Bull. Volcanol.,* **74**, 2383 (2012).

33. M. Wiegner, J. Gasteiger, S. Groß, F. Schnell, V. Freudenthaler and R. Forkel, *Phys. Chem. Earth.*, **45**, 79 (2012).

                    *A. Boselli et al.*

34. U. Schumann et al. *Atmos. Chem. Phys.*, **11**, 2245 (2011).

35. M. R. Perrone, F. De Tomasi, A. Stohl and N. I. Kristiansen, *Atmos. Chem. Phys.*,**12**, 10001 (2012).

36. C. Rolf, M. Krämer, C. Schiller, M. Hildebrandt and M. Riese, *Atmos. Chem. Phys.*, **12**, 10281 (2012).

37. G. Pappalardo, L. Mona, G. D'Amico, U. Wandinger, M. Adam, A. Amodeo, A. Ansmann, A. Apituley, L. Alados Arboledas, D. Balis, A. Boselli, J. A. Bravo-Aranda, A. Chaikovsky, A. Comeron, J. Cuesta, F. De Tomasi, V. Freudenthaler, M. Gausa, E. Giannakaki H. Giehl, A. Giunta, I. Grigorov, S. Groß, M. Haeffelin, A. Hiebsch, M. Iarlori, D. Lange, H. Linné, F. Madonna, I. Mattis, R.-E. Mamouri, M. A. P. McAuliffe, V. Mitev, F. Molero, F. Navas-Guzman, D. Nicolae, A. Papayannis, M. R. Perrone, C. Pietras, A. Pietruczuk, G. Pisani, J. Preißler, M. Pujadas, V. Rizi, A. A. Ruth, J. Schmidt, F. Schnell, P. Seifert, I. Serikov, M. Sicard, V. Simeonov, N. Spinelli, K. Stebel, M. Tesche, T. Trickl, X. Wang, F. Wagner, M. Wiegner and K. M. Wilson. *Atmos. Chem. Phys. Discuss.*, **12**, 30203 (2012).

38. M. A. Revuelta, M. Sastre, A. J. Fernández, L. Martín, R. García, F. J. Gómez-Moreno, B. Artíñano, M. Pujadas and F. Molero, *Atmos. Environ.*, **48**, 46 (2012).

39. K.N. Lio and Y. Takano, *Geophys. Res. Lett.,***29** (2002).

40. K.N. Liou, *Mon. Weather Rev.***114**, 1167-1199 (1986).

41. D. P. Wylie, W. P.Menzel,, H. M. Woolf and K. I. Strabala, *J. Climate*, **7**, 1972-1986 (1994).

42. R. M. Schotland, K. Sassen and R. J. Stone, *J. Appl. Meteor.*, **10**, 1011-1017 (1971).

43. C. M. R. Platt, *J. Atmos. Sci.*, **30**, 1191-1204 (1973).

44. C. M. R. Platt, J. C. Scott and A. C. Dilley, *J. Atmos. Sci.,* **44**, 729-747 (1987).

45. J. M. Comstock and K. Sassen, *J. Ocean. Atmos. Tech.*, **18**, 1658-1673 (2001).

46. Z. Wang and K. Sassen, *J. Appl. Meteor.*, **40**, 1665-1682 (2001).

47. CALIPSO mission: http://www.nasa.gov/mission_pages/calipso/main/index.html .

48. Y. Hu, S. Rodier, K. Xu, W. Sun, J. Huang, B. Lin, P. Zhai and D. Josset, *J. Geophys. Res.*, **115**, D00H34 (2011).

49. S.Ismail and E.V. Browell, *J. Atmos. Ocean. Technol.*, **11**, 76 (1992).

50. P. F. Ambrico, A. Amodeo, P. Di Girolamo and N. Spinelli, *Appl. Opt.*, **39**, 36, 6847 (2000).

51. S. Amoruso, A. Amodeo, M. Armenante, A. Boselli, L. Mona, M. Pandolfi, G. Pappalardo, R. Velotta, N. Spinelli and X. Wang, *Optics and Lasers in Engineering*, **37**, 5, 521 (2002).

52. C. Bellecci , M. Franucci, P. Gaudio, M. Gelfusa, S. Martellucci, M. Richetta and T. Lo Feudo, *Appl. Phys. B*, **87**, 373 (2007).

53. S. Kuang, M.J. Newchurch, J. Burris, L. Wang, P. I. Buckley, S. Johnson, K. Knupp, G. Huang, D. Phillips and W. Cantrell, *Atmos. Environ.,***45**, 6078 (2011).

54. R. Robinson, T. Gardiner, F. Innocenti, P. Woods and M. Coleman *J. Environ. Monit.*, **13**, 2213, (2011).

55. N. Cao, S. Li, T. Fukuchi, T. Fujii, RL. Collins, Z.wang and Z. Chen *Appl. Phys. B,* **85**, 163 (2006).

56. W. Baker, G. D. Emmitt, F. Robertson, R. Atlas, J. Molinari, D. Bowdle, J. Paegle, R. Hardesty, R. Menzies, T. Krishnamurti, R. Brown, M. Post, J. Anderson, A. Lorenc and J. McElroy, *Bull. Amer. Meteor. Soc.*76, 869-888 (1995).

57. A.Cress and W. Bergen, *Meteorologischrift,* **10**, 91 (2001).

58. L. P. Riishojgaard, R. Atlas and G. D. Emmitt, *J. App. Met.*43, 5 ,810-820 (2004).

59. C. Werner in *LIDAR Range-Resolved Optical Remote Sensing of the Atmosphere* , Ed. C. Weitkamp, Springer (2005).

60. J. T. Sroga, E. W. Eloranta, T. Barber, *J. App. Met.* **19**,598 (1980).

61. Y. Sasano, H. Hirohara, T. Yamazaki et al. *J. App. Met.* **21**, 1516 (1982).

62. N. Phillips, *Tellus,* **38A**, 314-322 (1986).

63. C. L. Korb, B. M. Gentry, S. X. Li and C. Flesia, *Appl. Opt.,* **37**, 3097, (1998).

64. M. J. McGill, W. D. Hart, J. A. McKay, J. D. Spinhirne, *Appl. Opt.,* **38**, 6388 (1999).

65. A. Souprayen, A. Garnier, A. Herzog, A. Hauchecorne and J. Porteneuve, *Appl. Opt.,* **38**, 2410 (1999)

66. A. Souprayen, A. Garnier and A.Herzog, *Appl. Opt.,*38, 2422 (1999).

67. V. A. Banakh, I. N. Smalikho, F. Kopp, C. Werner, *Appl. Opt.,* **34**, 2055 (1995).

68. D. G. H. Tan, E. Andersson, *Q. J. R. Meteorol. Soc.,* **131**, 1737-1757 (2005).

69. G. C. Guenther, A. G. Cunningham, P. E. LaRocque and D. J. Reid, *Proc. of EARSeL-SIG-Workshop LIDAR*, Dresden/FRG, June 16 – 17 (2000).

70. J. E. Tyler, *Limnol. Oceanogr.* **13**, 1-6 (1968).

71. G. C. Guenther, NOAA Professional Paper Series, National Ocean Service 1, *Nat. Oceanic and Atmosph. Admin.*, Rockville, MD, pp. 385 (1985).

72. N. G. Jerlov, *Marine Optics*, Elsevier Scientific Pub. Co., Amsterdam, pp. 231 (1976).

73. G. C. Guenther, P. E. LaRocque and W.J. Lillycrop, *Proc. SPIE Ocean Optics* XII, Vol. **2258**, 422-430 (1994).

74. G. C. Guenther and R. W. L Thomas, NOAA Tech. Rpt. OTES 01, *National Oceanic and Atmospheric Administration*, Washington, D.C., pp. 51 (1981).

75. G. C. Guenther and L. R. Goodman, *Proc.SPIE Ocean Optics* V, Vol. **160**, 174-183 (1978).

76. E. P. Baltsavias, *ISPRS J. of Photogrammetry & Remote Sens.,* **54**, 83–94 (1999).

77. F. Ackermann, *ISPRS J. of Photogrammetry & Remote Sens.,* **54**, 64–67 (1999).

78. M. Fraczek, A. Behrendt and N. Schmitt, *Appl. Opt.,* **51**, 2, 148-166 (2012).

79. A. Behrendt, in: *LIDAR Range-Resolved Optical Remote Sensing of the Atmosphere,* Ed. C. Weitkamp. Springer, NY, USA, pp. 19-42 (2005).

80. Y. Xie, Z. Sha and M. Yu, *J. of Plant Ecology,* 1, 9–23 (2008).

81. B. L. Becker, , D. P. Lusch and J. Qi, *Remote Sens. Env.,* **108**, 111–120 (2007).

82. M. Dalponte, L. Bruzzone, L. Vescovo and D. Gianelle, *Remote Sens. Env.,* **113**, 2345–2355 (2009).

83. E. Næsset, *ISPRS J. of Photogrammetry and Remote Sens.*, **52**, 49–56 (1997).

84. D. N. M. Donoghue, P. J. Watt, N. J. Cox and J. Wilson, *Remote Sens. Environment*, **110**, 509–522 (2007).

85. Ørka, H.O., Næsset, E., Bollandsås and O.M. *Remote Sens. Environ.*, **113**, 1163–1174, (2009).

86. A. H. Aldred and G. M. Bonnor, *Information Report PI-X-51, Petawawa National Forestry Institute* (1985)

87. E. Lindberg, K. Olofsson, J. Holmgren and H. Olsson, *Remote Sens. Environment*, **118**, 151–16, (2012).

88. S. W. MacFaden, J. P. M. O'Neil-Dunne, A. R. Royar, J. W.T. Lu and A. G. Rundle, *J. Appl. Remote Sens.*, **6**(1), 063567 (2012).

89. M. Dalponte, L. Bruzzone, D. Gianelle, *IEEE Transactions on Geoscience and Remote Sensing*, **46**(5), 1416–1427 (2008).

90. T. G. Jones, N. C. Coops, T. Sharma and T., Canada. *Remote Sens. Environ.*, **114**, 2841–2852 (2010).

91. I. Korpela, T. Tuomola, T. Tokola and B. Dahlin, *Silva Fennica*, **42**, 753–772 (2008).

92. H. K. Lichtenthaler, *J. Plant Physiol.* **131**, 101-110 (1987).

93. S. Svanberg, *Physica Scripta*, T58, 79-85 (1995).

94. N. D'Ambrosio, K. Szabò and H. K. Lichtenthaler, *Radiat. Environ. Biophys.* **31**, 51-62 (1992).

95. C. Buschmann, J. Schweiger, H. K. Lichtenthaler and P. Richter, *J. Plant Physiol.*, **148**, 548-554 (1996).

96. Y. Saito, R. Saito, T. D. Kawahara, A. Nomura and S. Takeda, *Forest Ecology and Management*, **128**, 129-137 (2000).

97. C. E. Brown and M. F. Fingas, *Marine Pollution Bulletin*, **47**, 477–484 (2003).

98. D. J. O'Connor, D. Iacopino, D. A. Healy, D. O'Sullivan and J. R. Sodeau, *Atmosph. Env.*, **45**, 6451-6458, (2011).

99. E. V. Browell, S. Ismail and W.B. Grant, *App. Phys. B*, **67**, 399–410 (1998).

100. M. J. McGill, D. L. Hlavka, W. D. Hart, E. J. Welton, J. R. Campbell, *J. of Geoph. Res.*, **108** (D13), 8493 (2003).

101. P. Chazette, J. Sanak and F. Dulac, *Anal. Environ. Sci. Technol.*, **41**, 8335–8341 (2007).

102. I. S. Stachlewska, R. Neuber, A. Lampert, C. Ritter and G. Wehrle, *Atmosph. Chem. and Phys.*, **10**, 2947–2963, (2010).

103. J. W. Hair, C. A. Hostetler, A. L. Cook, D. B. Harper, R. A. Ferrare, T. L. Mack, W. Welch, L. R. Izquierdo and F. E. Hovis, *Appl. Opt.*, **47**, 36, 6734-6752, (2008).

104. M. Esselborn, M. Wirth, A. Fix, M. Tesche and G. Ehret, *Appl. Opt.*, **47**, 3, 346-358 (2008).

105. S. F. J. De Wekker, K. S. Godwin, G. D. Emmitt and S. Greco, *J. Appl. Meteor. Climatol.*, **51**, 1558–1574 (2012).

106. W. H. Fuller Jr., S. Sokol and W. H. Hunt, *Science*, **216**, 4550, 1113-1115 (1982).

107. P. V. Hobbs, L. F. Radke, J. H. Lyons, R. J. Ferek, D. J. Coffman and T. J. Casadevall, *J. Geophys. Res.*, **96**, 18735-17552 (1991).

108. D. M. Winker and M. T. Osborne, *Geophys. Res. Lett*, **19**, 2, 167-170 (1992).

109. F. Marenco, B. Johnson, K. Turnbull, S. Newman, J. Haywood, H. Webster and H. Ricketts, *J. Geophys. Res.*, **116**, D21, (2011).

110. P. Chazette, A. Dabas, J. Sanak, M. Lardier and P. Royer, *Atmos. Chem. Phys.*, **12**, 7059–7072 (2012).

111. A. L. M. Grant, H. F. Dacre, D. J. Thomson and F. Marenco, *Atmos. Chem. Phys.*, **12**, 10145–10159 (2012).

112. http://airandspace.si.edu/collections/imagery/apollo/AS15/a15los.htm .

113. http://www.lpi.usra.edu/lunar/missions/apollo/apollo_17/ .

114. M. P. McCormick, D. M. Winker, E. V. Browell, J. A. Coakley, C. S. Gardner, R. M. Hoff, G. S. Kent, S. H. Melfi, R. T. Menzies, C. M. R. Platt, D. A. Randall and J. A. Reagan, *Bull. Amer. Meteor. Soc.*, **74**, 205–214 (1993).

115. D. M. Winker, R. H. Couch and M. P. McCormick, *Proc. IEEE 84*, 164-180 (1996).

116. C. M. R. Platt and D. M. Winker, in *SPIE Proceedings Vol. 2310 LIDAR Techniques for Remote Sensing*, C. Werner, Editors, **106**-35 (1994).

117. D. M. Winker and C. R. Trepte, *Geophys. Res. Lett.*, **25**, 17, 3351-3354, (1998).

118. H. Chepfer, J. Pelon, G. Brogniez, C. Flamant, V. Trouillet and P. H. Flamant, *Geophys. Res. Lett,*, **26**, 14, 2203-2206 (1999).

119. http://mola.gsfc.nasa.gov/.

120. http://nsidc.org/data/icesat/.

121. J. D. Spinhirne, S. P. Palm, W. D. Hart, *J. Geophys. Res.*, **32**, L22S03 (2005).

122. M. Simard, V. H. Rivera-Monroy, J. E. Mancera-Pineda, E. Castañeda-Moya and R. R. Twilley, *Remote Sens. Environ.*, **112**, 2131–2144 (2008).

123. http://www.nasa.gov/mission_pages/calipso/main/index.html .

124. D. M. Winker, J. R. Pelon and M. P. Cormick, *Proc. SPIE* **4893**, 1-11 (2003).

125. D. M. Winker, W. H. Hunt and M. J. McGill, *Geophys. Res. Lett.*, **34**, L19803, (2007).

126. X. Wang, M. G. Frontoso, G. Pisani and N. Spinelli, *Opt. Lett.*, **15**, 11, 6734 (2007).

127. Z. Tao, Z. Liu, D. Wu, M. P. McCormick, and J. Su, *Opt. Lett.*, **33**(24), 2986–2988 (2008).

128. L. Xiaomei, J. Yuesong, Z. Xuguo, W. Xuan, L. Nasti and N. Spinelli, *Opt. Express,* **19**, A72-A79 (2011).

129. European Space Agency, *ADM-Aeolus Science Report*, ESA SP-1311 (2008).

# NON INVASIVE TECHNIQUES FOR THE DIAGNOSIS OF AEROSPACE DEVICES


Federico De Filippis*, Luigi Savino, Alessio Cipullo and Elisa Marenna

*C.I.R.A, Centro Italiano Ricerche Aerospaziali*
*Via Maiorisi Capua, Italy*
**E-mail: f.defilippis@cira.it*


In aerospace applications the generation of gases at high temperature, pressure and velocity conditions is often necessary to simulate the atmospheric re-entry conditions. Experimental measurements on these gases cannot be realized using traditional intrusive sensors and represent a challenging target for developments of specific optical diagnostics. In this chapter important applications to the hypersonic facilities (Plasma Wind Tunnel (PWT)) are described.

The more powerful PWT in the world is operative in Italy at CIRA (Italian Centre for Aerospace Research). Measurements of temperature and gas species concentration are carried out by two optical non-intrusive techniques: Optical Emission Spectroscopy (OES) and Laser Induced Fluorescence (LIF). In the first part of this chapter an overview of the chemical-physical conditions of the analyzed gas is presented, with particular attention to the thermal and chemical non equilibrium due to high velocities and low pressures. In the second part, for the two previously mentioned optical diagnostic techniques, the experimental set-up, the data acquisition, the elaboration methods and the obtained results are described and discussed. Coherence between the obtained measurements and the theoretical-numerical predictions is continuously found to increase the confidence in the application of optical diagnostics for the characterization of this kind of aerospace devices.

## 1. Introduction

Origin of the current high interest for non-invasive measurement techniques in aerospace devices comes from the trans-atmospheric missions. These specially created an increasing interest in the field of the aero-thermo-dynamic design of hypersonic space vehicles.

These vehicles at high speed regime work at heavier conditions about heat flux and heat loads than those realized at supersonic or subsonic regimes. Non equilibrium hypersonic flows will wind these vehicles, affecting significantly aerodynamics, heat loads and aerodynamic efficiency.

Since the difficult to simulate these conditions numerically, the experimental simulation and characterization of the hypersonic regimes phenomena is fundamental in the design of atmosphere re-entering vehicles. These experiments are carried out in specific wind tunnels mentioned in the following as hypersonic facilities or Plasma Wind Tunnels (PWT). The high enthalpy regimes of the gases at hypersonic speed make almost impossible the use of conventional intrusive instruments for an experimental thermo-dynamic characterisation and require specific non-intrusive diagnostics developments.

A correct characterization of the flows generated by hypersonic facility allows the development of more sophisticated computational models useful for numerical simulations of hypersonic motion in reacting flows around models characterized by considerable geometric complexity.

At hypersonic flight conditions a vehicle travelling at high speed at high altitude can experiment thermal non equilibrium, since the low density reduces the frequency of the collision, and the high speed determines transition times of the molecules lower than those relive to low speeds. So a delay in the equilibrium between translational, rotational and vibrational modes can be revealed. At this conditions (thermal non equilibrium), the chemical model of the air requires a multi temperature approach, in contrast with classical formulation with single temperature, so a roto-translational temperature different from the vibrational one can be measured inside the hypersonic jet.

For a better understanding of the chemical effects in hypersonic aerodynamics, it is fundamental to know what happens to the chemical species of air (oxygen $O_2$ and nitrogen $N_2$) when temperature increases.

For this purpose Fig.1 below can be analyzed[1,2].

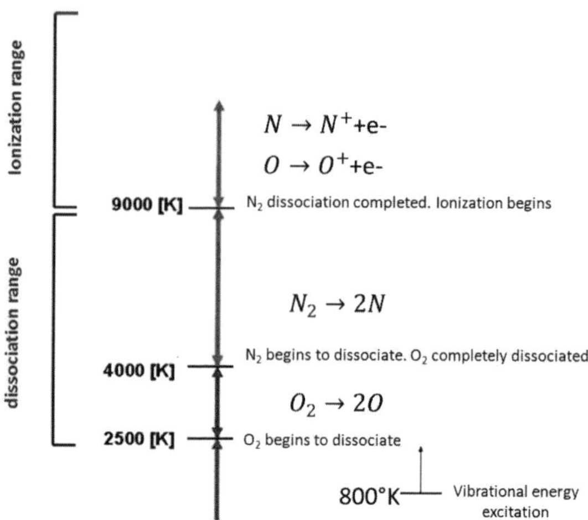


Fig. 1. Diagram representative of the temperature at which dissociation and ionization of the two main chemical species ($O_2$ and $N_2$) is verified.

It illustrates the dissociation and ionization ranges of oxygen and nitrogen at a pressure of 1 atm.

At a temperature of about 800 K, the vibrational energy of the molecules becomes considerable. When temperature reaches 2000 K, $O_2$ begins to dissociate. At about 4000 K, the oxygen dissociation is complete and so this species is present in atomic form (O). Near to this temperature, molecular nitrogen, $N_2$, begins to dissociate, completing this process around 9000 K. As a coincidence, this is the temperature at which oxygen and nitrogen ionization begins. So for temperature greater than 9000 K, air is composed by a mixture of O, $O^+$, N, $N^+$, and electrons.

## 2. Test Device: Hypersonic Facilities

For an efficient design of a hypersonic vehicle pressure, temperature over surfaces and heat load must be known[3]. To verify the accuracy of these design parameters, both test in wind tunnels and CFD (computations) analysis are necessary. To generate hypersonic free jet for test campaigns, these facilities must operate at high enthalpy and high pressures. For this type of experimental analysis, two kinds of plasma wind tunnels can be used: *impulse tunnels*[4] and *continuous tunnels*.

In the following applications the *continuous* ones are described.

Continuous arc heated tunnel are necessary to study radiative heating, problematic very important at high enthalpies and in particular during the interplanetary missions and to study the ablation phenomena, that can't be obtained with impulse facility.

The SCIROCCO Plasma Wind Tunnel[5-7] is an arc heated facility, capable of producing a low- pressure high-enthalpy hypersonic flow of large dimension on samples in scale up to 1:1 with test duration up to 30 minutes. Until now the facility's most important employ is testing and validating Thermal Protection System materials. This facility is operative in Capua (Italy) at CIRA, Centro Italiano Ricerche Aerospaziali.

The design and engineering of the SCIROCCO PWT represent the state of art for the arc jet technology. A facility schema is reported in Fig. 2. The arc heater of the PWT-SCIROCCO is a segmented constricted type with a bore diameter of 0.11 m and a length of 5.5 m. At the extremities of this tube there are 9 and 9 discs (cathode and anode), through which a confined electrical discharge is generated. Process air at a mass flow rate in the range of 0.3- 3.5 kg/s at high pressure of 87 bar is supplied to the arc heater, inside which temperatures in the range 2000-10000 K are achieved [5,6].

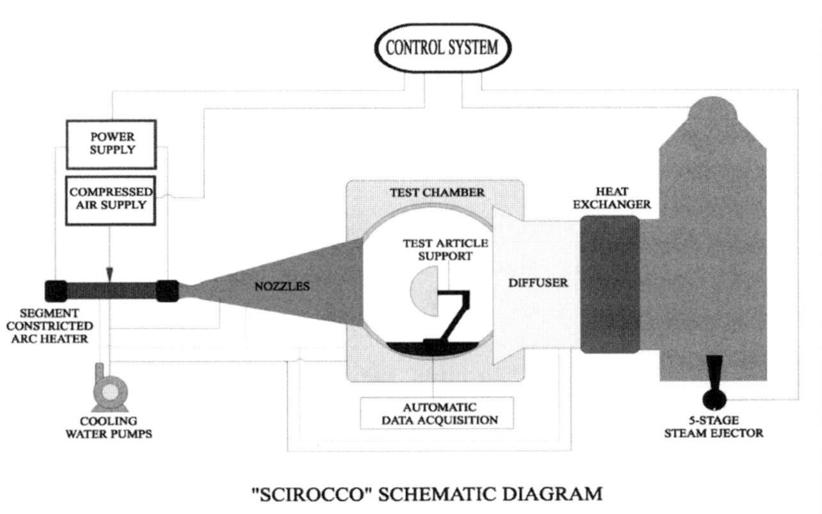


Fig. 2. Facility schematic and facility aerial view.

The heated compressed air is accelerated through a convergent-divergent conical nozzle. There are four different nozzle configurations in order to achieve the desired flow condition and to match any model size. A Model Support System inserts the test article in the plasma flow. The Test Chamber has a cylindrical vertical shape with an overall height of 9.2 m and an inner diameter of 5.17 m. After, the hypersonic jet is convoyed by a pick-up to a very long diffuser (50 m length and average diameter of 3 m).

The unique intrusive investigation of the plasma flow characteristics acting on the test model are performed using two Calibration Probes. They are two copper arms water cooled with a spherical nose of 100 mm in diameter. They support and locate instruments for stagnation heat flux and pressure measurements.


## 3. Experimental Diagnostic Techniques in Plasma Wind Tunnel

For a correct experimental characterization of the high enthalpy flow, the following non intrusive diagnostic technique are used:

1) Optical Emission Spectroscopy (OES);
2) Laser Induced Fluorescence (LIF).

Spectroscopy techniques represent useful tools to investigate the hypersonic jet features and to determine parameters like static temperature, vibrational temperature and molar fractions of the chemical species and flow velocity. In particular, OES, which is a spontaneous emission technique, can be used to determine the roto-translational temperature of the emitting species (in particular NO in the case of the SCIROCCO facility), which can be assumed equal to the static temperature of the plasma flow[8]. Whereas, fluorescence induced by laser (defined as Laser Induced Fluorescence) can be used to identify the species in free jet and determine their molar fraction. In the next paragraphs these non-intrusive diagnostic technique will be described more in details.

## 3.1. *Optical Emission Spectroscopy*

The following paragraph is dedicated to the application of the OES technique to the CIRA SCIROCCO facility. The first sub-paragraph describes the experimental setup, the second deals with the elaboration procedures applied on the emission spectra and the last presents the experimental results.

### 3.1.1. *Experimental Setup*

The set-up for high resolution OES measurements is sketched in Fig. 3.

The radiation is coupled to the entrance slit of the spectrograph (HR460, Horiba Jobin Yvon, Kyoto, Japan) by a UV objective (UV-Nikkor, Nikon, Tokyo, Japan) and a multi-mode silica optical fiber (200 – 1000 nm spectral range). The objective has a focal length of $f_l$ =105 mm and a maximum aperture of $f_l$/4.5. The special arrangement of the lenses eliminates the distortion and minimizes the chromatic aberration in the spectral range of interest for the investigations (200 – 400 nm). A circle having a diameter of $d$ = 20 cm is imaged at the jet centerline and the optical path is centered at a distance from the exit section of the nozzle of $x$ = 15 cm. Furthermore, a special anti-reflective (AR) coating minimizes reflections in the UV[9]. The objective is preferred to a classical achromatic doublet because in the wavelength range of interest, commercial UV doublets usually show unacceptably poor performances and much less depth of field.

The spectrograph has a plane holographic blazed 1200 grooves/mm grating with a ± 0.15 nm wavelength position accuracy and a total scanning range of 200-1300 nm with a slew rate of 45 nm/s. Light is then collected by a CCD camera (Symphony, Horiba Jobin Yvon, Kyoto, Japan) attached to the front exit of the spectrograph. The CCD matrix is made up of 1024×256 pixels, each pixel having a 26×26 μm size. It has a four stage thermo-electric cooling system (based on Peltier cells) to work at about 200 K. The CCD camera and the spectrograph have a remote control. The spectral coverage of single spectrograph scan is Δλ range = 50 nm and the entrance slit width is set to 10 μm for all the experiments.

The instrumental resolution is deduced for each test campaign by comparing the Full Width Half Maximum (FWHM) of several atomic lines of the experimental and theoretical spectra of a low pressure Hg lamp. The Hg source is also used to perform the wavelength calibration of the raw spectrum. The intensity calibration in the UV is performed in relative units, using a low power deuterium source (Avantes, AvaLight-DHc, Apeldoorn, The Netherlands). More details can be found in the contributions by Cipullo[10,11]. The measurement error on the emitted spectra is 10% rdg. and it has been estimated taking into account the superimposed noise on the raw signal (due to both the environmental conditions and the CCD), the optical components along the path, the calibration error and the digital-to-analog conversion error.

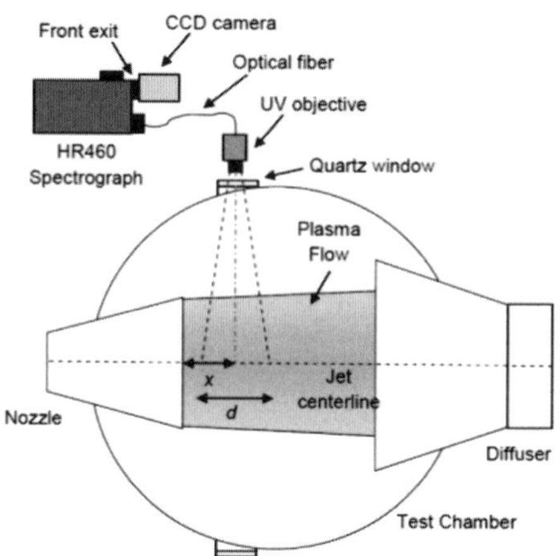



Fig. 3. Experimental setup for optical emission spectroscopy in free-stream. CIRA SCIROCCO facility.

### 3.1.2. *Elaboration Procedure*

The spectra acquired through the experimental setup described in the paragraph 3.3.1 are first of all calibrated and elaborated according to the procedures described in the contribution by Cipullo[11]. As already mentioned, the emitted spectrum is essentially characterized by the presence of NO emission. No other spectral signatures could be recorded using the present experimental set-up.

The rotational temperature and the vibrational temperature of the NO are obtained by a simple fitting procedure. The fitting algorithm is based on the work by Babou et al.[12], where the temperatures are determined through the minimization of the RMS Error (RMSE) function between experimental and predicted spectra:

$$RMSE(T_{rot}, T_{vib}) = \sqrt{\frac{1}{M} \sum_{i=1}^{M} \left( \overline{S_{exp}(i)} - \overline{S_{th}(i, T_{rot}, T_{vib})} \right)^2}$$ (1)

where, $M$ is the total number of spectral points of both normalized experimental spectrum $\overline{S_{exp}}$ and theoretical spectrum $\overline{S_{th}}$, $T_{rot}$ and $T_{vib}$ are the rotational and vibrational temperatures respectively. Obviously, both spectra must be interpolated to have the same wavelength vector.

The minimization of the RMSE (Eq. 1) is performed using the built-in optimization function *fminsearch* provided by *Matlab®*: *fminsearch* finds the minimum of a scalar function (in the specific case, this is Eq. 1) of several variables (two in the present case), starting from an initial estimate. This is generally referred to as unconstrained nonlinear optimization. For the application of the OES technique to the free-stream spectra produced by SCIROCCO, the optimization variables (or parameters) are $T_{rot}$ and $T_{vib}$. The theoretical spectrum is generated using the spectral simulation software Specair 2.2[13].

The RMSE, determined at each iteration from Eq. 1, is an important indication of the quality of the fitting. The algorithm is iterated until the termination criteria have been fulfilled. In particular, the tolerance on the temperature is set to 10 K and the one on the RMSE is set to 1e-4. The validation of the fitting procedure is described in a previous work[14].


### 3.1.3. *Experimental Results*

Results of the application of the OES to the plasma acquired from the free-stream flow are reported in Table 1. Test 2 and 2bis are performed during the same run with same nominal operating conditions within 2 minutes.

Table 1. Experimental Results of the Optical Emission Spectroscopy in Free-stream. CIRA SCIROCCO Facility.

| Test No. | Itot [A] | mair [kg/s] | Trot [K] | $\Delta$Trot [K] | Tvib [K] | $\Delta$Tvib [K] | RMSE [a.u.] | $\Delta$RMSE [a.u.] |
|---|---|---|---|---|---|---|---|---|
| 1 | 2902 | 1.29 | 631 | 17 | 1137 | 9 | 0.0134 | 6.774x10-5 |
| 2 | 4001 | 1.31 | 634 | 25 | 1100 | 15 | 0.0161 | 5.189x10-5 |
| 2bis | 4001 | 1.31 | 655 | 23 | 1100 | 16 | 0.0175 | 5.348x10-5 |
| 3 | 3992 | 1.30 | 686 | 13 | 1134 | 8 | 0.0108 | 4.622x10-5 |

The spectral fitting is performed between 232 nm and 265 nm and the result is reported in Figure 4. The (0,1), (0,2) and (0,3) vibrational bands of the $NO - \gamma$ collection and the (0,4), (0,5) and (0,6) of the $NO - \delta$ collection are clearly observable. The quality of the fitting is satisfactory, meaning that the two-temperature approach may be sufficient to model the behavior of the free-stream plasma flow produced by the CIRA SCIROCCO facility. The resulting temperatures have been compared to the results of the simulations performed by DART code, developed at CIRA. The rotational temperature resulted in partial agreement with the experimental value. Conversely, the vibrational temperature of the NO showed a significant discrepancy.

The estimation of the measurement error reported in Table 1 is performed using a simple procedure, based on the propagation of the error on the equation of the RMSE[12].

The integral of the two spectra of Test 2 and 2bis in the range of acquisition is in agreement within 0.5 %, meaning that there is good short term repeatability (within 2 minutes) for the emission intensity.

The results of the spectral fitting, performed between 232 nm and 265 nm, are reported in Fig. 5. Again, the quality of the fitting is satisfactory for both fittings. As it is possible to note from Table 1, the rotational temperature values are in agreement within 3% (within the error bars) and the estimated vibrational temperature are coincident.

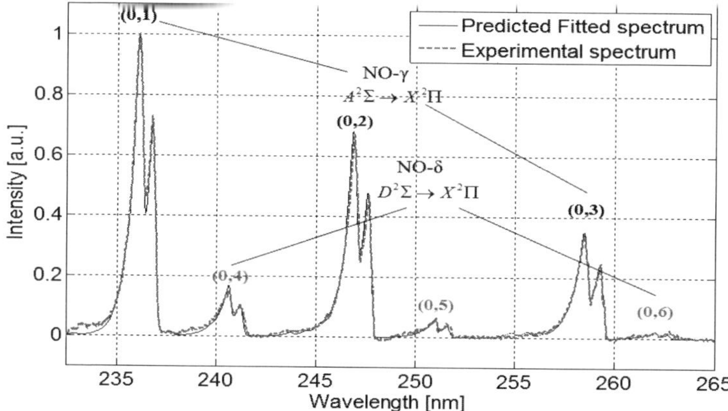

Fig. 4. Result of the fitting procedure on the spectrum produced by the free-stream plasma flow of the CIRA SCIROCCO Facility. Test conditions: $I_{tot}$ = 2902 A, $m_{air}$ = 1.29 kg/s (Test 1, Table 1).

Therefore, the thermodynamic condition of the plasma flow produced by CIRA SCIROCCO can be considered repeatable in the short term. It's interesting to note that both rotational and vibrational temperatures are not changing significantly between Test 1 and Tests 2 and 2bis, even though the mass-averaged total enthalpy changes significantly. This could be explained by the low sensitivity of the spectral shape to the temperature, and the relatively low resolution of the experimental setup for the OES technique can play a significant role.

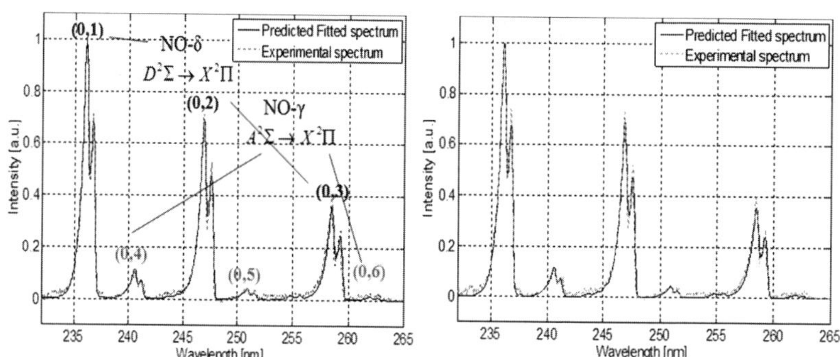

Fig. 5. Result of the fitting procedure on the spectrum produced by the free-stream plasma flow of the CIRA SCIROCCO Facility for Test 2 (left) and Test 2bis (right). Test conditions: $I_{tot}$ = 4001 A, $m_{air}$ = 1.31 kg/s.

The fitting result for Test 3 is not reported here. However it shows a similar fitting performed with respect to the other results. Finally, for each testing condition reported in Table 1, the spectral fitting was also performed in different wavelength ranges, either including single vibrational bands or multiple bands. The resulting rotational temperature does not show any noticeable difference with the ones reported in the Table. Conversely, the vibrational temperature shows significant differences when single vibrational bands are considered. The reason is that the vibrational temperature mainly influences the relative strength of different vibrationals bands of $NO$, belonging to the same electronic transition ($NO - \gamma$ or $NO - \delta$) or not. Therefore, taking a single band the sensitivity of the spectral shape to the vibrational temperature experiences a strong decay.

Additional experiments and results obtained through the OES technique are described in details in the contribution by Cipullo[12].

## 3.2. *Laser Induced Fluorescence*

The following paragraph is focused on the application of Laser Induced Fluorescence (LIF) technique to the CIRA Scirocco facility. The first sub-paragraph describes the basics of the technique and the second one the experimental set-up and the preliminary experimental results.

### 3.2.1. *Basics of LIF Technique*

The LIF is a non-intrusive technique with great potentiality for the determination of temperature, concentration and velocity of chemical species. In particular, this methodology is based on the excitation of atoms/molecules to higher energy states by absorption of a laser radiation and subsequent emission of fluorescence by excited species at a different wavelength. The LIF technique has the advantages of high sensitivity and selectivity for species analysis. However, the large number of optical components, such as laser source, detection unit and synchronization system, requires a complex optimization activity. In this case, the LIF system implemented to Scirocco facility is used to detect atomic oxygen in the free-jet since it is one of the most important constituents of the hypersonic jet at the nozzle exit.


### 3.2.2. *Experiment and Preliminary Results*

The basic approach used to detect oxygen atoms in the free-jet of Scirocco facility using LIF is based on the energy level diagram showed in Fig. 6. Oxygen atoms in the ground electronic state are excited by the $2p^3P \rightarrow 3p^3P$ two-photon transition at about 226 nm, and detected by the $3p^3P \rightarrow 3s^3S$ fluorescence transition at about 845 nm [15-17].

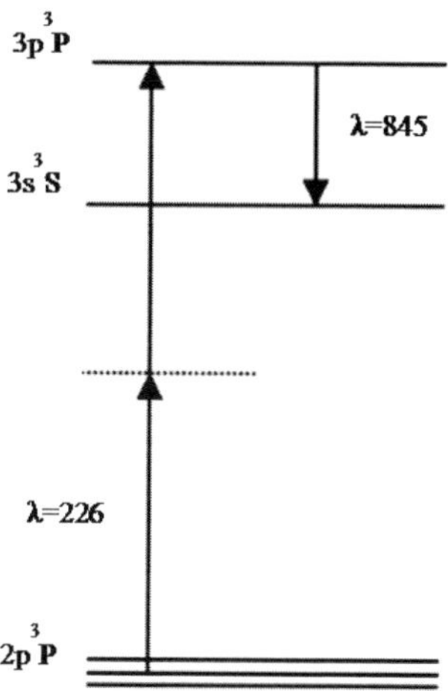


Fig. 6. Energy level diagram for atomic oxygen.


A sketch of the LIF experiment is showed in Fig. 7. A tunable pulsed Nd:YAG laser (Continuum) is used for the detection of atomic oxygen.

The laser system is composed of three units: the first one (Powerlite) generates a signal at 355 nm that pumps the second unit (Sunlite, single-mode OPO) and allows to obtain a new signal in the range 400-700 nm.

The third unit (FX-UV), pumped by the laser beam at 355 nm of the Powerlite and by the new signal generated in Sunlite, allows to obtain a signal in the range 205-230 nm. The main advantage of this LIF system is the possibility to synchronize the laser with 10 Hz repetition rate to the

                    *F. De Filippis et al.*

Scirocco facility jet that generates a continuous flow with a test time duration up to 30 min. The laser beam is adjusted to reach the requested measurement location on the hypersonic flow by a mirror system (periscope) that allows to correct the altitude of the laser beam and to direct it into test chamber. Specifically, the experimental investigation is performed at a location of about 30 cm from the nozzle exit along the flow centerline. For our application the output laser beam is fixed at 226 nm and the main output energy per pulse at 3.5 mJ. This energy value is chosen to take in account both the loss of the signal intensity due to periscope and the not perfect UV transparency of Scirocco window. In fact, the total energy loss is about 40% of the overall laser energy sent into test chamber.

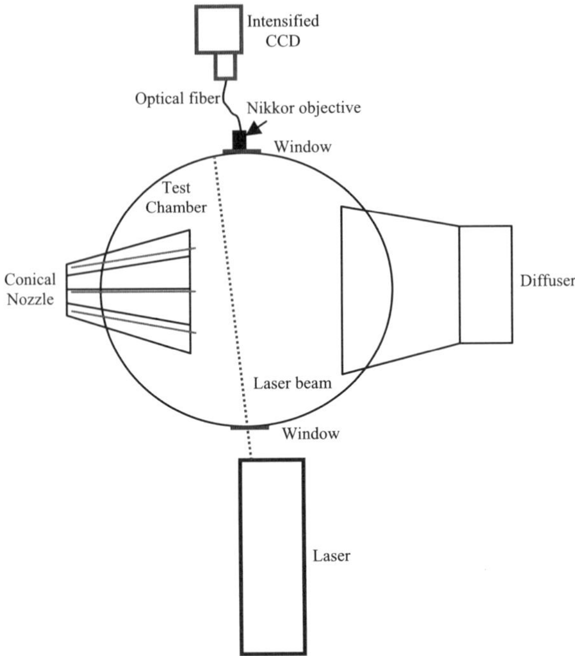


Fig. 7. Sketch of the optical set-up for LIF measurements in the Scirocco facility.

LIF from atomic oxygen is detected using a time-resolved system constituted by a Nikkor objective, a spectrograph (1200 grooves/mm) and a cooled intensified CCD camera (PI-MAX 3 Princeton Instruments) with a 1024x1024 pixels array and a sustained gating repetition rate of 1 MHz. For this experiment, the camera is synchronized by gating the intensified CCD camera at incremental delays after the excitation source, allowing the lifetime of the oxygen atoms to be measured from the intensity vs. time profile.

A precise synchronization of the pulse laser and camera illumination is necessary because the fluorescence emission lasts few nanoseconds. Synchronization is performed with a photodiode and controlled with an oscilloscope. The gate of the camera is set at 300 ns to collect almost all fluorescence radiation from the considered oxygen excited level. The delay increment is fixed at 33 ns and the number of spectra to be acquired at 50. Fig. 8 shows all the 3-D spectra acquired in the range 830–860 nm, where the $3p^3P \rightarrow 3s^3S$ fluorescence transition of atomic oxygen is expected. These 3-D spectra are reported as emission radiation intensity vs. wavelength vs. time.

Among all registered spectra, in the spectrum number 33 a narrow and intense peak located at 845 nm can be observed. The peak occurs about 150 ns after the laser pulse and it is reasonable to relate it to the fluorescence of excited oxygen atoms.

Future work will be aimed at improving the LIF experimental set-up to convert the measured oxygen atom LIF intensity into absolute number density.


## 4. Conclusions

This chapter has outlined the experimental necessities of optical diagnostics to analyse the high enthalpy condition gases in aerospace devices as the hypersonic facilities.

Affordable results obtained using the Optical Emission Spectroscopy technique have been discussed, in particular for temperatures measurements in thermal non-equilibrium conditions.

Fig. 8. 3D Spectra acquired in the range 830–860 nm as emission radiation intensity vs. wavelength vs. time.

The optical set-up and the preliminary results obtained by application of Laser Induced Fluorescence technique on the analysis of chemical species have been reported.

The results of illustrated activities underline the importance to increment the links between the optics and photonics research community about the aerospace applications to reach the goal of an experimental and complete description of the high enthalpy gas behaviour. It's a starting point for an affordable atmosphere re-entering space vehicle design activity.

## References


1. S. P. Sharma, "Study of Non-Equilibrium Transport Phenomena", NASA Contractor Report -180441(1984).
2. J. D. Anderson "Hypersonic and High Temperature Gas Dynamics", 2nd Ed. AIAA Education Series (2006).
3. J. D. Jr. Anderson, "*A History of Aerodynamics*" Cambridge, England: Cambridge University Press (1997).
4. C. Park, *J. of Thermophysics and Heat Transfer*, Vol. **11** No. 1 (1997).

5. F. De Filippis, A. Del Vecchio, S. Caristia and E. Graps, "SCIROCCO arc-jet facility for large scale spacecraft TPS verification", *Proceedings of 55th International Astronautical Congress*, 4-8 October 2004, Vancouver (Canada).

6. G. Russo, F. De Filippis, S. Borrelli, M. Marini and S. Caristia, "SCIROCCO 70-MW Plasma Wind Tunnel: A New Hypersonic Capability", in *AIAA 'Advanced Hypersonic Test Facilities'* Chapter 11, pp. 313-351 (2002).

7. F. De Filippis, S. Caristia, A. Del Vecchio and C. Purpura, "The SCIROCCO PWT Facility calibration activities" *Proceedings of the 3rd International Symposium on Atmospheric Re-entry Vehicles and Systems*, Arcachon (France), 24 - 27 March, 2003.

8. L. Savino, F. De Filippis, E. Marenna and A. Cipullo, "Experimental Characterization of the Cira Plasma Wind Tunnel Flow by Optical Emission Spectroscopy", 17th AIAA International Space Planes and Hypersonic Systems and Technologies Conference San Francisco (CA), 11-14 April, 2011.

9. Nippon Kogaku K.K., "Uv-Nikkor 105 mm f/4.5. Nikon UV Objective Instruction Manual", Tokyo, Japan (2000).

10. A. Cipullo, L. Savino, E. Marenna and F. De Filippis, *Aerospace Science and Technology*, Vol. **23**, No. 1 (2012).

11. A. Cipullo, "Ground Testing Methodologies Improvement in Plasma Wind Tunnels using Optical Diagnostics", PhD thesis, Department of Information Engineering, Second University of Naples, November 2012.

12. Y. Babou, Ph. Rivière, M. Perrin and A. Soufiani, *Plasma Sources Science & Technology*, Vol. **17**, No. 4 (2008).

13. C. O. Laux, "Radiation and Nonequilibrium Collisional radiative Models. Physicochemical Modeling of High Enthalpy and Plasma Flows", von Karman Institute Lecture Series 2002-07, D. G. Fletcher, J.-M. Charbonnier, G. S. R. Sarma, T. Magin (Eds.), Rhode-Saint-Genèse, Belgium (2002).

14. A. Cipullo, "Plasma Flow Characterization by means of Optical Emission Diagnostics", DC Report, von Karman Institute for Fluid Dynamics, Rhode-Saint-Genèse, Belgium (2010).

15. D. J. Bamford et al., *J. of Thermophysics and Heat Transfer*, Vol. **9**, No. 1 (1995).

16. S. Lohle et al., *J. of Thermophysics and Heat Transfer*, Vol. **21**, No. 3 (2007).

17. U. Koch et al., Two Dimensional Spatially Resolved Two Photon Oxygen Atom Laser Induced Fluorescence Measurements in the Flow Field of the Arc Heated Facility L3K, *Proceedings of the 5th European Symposium on Aerothermodynamics for Space Vehicles* (ESA SP-563). 8-11 November 2004, Cologne, Germany.

# NIGHT VISION


Carlo Corsi

*Consorzio C.R.E.O. Centro Ricerche Elettro Ottiche*
*SS. 17 Località Boschetto 67100 L'Aquila, Italy*
*E-mail: corsi@romaricerche.it*


"Night Vision" is a fascinating word-term to identify devices, systems and techniques that are involved in "Seeing in the Dark" or, even more fascinating, in "Seeing the invisible". Devices and instruments able to see in the dark have been tried even in old times, but only since the discovery of "Infrared" Science and Technology was possible "to see outside the part of e.m. spectrum that is visible by human eye". In fact in 1900's years, almost a century after the discovery of the "Infrared", the first prototypes of devices and instruments capable of seeing in the darkness were developed with a great impulse originated by World War: those military applications have deeply conditioned the development of the "Night Vision".

Recently, the development of bidimensional mosaic sensors FPAs (Focal Plane Arrays) and IR smaller and lighter systems has allowed a wider use in civil applications thanks to the reduction in costs. This was due to the elimination of the mechanical scanning systems and the possibility of contained cooling and even of room working temperature in the case of high sensitivity Microbolometers FPAs.

A review of Science and Technology of Infrared Sensors and their applications to Night Vision in the electro-magnetic bands of middle and far infrared will be illustrated and discussed.

## 1. Historical Introduction

The history of IR radiation sensors is starting in 1800 when the astronomer, William Herschel, discovered the existence of infrared radiation, a form of energy in the light beyond the "red" (from the Latin "*infra*"-below),by trying to measure the heat of the separate colors of the rainbow spectrum cast on a table in a darkened room[1] (Figure 1).

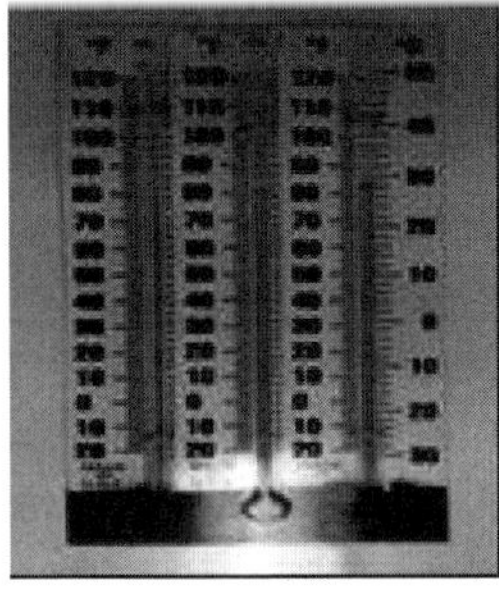

Fig. 1. Herschel's original prism and mirror (Museum of Science and Industry London).

In 1884, many years after the discovery of IR radiation, L.E. Boltzmann derived, from the physical principles of thermodynamics, the theoretical formula of $T^4$ Black Body Radiation Law, stated empirically in 1879 by the Stefan-Boltzmann's Law:

$$W = \sigma \cdot T^4 \tag{1}$$

where $W$ is the Radiation Power, $T$ is the Absolute Temperature and $\sigma$ is the Stefan-Boltzmann's Constant.

In 1901, Nobel Prize Max Karl Ernst Ludwig Planck developed the Planck's law which stated that the radiation from a blackbody at a specific wavelength can be calculated from

$$I(v)dv = \frac{2hv^3}{c^2} \cdot \frac{1}{\exp\left(\dfrac{hv}{kT}\right) - 1} dv \tag{2}$$

where $I(v)\delta v$ is the radiation power emitted per unit of surface and solid angle unit, in the frequency interval $(v-v+\delta v)$, T is the Absolute Temperature, c is the speed of light and h is Plank's constant.

Soon after, Wilhelm Wien (Nobel prize 1911) established the Wien's displacement law taking the derivative of the Plank's law equation to find the wavelength for maximum spectral radiance at any given temperature

$$\lambda_{Max} \cdot T = 2897.8 \mu m \cdot k \tag{3}$$

## 2. Night Vision Applications

As written over, Infrared is a Science/Technology originated as Radiation measurement only in 1800 with Herschel's discovery with the main application of "seeing the invisible", that is to see "after/below ("*infra*" in Latin) the visible red, that is in the e.m. region between the visible and the sub-millimeter/microwaves bands (Figure 2).

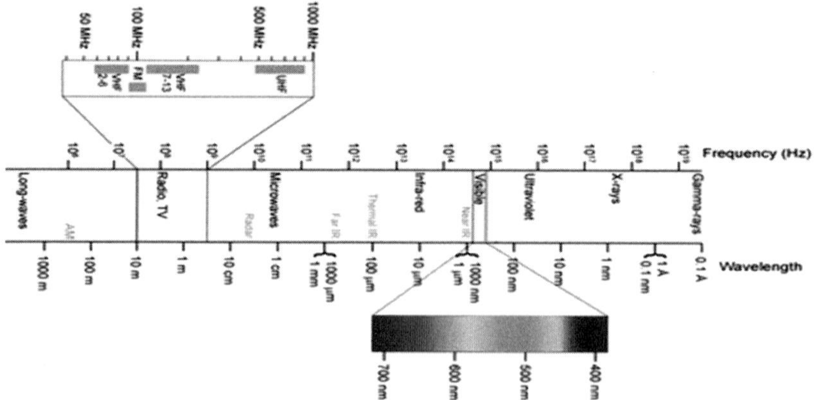


Fig. 2. Electromagnetic (EM) Spectrum main bands (in Frequency and Wavelength).

We shall confine the analysis to the "Night Vision Applications" as capability of seeing and/or detecting specific targets mainly for Security, just mentioning that historically, since Herschel's discovery, one of the main application of Infrared Imaging has been Astronomy, even if infrared astronomy achieved a real breakthrough in the 1960's thanks to the development of new, sensitive infrared detectors [2,3].

Before going to specific "Night Vision" where" Imaging" is the main task (even if sometimes with a strong characterization in its functional interpretation), it is important to mention important applications of Infrared Sensing , that is Security applications with specific "Targets Detection" as the main task. Thermal detectors development, even after the discovery of Infrared Radiation by Herschel in 1798, were mainly based on the use of thermometers which dominated IR applications till the 1st World War, although in 1829 L. Nobili had fabricated the first thermocouple, allowing in 1833 the multielement thermopile development by Macedonio Melloni with the first detection of a person at 10 meters. In the late 1870s, S. P. Langley developed the "bolometer", a device capable of measuring thermal radiation, so sensitive that it could detect the thermal radiation emitted by a cow from 400 meters away [2,4].

In fact, after almost a century of limited applications of infrared, at the beginning of 1900, important patents based on infrared technology were issued for Security applications devices to detect personnel, artillery, aircrafts and ships. Really, the first advanced applications of IR technology to security were in civil area, when (probably developed in 1910 and proposed for Patent in 1913, shortly after the Titanic tragedy in April 1912) L. Bellingham, by using a mirror and the original Nobili thermopile/Langley bolometer, patented a device to detect the presence of icebergs and steamships (Figure 3).

From Bellingham's Patent[5]: *"The infra-red eye l which in this case is a thermopile is mounted laterally in a tube, the toroidal surface being of such curvatures in the vertical and horizontal planes respectively as to produce. on the sensitive surface of the thermopile a sharply localized images in a horizontal plane, the rays in the vertical plane on the other hand being distributed over an angle of from several degrees below to several degrees above the horizontal, the object being to prevent loss of the image by pitching of the ship when the apparatus is in use at sea."*

Similarly, R.D Parker in the same year was patenting an advanced sensor using the high sensitivity of "Delta zero measurement" by means of a Wheatstone bridge. From Parker's Patent, Thermic Balance or Radiometer Application Filed July 31, 1912,- Patented June 9, 1914[6]:

*"This invention relates to thermic balances or radiometers and has for its object an improved capable of detecting the presence of a body by its sensitiveness to the ethereal radiation produced by that body consists in improvements in radiometers as well as in a novel adaptation thereof to produce an instrument sensitive only to the energy radiated by bodies, and also rugged enough to be used as a commercial instrument with particular reference to its use as a detector of the presence of cold bodies, as icebergs. I accomplish this and the other objects by the apparatus herein after described".*

 *C. Corsi*

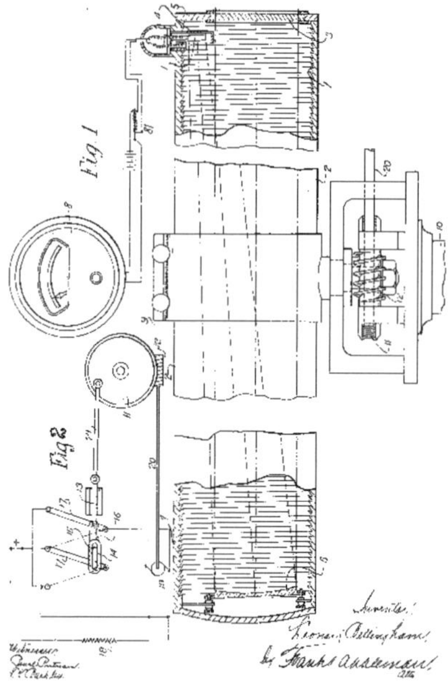

Fig. 3. Bellingham. Means for detecting the presence at a distance of icebergs, steamships, and other cool or hot objects.

So, between the years 1900 and 1920, many inventions in the world were based on the infrared with patents issued for devices to detect personnel, artillery, aircraft, ships and even icebergs. The first operating systems, in the modern sense, began to be developed during the 1914-18 war, when both sides had research programs devoted to the military exploitation of the infrared. These programs included experimental systems for enemy intrusion/detection, remote temperature sensing, secure communications, and 'flying torpedo' guidance. An infrared search system tested during this period was able to detect an approaching airplane at a distance of 1.5 km, or a person more than 300 meters away.

The most sensitive systems up to this time were all based upon variations of the bolometer idea, but the period between the two World Wars saw the development of two revolutionary new infrared detectors: the image converter and the photon detector. In fact early thermal detectors, mainly thermocouples and bolometers, were sensitive to all infrared wavelengths and operating at room temperature and normally, until few years ago, they were relatively low sensitivity and slow response time.

The first photon detectors (based on photoconductive effect discovered by Smith[7] in 1873 in Selenium and, later on, by Bose in photovoltaic lead sulphide, but not applied for many years) were developed by Case in 1917[8]. In 1933 Kutzscher developed IR PbS detectors (using natural galena found in Sardinia): these sensors were widely used during the 2nd War. These detectors have been extensively developed since the 1940's. Lead sulfide (PbS) was the first practical IR detector, sensitive to infrared wavelengths up to ~ 3 µm. In the meantime Cashman developed TaS, PbSe and PbTe IR detectors with high performances supporting the developments in England and US.

High level results were achieved in the 40's, especially in lead salts (PbSe and PbTe good stable cells were developed by OSRD -Office of Scientific Research and Development in co-operation with the MIT, Harvard and the British TRE Telecommunication Research Establishment-Great Malvern). The history of IR detector developments has been therefore almost coincident with optoelectronics for military applications for many decades, strongly conditioning the cultural behaviour of the IR industry and in some way of R&D labs [9].

## 3.  History of Military Night Vision

### 3.1. *Night Vision Technologies*

Night vision, the capability to see in the complete dark or in very low light conditions by technological means, can be divided into three main technologies[10]:

### 3.1.1. *Image Intensifiers*

At first, the image converter received the greatest attention by the military, because it enabled an observer to literally 'see in the dark'. However, the sensitivity of the image converter was limited to the near infrared wavelengths and the enemy military targets had to be illuminated by infrared search beams. (Since this involved the risk of giving away the observer's position to a similarly-equipped enemy observer, it is understandable that military interest in the image converter eventually faded). For this reason it became vital that a decisive advantage was to be achieved by gaining the ability to see at night without being seen mage intensification technologies work on the principle of magnifying the amount of received photons from various natural sources such as starlight or moonlight. The image intensifier is a vacuum-tube based device that converts visible light from an image so that a dimly lit scene can be viewable by a camera or open eye. When light strikes a charged photo cathode plate, electrons are emitted through a vacuum tube that strike the micro channel plate that cause the image screen to illuminate with a picture in the same pattern as the light that strikes the photo cathode, and is on a frequency that the human eye can see. The image is said to become "intensified" because the output visible light is brighter than the incoming IR light. A Night Vision Device (NVD) is a device comprising an image intensifier tube in a rigid casing, commonly used by military forces and recently has become more widely available for civilian use, for example, EVS (Enhanced Vision Systems), which are installed in the latest avionics instrumentation to help pilots to avoid accidents. The Night Vision Goggle (NVG) is a night vision device with limited performances (hundred meters) and is strongly conditioned by environmental lightness.

### 3.1.2. *Active Infrared*

For improving the performances on NVG, active illumination technologies in the Near Infrared (NIR) or Shortwave Infrared (SWIR) bands have been added.

Examples of such technologies include low light camera. Active infrared night vision combines infrared illumination of spectral range 700-1000 nm (just below the visible spectrum of the human eye) with CCD cameras sensitive to this light.

The resulting scene, which is apparently dark to a human observer, appears as a monochrome image on a normal display device. Because active infrared night vision systems can incorporate illuminators that produce high levels of infrared light, the resulting images are typically higher resolution than other night vision technologies. Active infrared night vision is now commonly found in commercial, residential and government security applications, where it enables effective night time imaging under low light conditions. Recently, Laser Illuminator, mainly for range gated imaging, has been developed as a new form of active night vision which utilizes a high powered pulsed light source for illumination and imaging.

### 3.1.3. *Passive Infrared/Thermal Imaging*

Thermal imaging technologies work by detecting the temperature difference between the background and the foreground objects thanks to sensors sensitive in the Middle and Far Infrared, that is the "thermal region of e.m. spectrum.

Thermal imaging cameras are excellent tools for night vision. They detect thermal radiation and do not need a source of illumination. They produce an image in the darkest of nights and can see through light fog, rain and smoke. Thermal imaging cameras make small temperature differences visible. Thermal imaging cameras are widely used to complement new or existing security networks.

The breakthrough came about in 1936 when the first Active Infrared system was developed using a silver photocathode. At first, the image converter received the greatest attention by the military, because it enabled an observer to literally 'see in the dark'. During World War 2, the United States, UK and Germany developed a rudimentary infrared sniper scope that used near-infrared cathodes coupled to visible phosphors to provide a near-infrared image converter to be used in night fighting, but with a limited range of less than 100 meters, they could only provide perimeter defense, moreover rifle-mounted scopes were requiring too heavy batteries and active IR searchlights so large that they had to be mounted on flatbed trucks. Moreover, the sensitivity of the image converter was limited to the near infrared wavelengths, and the enemy military targets had to be illuminated by infrared search beams.

Fig. 4. Infrared Military Applications 2° War: (a) German Infrared Night-Vision Devices, (b) 1940s Sniperscope ZG 1229, (c) 1940s Metascope, (d) Vampir Panzerkampfwagen with Sperber/FG 1250.

Despite its inadequacies, this infrared Sniperscope initiated R&D into advanced night vision technology by developing passive systems without IR searchlights to avoid to betray soldier's position [10, 11] (Fig. 4).

At first, the image converter received the greatest attention by the military, because it enabled an observer to literally 'see in the dark'. However, the sensitivity of the image converter was limited to the near infrared wavelengths, and the most interesting military targets had to be illuminated by infrared search beams.

Since this involved the risk of giving away the observer's position to a similarly-equipped enemy observer, it is understandable that military interest in the image converter eventually faded. The tactical military disadvantages of so-called 'active' IR imaging systems generated military IR-research programs for developing 'passive' (no search beam) systems based on highly sensitive photon detectors. During this period, military secrecy regulations completely prevented disclosure of the status of IR-imaging technology. This secrecy only began to be lifted in the middle of the 1950's and adequate thermal imaging devices began to be available to civilian science and industry. High level results were achieved in the 40's, especially in lead salts (PbSe and PbTe good stable cells were developed by OSRD - Office of Scientific Research and Development in co-operation with the MIT, Harvard and the British TRE - Great Malvern). The history of IR detector developments has been therefore almost coincident with optoelectronics for military applications for many decades, strongly conditioning the cultural behavior of the IR industry and of R&D labs[12,13].

## 4. Blackbody Radiation, Atmospheric Transmission, Night Vision

The physical Laws of Radiation (Planck's and Wien Displacement's laws) correlated to the atmosphere transmission are imposing some key parameters for the development of high performances Night Vision. Moreover the applications of Infrared Night Vision in military environments, with the main task of detecting specific tasks at long distance, were enhancing the importance of atmosphere transmission and the solid state detectors best matched ,therefore creating the well-known two bands (3-5 and 8-12 microns ) infrared FLIR (Forward Looking IR).

More specifically the 3 to 5 micron band was mainly used for detecting warmer targets, while 8 to 12 microns has been applied to thermal imaging of targets (mainly human beings) close to room temperature. Ground or sea-based long range IR imaging is strongly attenuated by the atmosphere excluding the 3-5 and 8-12 microns bands (Fig. 5a, 5b, 5c).

Near Infrared (NIR) and Short-Wave Infrared (SWIR), on the other hand, have a behavior quite similar to more familiar visible light.

Both NIR and SWIR systems can take advantage of moonlight, starlight and an atmospheric phenomenon called "nightglow," but typically requires some type of artificial illumination at night.

Active Infrared systems continued in use until the late 1970's in some countries (NATO was phasing them out by the late 1960's).

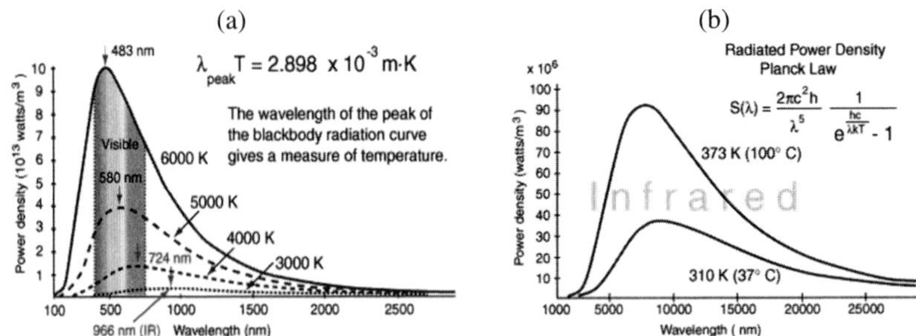



Fig. 5. Blackbody Planck's (a) and Wien Displacement's (b) laws vs. temperature.

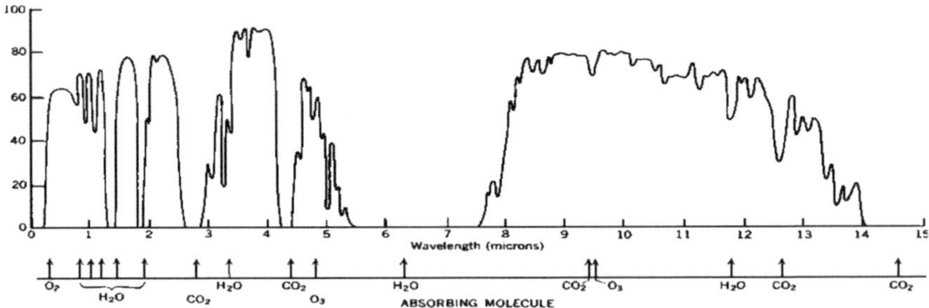



Fig.5(c). Atmospheric Transmission/Absorbing Gases  vs. wavelength.

## 4.1. *Vision Systems: the IR Focal Plane Arrays (FPA)*

The most important application of IR detectors is Thermovision which is allowing to see the thermal emission of the scene. The "ThermoVision" systems as "Night Vision" have strongly evolved in time  since the first thermovision systems developed during the Second  World War ,based on a simple opto-mechanical scanning focusing scene onto a single detector (Figure 6-a).

Such type of image reconstruction, based on a serial scanning of the image points (pixels), has been improved in the time, mainly by enhancing the signal-to-noise ratio.

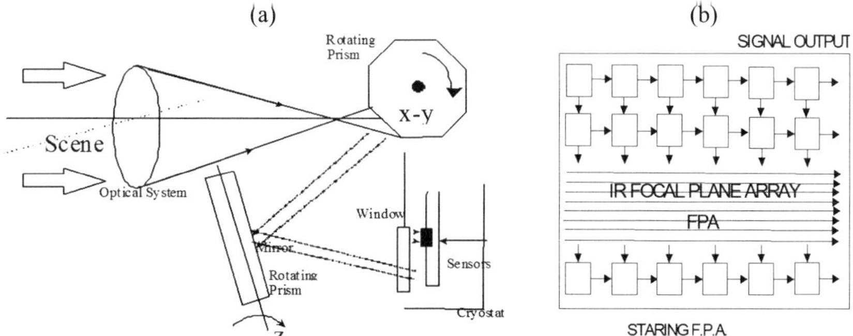


Fig. 6. (a) Optomechanical Scanning, (b) FPA Electronic Scanning.

This was made possible thanks to the increase of the sensors number, which has been achieved firstly, by using linear detectors arrays, with a serial read-out integrated in time (Time Delay Integration TDI) or by using a parallel structure, allowing to read simultaneously more rows of the scanned image and then by integrating them in the image reconstruction. It is evident the importance of the development of FPA detectors working in the so called "staring" mode, that is capable of seeing the image by a simultaneous vision of the scene thanks to a mosaic sensors structure positioned in the focal plane of the image avoiding the use of any opto-mechanical scanning (Figure 6-b) [9].

The readout have been continuously evolving till 1970 when the new charge-coupled- device CCD read-out based on electronic charge transfer has allowed a completely integrated read-out, with the development of FPA sensors with a high number of detectors (close to 1 million pixels).

In all the latest developments the really driving key technology has been the integration of the IR technology with silicon microelectronics and it was, more and more, emerging the importance to free IR from the constraints of the cooling requirements due to its high cost (almost 1/3 of the total cost) and low reliability and heavy need for maintenance. For the above reasons, work on uncooled infrared detectors has shown an impressive growth since the first developments, allowing the real expectation for a production of low cost, high performance detector arrays which finally should follow the rules of a real global market, opening a real market for civil applications following the winning rules of silicon microelectronics. For these reasons the emerging room temperature detectors in the '70 by the use of pyroelectric materials[14], which shows the limitations of not being fully monolithic, but the more innovative room temperature silicon microbolometers [15] appearing on the IR scene in the 1990, appear to be a real breakthrough for future IR sensors.


## 5.  Future Infrared Detectors

Electro-Optics (E.O.) is continuously growing in complexity and difficulty in science and technology developments: particularly IR sensors might be considered at the frontier of solid state technology[16-19].

Electro-Optics has shown winning performances thanks to the reliability and cost achievable by integration with advanced silicon microcircuits. The exploding market of optical telecommunication has, in some way, concentrated the main interest to the near infrared bands pushing the Infrared technology (especially the long wavelengths) towards thermovision and surveillance/ Security applications. This gave an impressive impulse to the development of FPA arrays with an ever growing performances and tasks of these advanced high–tech components and systems. The main effort in nowadays IR detectors developments are oriented towards the highest number of pixels (of the order of $10^6$) with a Focal Plane Array (FPA) structure, with an integrated electronics for signal read-out and elaboration and with the highest possible working temperature.

Key parameters of a single pixel sensor such as the ultimate sensitivity (measured by NETD), response time and working temperature are integrated more and more by key parameters of FPAs as number of pixels, uniformity, reliability and cost. All these requirements will be strongly conditioned by the complete integrability with silicon microelectronics technologies[9].

The competition among various technologies and "technical schools" was strong with unforeseeable new actors emerging in the last years (overall, Room Temperature Microbolometers for future and extremely valid applications in the civil field.). Operational requirements (mainly of maintenance and reliability) were pushing IR science to look for new advanced sensors which could avoid the cryogenic needs. After a serious, but partially successful effort by using pyroelectric effect, limited by low sensitivity and chopper requirement, the new microbolometers technology, due to the microsize of thin films bolometers, completely integrated with silicon technology and therefore often named silicon microbolometers, have been emerging in the last years with a very high promising future for IR sensors market growth.

So, for the first time, thanks to the elimination of cryogenic cooling wide use of I. R., Smart Sensors are emerging on the international market, becoming strategic components for the most important areas, like transports, (especially cars, aircrafts and helicopters), environment and territory control, biomedicine and helps to "human beings better life" (intelligent building, energetic control, thermo-mechanical structuring, auxiliaries to handicapped people etc).

The pixel size is conditioning the achievable number of pixels achievable in a feasible FPA chip size conditioned by the size and the cost of optics. And overall by the fundamental limit on the pixel size determined by diffraction law[13, 22] In fact the size of the diffraction-limited optical spot, or Airy disk, is given by:

$$d = 2.44 \cdot \lambda \cdot F \qquad (4)$$

where $d$ is the diameter of the spot, $\lambda$ is the wavelength, and $F$ is the f number of the focusing lens (for high luminosity F/1 optics at 10 μm wavelength, the diffraction limited spot size is ~ 25 μm).

Therefore, in future a reduction of the pixel size, mainly thanks to oversampling (up to a factor of 4), could be achieved for applications requiring high spatial resolution.

The LWIR pixels could reach the limit of ~ 5 μm and SWIR pixel sizes could shrink to correspondingly smaller dimensions. Therefore the pixel size, unlike the size of a DRAM memory cell, cannot be significantly reduced. So, in order to increase the number of pixels in IR FPAs, the chip size should grow with a high cost and limited handling.

This a further reason for which room temperature IR sensor technologies allowing monolithic integration with silicon microelectronics are favorite for future developments[9].

New requirements in most advanced applications such as in military surveillance, astronomy, medical and environment fields are pushing the request, as strategic parameter, of the multi-spectral operability (for some applications even hyperspectral resolving detection is required). This can allow to foresee either high sensitivity tunable bandgap, photon detectors either wide-band, microbolometers with integrated coating multispectral filters.

The cost of an FPA depends on type and maturity of the technology with different reduction factors in case of mass quantity production (again productions based on silicon technology is more competitive)[38]. The cost of an IR sensors array can be shared in the following main blocks: a) IR FPA physical detectors, including read-out electronics, b) assembling and testing, c) optics with mechanical structure, d) cryogenics in case of cooled detectors.

Up to few years ago there was the thumb's evaluation of almost 25% for each part, although, for high performances photon cooled hybrid detectors, more than 50% was due to the sensors chip and a significant cost improvement in the of cryogenic cooler was to be taken in account in case of working temperature lower than 77 K. This explains why PtSi and QWIP detectors requiring respectively wider luminosity optics and lower operating temperatures have a cost comparable to other photon detectors even though the raw material (Si or GaAs) and fabrication technologies are much cheaper than for HgCdTe.

IR imagers, mainly for military applications, using cryogenic or thermoelectric coolers, complex IR optics and expensive sensor materials, have typical costs of some tens thousands Euro, while IR cameras, based on emerging micromachined silicon bolometer arrays with NEDT close to 10 mK and costs of few thousands Euro, are expected in the civil market in the near future.

This cost, in case of mass production, as it is expected for collision avoidance and guidance assistance in automobile production, could be reduced down to less than one thousand Euros even to few hundreds euro with plastic optics.

The actual explosions of Microsystems Technologies (sensors, control and actuators), especially in automotive applications, is moreover increasing this positive growth of IR market with a real possibility of high level products at contained cost. New markets (automotive, law enforcement, intelligent building, environmental control) and old ones (biomedical and medical, industrial process monitoring, energy control, surveillance and warning) will get strong benefits from the new technology feeding also new military market applications (especially portable equipments, soldier of 2000)[13].

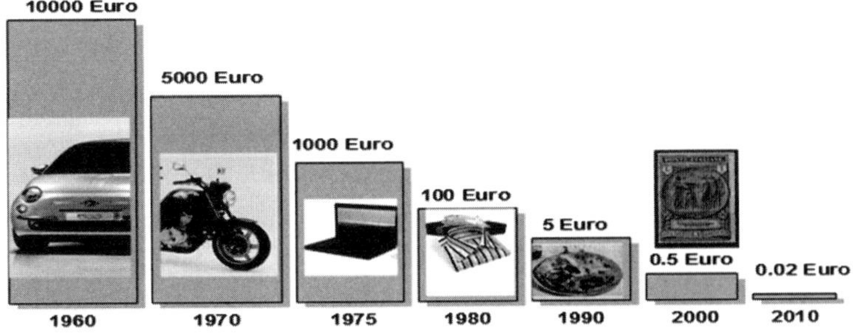


Fig. 7. Single Sensor Pixel Cost Evolution.

In Figure 7, the single pixel detector cost in the last 50 years (at actual value of the year 2000) is shown with a reduction of more than $10^6$. This is comparable only to the cost reduction for 1 Mb memory chip for Computers obtained by Silicon Microelectronics, that is the most advanced technology developed up to now (this is another factor demonstrating the strong liaison between IR FPAs and Silicon Microelectronics).

In conclusion, the cost of IR sensors strongly reduced for single pixel's cost, is however still high because of the improved performances of the new high quality thermal camera. Similar analysis can be done for PC where the cost of memory bit has been reduced by a factor of $10^8$, but the cost of the top Computer is reduced only of a factor of 10.

A strong reduction of the cost, even for high pixels number IR FPAs, is expected, thanks to silicon technology, which, in the case of silicon microbolometer, is avoiding the high cost of cryogenic components.

## 6.  Future Infrared Detectors for Security

For a long time infrared was confined to military applications, especially based on thermovision, technology that is "Infrared: to see the invisible" (Figure 8).

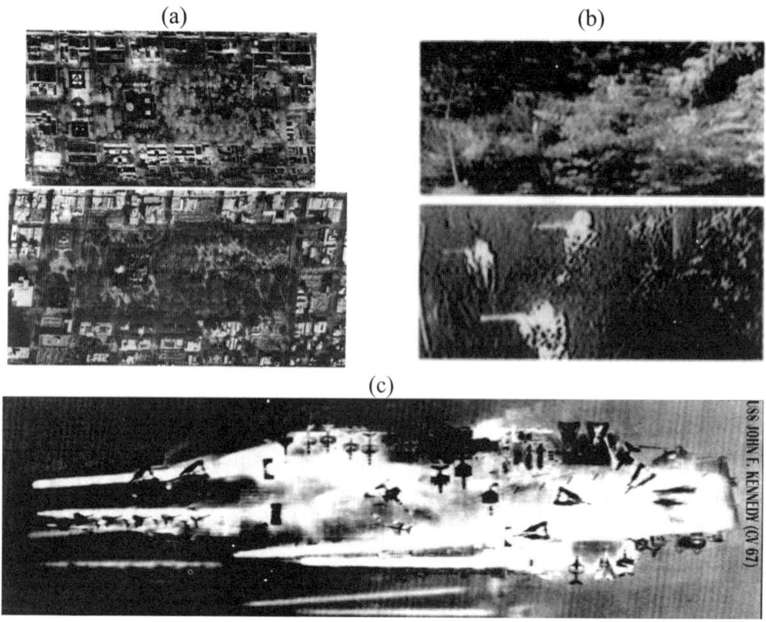



Fig. 8. (a) High Resolution Visible and Infrared Aerial Photos. (b) Visible and Infrared Photos of a Wood (FLIR System Inc.) Photos show evidence that targets in the infrared photo are not detectable in the visible. (c) Aerial Infrared photo of "Kennedy Aircraft Carrier": aircrafts parked or just landed can be recognized.

### 6.1. *Surveillance and Detection and Warning*

The main application for IR technology was in the past and will be in the future and it is Surveillance and Warning, more specifically for military applications that conditioned for a long time the development of IR devices and systems. The highlights of this application to security field are quite well known and have allowed to develop an impressive know-how either in the systems performances.

The first historical lesson, missed for lack of knowledge of the users, was the underestimation of the strategic value of IR surveillance systems: at that time the RADAR was not yet operating although in 1904 Christian Hülsmeyer had used radio waves for detecting "the presence of metallic distant objects".

Only in 1922 Guglielmo Marconi proposed the idea of a RadioTelemeter for localizing metallic objects at distance and therefore the remote sensing was mainly just optical[23,24].

In fact in 1910 (later patented in 1913), L. Bellingham had presented a method to detect the presence of icebergs and steamships by using a mirror and the original thermopile. His infrared radiometer's primary advantage over the disappearing-filament optical pyrometer was that it was able to detect temperatures substantially lower than environmental.

If this device was installed on the Titanic ship avoiding that grave tragedy, probably the efforts in developing IR surveillance systems would have been much greater. During the second World War great efforts have been dedicated to the development of IR surveillance systems especially in the Army with both parties capable of IR detection of enemy targets and support to night moving.

After the 70's, R&D developments of IR surveillance systems, especially for navy applications, were done in advanced Countries (especially US, UK, France and Italy, where the first Modular FPAs Staring Omni-Directional Surveillance System prototype was realized)[12,25].

In the 80's, the SDI Programme for Ballistic Missile Defence by US was originating highly advanced E.O. Surveillance System with performances close at BLIP limits. Nowadays the main efforts are dedicated to the multispectral detection capability. In the end of 90's the crisis originated by the solving of Soviet Union and the consequent end of "Cold War" was pushing the use of infrared towards civil applications (the famous slogan "from military to civil" and "the dual- use mode" with the explosion of thermovision technology applied to environmental control, building /art masterpieces analysis, medical digital diagnostics and recently car guidance/collision avoidance systems) [26-40]. In fact, wide application of IR warning is expected in automotive for Smart Collision Avoidance Systems in poor visibility conditions: room temperature Smart IR Receivers could be installed with high reliability and simple, immediate man-machine interface in any type of motor-vehicles[41, 42].

The developments were so great that, thanks to simultaneous lowering of costs due to mass production of advanced room temperature FPAs sensors, a dual way use from civil to military markets was emerging [16, 21].

## 6.2. *Smart Sensors for Collision Avoidance in the Fog*

An important application of the IR Smart Sensors for civil security is an IR system for driving assistance in low visibility due to fog/smokes as "Collision avoidance in low visibility". In fact, thanks to the better visibility trough fog in IR field with respect to visible, car producers have proposed many systems for the use of thermal viewers to be installed on board. These IR systems up to now have shown heavy limits for the sensible cost even if using the IR room temperature microbolometers. Other limitations include maintenance and reliability and, over all, "man interface" (few car drivers can use an helmet type display or can have enough skill to look at a display while driving in very low visibility).

For these reasons a new generation of simple, reliable, Smart Sensors, operating at room temperature with no costly thermo-control, supplying a sound and light alarm in case of presence of an obstacle on the road could be a winning solution. In the presence of fog, air is more transparent to IR radiation than to visible light.

A low cost smart sensor, based on a 32 x 2 bi-linear micro-bolometer array, to be fixed close to the car headlamps has been developed[43]. This sensor is capable to perform radiometric identification of hot objects and deliver an audible alarm to the driver.

Figure 9-a shows a high resolution thermal image of two possible obstacles in a winter environment: a car parked for more than 10 minutes and a running car.

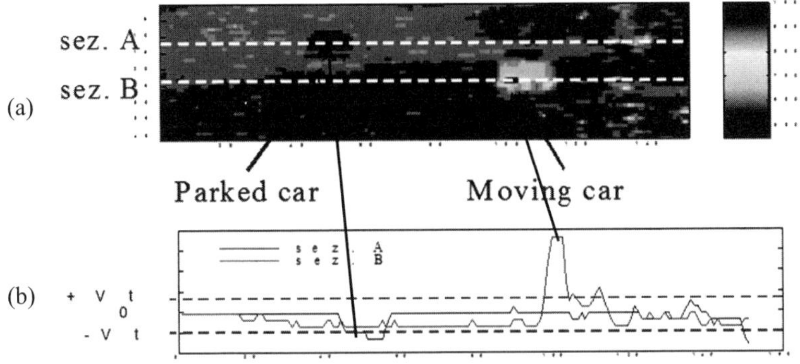



Fig. 9. (a) Thermal imaging; (b) Smart Sensors Signals.

In Fig. 9-b the signal detected by the IR Smart Sensor structure, supplying automatic alarm in both cases, is shown.

## 6.3. *Toxic Gas Sensing*

Owing to the increasing demand for security in civilian crowded areas, both for the risk of accidents in dangerous materials transportation and in industrial toxic substances processing and recently of terrorist threat, the development of novel large scale applicable technologies and methodologies for quick detection and identification of extremely toxic compounds is strongly advised.

Particularly, in military battlefield operations or in case of terrorist attacks, high detection capability coupled to low false alarm rate and identification capability, at least for classes, are required. Moreover, most of the CWA (Chemical Warfare Agents) are heavier than air and are spreading in the atmosphere when there is ground movement: therefore it a point stand-off detection system is needed to detect the presence of toxic gases. Last, but not least, the system could be unattended without the need of any intervention by human operators.

Existing technologies are based on two main classes of sensors: one based on simple physical-chemical sensors which are modifying their electronic properties in presence of some gases; the other class based on complex sensors subsystems which are correlating the physical-chemical properties detected by specific measurements.

Both sensors may be highly sensitive although the first class type is confined to be used only for specific gases and with performances changing with time and up to now unable to supply gas identification.

The second class type (e.g. surface acoustic wave, ion mobility spectrometry, and mass spectrometry), can achieve high sensitivity and selectivity, but most of these technologies cannot give both at the same time and not in unattended way, needing human intervention in their use.

Best performing systems usually require costly and sophisticated laboratory equipments (e.g. FTIR spectrometry, gas chromatography and mass spectrometry[40-42]. Furthermore, they often require ancillary operations, such as sample preparation and pre-concentration, thus requiring long time for analysis which is not operationally acceptable.

### 6.3.1. *Toxic Gas Sensing based on NDIR Absorption Spectroscopy*

As mentioned above, best performing systems such as FTIR spectrometry, gas chromatography and mass spectrometry[44-46] are recently substituted by innovative systems, based on consolidated technique of Non Dispersive Infrared Spectroscopy (NDIR).

Innovative systems, performing a fast multi-spectral measurement of the radiation emitted from a thermal blackbody source in the IR spectral range, particularly between 8 and 12 micron wavelength region, have been recently developed with Bidimensional Multi-paths Multi-Spectral Staring Smart Sensors in the IR (3÷5 and 8÷12 micron) using a MEMS Multi-element Blackbody Source optically coupled to a Multi-pass Cell /Spectrally Linear Variable Filter.

A new Sensor Staring Array Structure, designed to perform Multi-spectral measurements for Gas Detection, Identification and Alarm, has been developed. Spectral resolution of 0.3 microns allows not only to detect but even to identify, at least for classes, even a small amount (down to few parts per million) of toxic gases operating at room temperature and pressure, even if unattended[47].

The detection and identification of absorbing species is achieved by analyzing the spectroscopic signal variation. The novel technical approach consists in implementing a low cost source and a microbolometer room temperature sensor array. This consists of a smart architecture coupled to a spectrally linear variable filter by using a high throughput efficiency optical system to increase the path length and make simultaneous acquisition of several spectral bands within a reasonably compact size. The linear variable filter is a narrow band-pass filter in which the wavelength of peak transmittance changes linearly along one direction of the filter surface. The optical resolution of each detection channel is optimized in width to be matched at the best between resolving power and integrating absorption signal in the widest band as possible. Therefore, it has to be not too narrow and not too large.

Also, considering technological constrains associated to the filter feasibility, a good trade-off is around 0.1 micron bandwidth. The systems are working at room temperature, they are small, portable, low cost and low power consumption, in order to allow easy operation and large scale distribution both for environmental and security applications.

The sensor is based on Active Multispectral Infrared Absorption Spectroscopy in the gas phase and uses 42 spectral channels to represent molecular fingerprints across the Medium and Long Wave IR spectrum (MWIR and LWIR). The heart of the sensor is an advanced detector device operability (self-calibration capability); the system is showing sensitivity for most Toxic Industrial Compounds and Chemical Warfare Agents with a limit of Detection to Nerve Agents around 500 ppb. Resolution of 0.1 microns spectral resolution allows not only to detect but even to identify, at least for classes, even a small amount (down to few parts per million) of toxic gases operating at room temperature and pressure, even if unattended [48].

The NDIR system matches very well the requirements of an early warning system for toxic gases, that could be implemented both as an autonomous unit and as the node of a sensor network.

Key assets are its ability to detect and correctly identify a huge number of chemicals of different nature and class of risk (CWAs, TICs, solvents, perfumes..), with minimal false alarm rate; other crucial assets are its compact and simple design, cost-effectiveness, unattended and continuous operability (self-calibration capability). Moreover it shows identification capacity and high rejection of interfering substances and it is suitable for surveillance and warning in the presence of toxic gases intentionally or accidentally released in the air.

## 6.3.2. *Stand-Off Detection*

Infrared /microwaves techniques and Imaging Hyperspectral Sensors are mostly developed for their selectivity and specificity to detect differences in the spectra of two or more species of chemical warfare agents (CWA) allowing detection of explosives with high sensitivity and very low false alarm rate. Infrared transmission spectroscopy is a analytical tool to detect and identify chemicals in the gas phase.

Open-path Fourier transform infrared (FTIR) spectroscopy is an established method to monitor environmental pollutants 1), e.g., aircraft exhaust gases 2), detect toxic gas emissions at industrial plants 3) and even detect chemical warfare agents (CWA).

Thus, open-path FTIR is a potential method for explosive detection too. A major drawback of open-path FTIR is the long integration time needed to collect a spectrum because of the low power of the thermal radiation sources usually employed. This is especially the case if trace compounds have to be detected, which only show up in tiny spectral features in the spectra. The sensitivity of infrared spectroscopy is linked to the measurement of small relative transmission changes of, e.g., less than $10^{-3}$. Selectivity and specificity are linked to the ability to detect differences in the spectra of two or more species. If spectral features of two compounds overlap at one wavelength, at another wavelength a difference may occur which could be used to discern between the two species.

Recently, integration of multispectral IR spectroscopy with quantum cascade lasers and hollow-waveguides is promising improved sensitivity and area coverage allowing fast and effective detection at public places like airports, railway or coach stations[49]. In fact, infrared laser spectroscopy, an established method to measure gas concentrations, due to the high power of the laser and its higher optical collimation compared to the thermal emitters used in FTIR equipment sensitive, allow making long distance measurements. Especially for small molecules with singe rotational-vibrational lines in the infrared laser spectroscopy methods are extremely sensitive and relative transmission changes of $10^{-4}$ and below have been measured even in industrial in-situ applications. The stronger and more specific infrared absorptions of most chemicals/toxic gases are in the mid/far infrared range between 3µm and 20 µm wavelength, making this region interesting for Infrared spectroscopy, for this reason, although most work on laser spectroscopy is performed in the near infrared, where suitable laser sources thanks to wide use in telecommunications wavelength bands, the quantum cascade lasers (QCL) in the middle/far IR are the ideal source for IR laser spectroscopy.

After the first QCL spectroscopic applications made with pulsed QCL operation, most of the developments are using CW sources at room temperature, nowadays available for many wavelengths in the mid/far infrared. Recently, QCL systems for gas detection operating in the pulse mode have been developed. QCL devices are driven by a short pulse in the 10- 100 ns range with pulse repetition rates in the low 10-100 kHz range allowing enough time to get consistent absorption features of the sample gas within the pulse duration. Advanced QCL laser systems have detection capability of hidden explosives and other items on personnel.

As is the case with chemical warfare agents (CWA) for detection of explosives, very good sensitivity and a very low false alarm rate, i.e. specificity, are a must. Fast and effective detection of explosives is a key security issue against possible terrorist attacks. Especially at public places like airports, railway or coach stations efficient detection systems are needed. Usual methods at airport security checks are wiping carry-on baggage / laptop computers etc. to collect samples which are analyzed subsequently by e.g. gas chromatography mass spectroscopy (GC-MS), gas chromatography chemo luminescence (GC-CL) or ion mobility spectrometers (IMS)[50-55].


## 7. Security Applications of THz Technology

Recent terroristic events, based on new types of threats and explosives, have pushed towards the developments of new techniques of detection and alarm employing different parts of the electromagnetic spectrum, particularly extending from infrared to terahertz radiation, that is the e.m. band between infrared and microwave. Also in these new developments there is a priority to the developing of imaging systems, at least for the know-how originated in infrared technologies, while spectroscopic detection is mostly developed by microwaves techniques.

Terahertz (THz) technology is a promising, emerging candidate. The "Terahertz gap" due to the fact that until recently there was a lack of high power sources and  high sensitivity detectors can be allocated between 100 GHz ( 3 mm) and 10 THz (30 μm) that is between the millimeter -microwave part of the e.m. spectrum and the far-infrared.

Concerning microwave sources till now few sources (recently solid state) are capable of generating enough power radiation efficiently produced at frequencies above undred gigahertz, whereas solid state laser sources have been limited by thermal effects in their performances in the Far Infrared region. However, in recent years, several approaches have been developed that enable the efficient generation and detection of terahertz radiation truly commercially viable. The most mature technology uses ultra-fast pulsed laser technology, and produces very short terahertz pulses. As a pulsed technique, with picoseconds timescales, the method is intrinsically broadband. Radiation at terahertz frequencies has unique properties that may be advantageous for security applications. It penetrates many non-conducting materials, but unlike X-rays is non-ionizing. The short pulses produced by laser techniques also allow radar-like imaging in three dimensions, as well as the simultaneous collection of spectroscopic information as in magnetic resonance imaging (MRI) or optical spectroscopy. This is important because many substances have characteristic intermolecular vibrations at far infrared/ terahertz frequencies that can be used to characterize them as molecules.[56-58]

Radiation at terahertz frequencies has unique properties that may be advantageous for security applications, in fact, can penetrate many non-conducting materials, but unlike X-rays is non-ionizing, and can allow radar-like imaging in three dimensions thanks to the extremely short pulses used in pulsed terahertz techniques, as well as the simultaneous collection of spectroscopic information like Infrared. This is important because many substances have characteristic intermolecular vibrations at Far IR-THz frequencies that can be used to characterize them as molecules like IR spectroscopy can detect bond vibrations of single molecules.

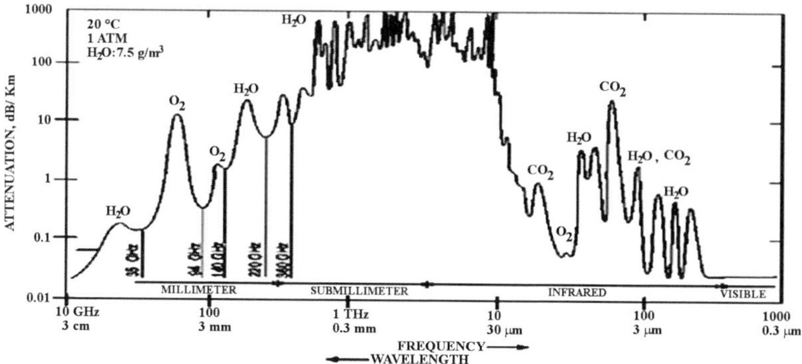


Fig. 10. Absorption Bands in Atmospheric Transmission.

In synthesis Multispectral (IR-THz) Systems for Security applications will have high growth especially coupled to new solid state tunable Laser Sources with innovative, new technologies for Multispectral Pluridomain Smart Sensors.

## 8. Conclusions and Outlook

The first historical lesson, missed for lack of knowledge of the Users, was the underestimation of the strategic value of IR surveillance systems (at that time the RADAR was not yet operating and therefore the remote sensing was mainly just optical). In fact around 1910 an infrared radiometer had been presented for detecting the presence of icebergs and steamships by using a mirror and the original thermopile to detect. If this device was installed on the Titanic ship avoiding that grave tragedy, probably the efforts in developing IR surveillance systems would have been much greater.

During the 2$^{nd}$ World War, great efforts have been dedicated to the development of IR surveillance systems especially in the Army with both parties capable of IR detection of enemy's tanks and support to night moving. After the 70's, R&D developments of IR surveillance systems, especially for navy applications, were done in most advanced Countries.

In the 80's, the SDI Programme for Ballistic Missile Defense by US was originating highly advanced E.O. Surveillance System with performances close at BLIP limits. Terroristic events, based on new types of threats and explosives, have pushed towards the developments of new techniques of detection and alarm employing different parts of the electromagnetic spectrum, particularly extending from infrared to terahertz radiation, that is the e.m. band between infrared and microwave. Also in these new developments there is a priority to the utilization of imaging systems, at least for the know-how originated in infrared technologies, while spectroscopic detection is mostly developed by microwaves techniques.

Recently, emerging Terahertz technologies, thanks to their non ionizing radiation and their detection capability of hidden objects in clothing and in packaging containers and luggage, coupled to the spectroscopic detection of plastic explosives and other chemical and biological agents are emerging as very promising technologies for integrated, efficient systems for security screening and counter-terrorism. Far-infrared /microwaves techniques and Imaging Hyperspectral Sensors will be mostly developed for their selectivity and specificity to detect differences in the spectra of two or more species of chemical warfare agents (CWA) allowing detection of explosives with high sensitivity and very low false alarm rate.

Recently, integration of multispectral IR spectroscopy with quantum cascade lasers and hollow-waveguides is improving sensitivity and area coverage allowing fast and effective detection at public places like airports, railway or coach stations. In future, an evolution from Infrared Night Vision to Digital Functional Thermography, and from Imaging to Pattern Analysis, and from Detection to Smart Sensors Alarm Systems is expected. Multispectral (IR-THz) Systems for Security applications will have high growth especially coupled to new solid state tunable IR Laser Sources with innovative, new technologies for Multi-Spectral Multi-Domain Smart Sensors.

# References


1. W. Herschel, "Experiments on the Refrangibility of the visible Rays of the Sun" Philosophical Transactions of the Royal Society of London Vol. **90**, 284-292 (1800).
2. E. S. Barr, Amer. J. Phys. **28**, 42-54 (1960).
3. www.ipac.caltech.edu/outreach/Edu/irtech.html .
4. S. P. Langley, "The bolometer," Vallegheny Observatory (Dec.1880), New York: Publ. by the Soc. Gregory Bros. Printers (1881).
5. L. Bellingham, "Means for Detecting the presence at a Distance of icebergs, steamships, and other cool or hot objects" U.S. Patent n. 1,158, Pat. Nov. 2, 1915.
6. R. D. Parker "Thermic Balance or Radiometer" U.S. Patent n. 1.099,199, Pat. June 9, 1914.
7. W. Smith, Nature 7, 303 (20 Feb. 1873)
8. T. W. Case, Phys. Rev. **9**, 305 (1917).
9. C. Corsi "Infrared Detectors" –"An Introduction to Optoelectronic Sensors". Vol.7 Ed. G. C. Righini, A. Tajani and A. Cutolo, World Scientific Publisher Co. Singapore, (2009).
10. www.nvl.army.mil/.
11. www.achtungpanzer.com/german-infrared-nig.
12. C. Corsi, "International NATO Electronics Warfare Conference" (1978).
13. C. Corsi, *J. Modern Optics*, Special Issue Infrared Technologies, Vol. **57**, Issue 18, pp.1663-1686 (2010).
14. R. Watton, *Ferroelectrics*, **91**, 87 (1989).
15. R. A. Wood et al., IEEE, Solid State Sensors & Actuators Workshop, June 1992, USA.
16. C. Corsi "History lessons and future trends in advanced IR detectors" Giorgio Ronchi Foundation, Atti LIII, pp.11- 20 (1998).
17. M. Razeghi, *Opto-Electr.*, Vol. **6** , 155-194 (1998).
18. P. Norton et al., "3rd infrared imagers," *Proc. SPIE* 4130, 226 - 236 (2000).
19. A. Rogalski "IR detectors: the next millennium", *Proc. SPIE* VI 4413, (2001).
20. A. Rogalski, "IR detectors: status trends", *Prog. Quantum Electr.*, 27, 59-210 (2003).
21. C. Corsi, "Infrared and the Others", Giorgio Ronchi Foundation, Atti LVII, pp.363–369 (2002).
22. L. J. Kozlowski and W. F. Kosonocky, "Infrared detector arrays," Chap. 23 Handbook of Optics, M. Bass, Williams, and W. L. Wolfe, McGraw-Hill (1995).
23. "The Telemobiloscope," Electrical Magazine (London) Vol. **2**, p. 388 (1904).
24. G. Marconi Conference "American Institute of Electrical Engineers/Institute of Radio Engineers", 20 June 1922.
25. C. Corsi et tl. Nat. Patent n.47722°/80, Tech. Rep. PT-79 Elettronica S.pA. (1979).
26. http://www.reddingthermography.com/ .
27. E. F. J. Ring, "The History of Thermal Imaging" in the "Thermal Image in Medicine and Biology" eds. K. Amner and E. F J. Ring, pp.13-20, Uhlen-Verlag Vienna (1995).
28. M. Gauthrie, *Biomedical Thermology*, Alan R. Liss, Inc., New York, NY, pp. 897-905(1982).

29. C. Corsi, *Proc. Biomedical International Conference*, S.Elpidio Italy, Ed. Elettronica SpA, pp. 3-9 (1981).
30. R. Amalric, D.Giraud, et al., *Acta Thermographica*, **8**, pp. 21-24 (1984).
31. J. Spitalier, D. Amalric, et al. *Proceedings International Conference*, MTP Press Ltd., pp.173-179 (1983).
32. C. Corsi, *Acta Thermographica*, **7**, pp. 33-39 (1983).
33. X. Maldague, ASNT Handbook 3$^{rd}$ Edition, Vol. **3** "Infrared and Thermal testing" Maldague, Editor: Patrick O. Moores (2001).
34. P. Bison, F. Cernuschi and E. Grinzato, *Int. J. of Thermophysics* 29:2149-2161 DOI 10.1007/s10765-008-0421-1 (2008).
35. V. Vavilov, T. Kauppinen and E. Grinzato, "Research in non-destructive evaluation", Springer-Verlag, Vol. **9** No. 4, pp. 181-200, New York (1997).
36. E. Grinzato, P. G. Bison and S. Marinetti, *J. of Cultural Heritage*, Elsevier, **3**, pp. 21-29 (2002).
37. G. M. Carlomagno, et alt, *Thermography Rev. Gen. Therm.*, **37**, pp. 644-652, Elsevier Science, Paris (1998).
38. D. W. Banks et al. , Technical report  2001-210848, NASA Dryden Flight Res. Center Edwards CA. (2001).
39. T. Astarita, G. Cardone, G. M. Carlomagno and C. Meola, *Optics & Laser Technology*, Elsevier Science (2001).
40. C. Corsi, *Microsystem Technologies*, pp. 149-154 (1995).
41. C. Corsi "Smart Sensors" 8$^{th}$ AITA -Infrared Physics Vol. **49,** n. 3, 192-197 (2007).
42. C. Corsi "Systems for avoiding collision of vehicles in low visibility conditions" European Patent WO/2000/022596
43. C. Corsi European Patent PCT/IT99/00327-Int. Publ. N. WO 00596.
44. S. Harden, "Ion Mobility Spectrometry for Detection of Chemical Warfare Agents 1960's to the Present", Proceedings 13$^{th}$ Int. Conference on Ion Mobility Spectrometry, Gatlinburg, Tennessee, USA, July 25-29 (2004).
45. J. Xu, W. B. Whitten and J. M Ramsey, *IJIMS* **5**, 207-214 (2002) 2.
46. J. Stach and J. I. Baumbach, *IJIMS* **5**, 1-21(2002) 1.
47. C. Corsi, N. Liberatore, S. Mengali, A. Mercuri, R. Viola and D. Zintu, *Proc. SPIE* **6739**, (2007) 673915, pp. 1-10 (2002)
48. C. Corsi, A. Dundee, P. Laurenzi, N. Liberatore, D. Luciani, S. Mengali, A. Mercuri, A. Pifferi, M. Simeoni, G. Tosone, R. Viola and D. Zintu, *Advances in Optical technologies*, **12**, Article ID 808541 (2012).
49. S. Mengali, N. Liberatore and R. Viola "DIRAC" project, *Lecture Notes in Electrical Engineering*, Vol. **162**, in *Proc. of the First National Conference on Sensors*, Rome, 15-17 February 2011, F. Baldini, A. D'Amico, C. Di Natale, P. Siciliano, R. Seeber, L. De Stefano, R. Bizzarri and B. Andò (Eds.), Springer (2012).
50. H. Sekiguchi, K. Matsushita, S. Yamashiro, Y. Sano, Y. Seto and T. Okuda, Forensic Toxicology 24:17 – 22 (2006).
51. J. Hildenbrand, J. Herbst, J. Wöllenstein and A. Lambrecht, "Explosive detection using infrared laser spectroscopy", Fraunhofer, Institute for Physical Measurement Techniques (IPM), Heidenhofstr. 8, D-79110 Germany.
52. J. Faist, F. Capasso, D. L. Sivco, C. Sirtori, A. L. Hutchinson and A. Y. Cho, *Science* **264**, 553–556 (1994).
53. A. A. Kosterev, F. K. Tittel, *IEEE J. Quantum Electronics* **38**(6), 582-590 (2002).

54. B. G. Lee, M. A. Belkin, R. Audet, J. MacArthur, L. Diehl, C. Pflügl, F. Capasso, D.C. Oakley, D. Chapman, A. Napoleone, D. Bour, S. Corzine, G. Höfler and J. Faist, *Appl. Phys. Lett.*, **91**, 231101 (2007).
55. A. Lambrecht, S. Hartwig, J. Herbst and J. Wöllenstein, "Hollow fibers for compact infrared gas sensors ", Proc. SPIE, 69010V, (2008).
56. M. C. Kemp, P. F. Taday, B. E. Cole, J. A. Cluff, A. J. Fitzgerald, and W. R. Tribe, "Security applications of terahertz technology", Proc. SPIE, 5070, 44-52 (2003).
57. K. Kawase, Y. Ogawa, Y. Watanabe and H. Inoue, *Opt. Express*, **11**, 2549-2554 (2003).
58. I. S. Gregory, W. R. Tribe, B. E. Cole, C. Baker, M. H. Evans, I. V. Bradley, E. H. Linfield, A. G. Davies and M. Missous, "Continuous-wave Thz imaging using diode lasers, in W. R Tribe, D. A. Newnham, P. F. Taday, and M. C. Kemp Eds. *Proc. SPIE*, Vol. **5354**, pp. 139-150 (2004).

# QUANTUM CRYPTOGRAPHY: A NOVEL APPROACH TO COMMUNICATION SECURITY


Alberto Porzio

*CNR - SPIN*
*Complesso Universitario di Monte Sant'Angelo*
*Via Cintia, 80126 Napoli - Italy*
*E-mail: alberto.porzio@spin.cnr.it*



Quantum Cryptography aims at solving the everlasting problem of unconditional security in private communication. Every time we send personal information over a telecom channel, like when we buy something on the web, a sophisticated protocol protects our privacy making our data unintelligible to unauthorised receivers. These protocols resulted from the long history of cryptography. The security of modern cryptographic systems is guaranteed by complexity: the computational power that would be needed for gaining info on the code key largely exceeds the available one. Security of actual crypto systems is not "by principle" but "practical". On the contrary, quantum technology promises to make possible to realise provably secure protocols. Quantum cryptology exploits paradigmatic aspects of quantum mechanics, like superposition principle and uncertainty relations, for realising protocols whose security is intrinsic. In this contribution, after a brief historical introduction, we aim to provide a survey on the physical principles underlying the quantum approach to cryptography. Then, in the second part, we analyse in more detail two of the possible protocols and glance over the state of art of the research in the field.


## 1. Introduction

Ever since its advent in the first decades of the $20^{th}$ century, quantum mechanics (QM) has been recognized as a scientific revolution. Its impact is nowadays under everyone's eyes. As a matter of fact, lasers and silicon electronics have their foundations in the quantum world. So that, every time we use electronics devices, as dishwashers, vacuum cleaners, cellular phones or we transmit and receive information we unwittingly exploit our knowledge of the quantum nature.

At the same time, two of the most intriguing aspects of QM, the existence of single quanta and the possibility of generating entangled states, are, so far, mostly unexplored in everyday life.

On one hand, the *quantum* idea[a] is the core of the Planck[1] hypothesis that gave birth to the quantum revolution. On the other hand, the nature of entanglement has been disputed, in the scientific community, for a long time[2]. So that, this concept is not part of the common knowledge, contrarily to gravity, or interference, or many old physical concepts.

*Entanglement*[b] was firstly mentioned by Erwin Scrödinger[3] in replying to the famous 1935 EPR paper[4]. In the latter the classically counter-intuitive result of a *spooky action at distance* in a bipartite quantum system was indicated as the proof for the failure of the quantum theory. Schrödinger, taking for granted the quantum view, applied the superposition principle to multi-partite systems to explain that paradox[c].

In doing so he found that, in the quantum realm, correlation is a non-local property. That is, space-time separated quantum objects are interconnected in a way that, if seen with classically eyes, can contradict causality. So that, in a thought experiment, the result of a measurement on one sub-system instantaneously gives information on the result attainable on the other.

After nearly 50 years of theoretical discussions, since the early 80's of last century, scientists have recognized *entanglement* as a possible quantum resource for realizing communicational and computational protocols that would be forbidden in a classical world.

---

[a] The word *quantum* in this context indicates a single quanta.

[b] The word *entanglement* is a rather literal translation of the German word *Verschränkung* originally used by Schrödinger.

[c] Since then named as the *EPR paradox*.

The need for a quantum approach to computation was brought to the attention by Feynman[d] who recognized that the quantum realm asks a quantum computer, *i.e.* a computer obeying quantum mechanical rules, for being fully understood. Moreover, it is well known that the increase in performances in modern computers goes hand in hand with size reduction[e]. So that, soon or later, a single transistor will be so small that it will be necessary to account for quantum effects to fully understand and deterministically predict its behaviour.

Both entanglement and single quanta lay at the core of Quantum Cryptography (QC) the art of secret sharing in a quantum world.


## 2.   A Bit of History

The first documented use of a trick for a secret communication dates back to the ancient Greeks. In his Histories[f], Herodotus tells of a messenger that had a message imprinted on the scalp under his hair. He was then sent over to an allied king with the only instruction of asking to be shaved once there. Even if this represents an example of steganography[g] rather than cryptography, it proves that the quest for secret *unbreakable* communication roots at the very beginning of human history. Probably, the first attempts are as old as the human ability of writing and methods of secret communication were developed by many

---

[d] The idea was one of the core concepts during his famous speech at the *First Conference on the Physics of Computation*, MIT, Boston (USA), 1981.

[e] The empirical law on size reduction vs. time is known as the Moore's law (http://en.wikipedia.org/wiki/Moore%27s_law).

[f] " ... the man with the marked head came from Susa, - ... - Histiaeus, - ... - could find but one safe way, as the roads were guarded, of making his wishes known; which was by taking the trustiest of his slaves, shaving all the hair from off his head, and then pricking letters upon the skin -... - and as soon as ever the hair was grown, he despatched the man to Miletus, giving him no other message than this - When  thou art come to Miletus, bid Aristagoras shave thy head, and look thereon. " (Herodotus, Histories, book V, 440 B.C.).

[g] Steganography is the art of hiding messages in such a way that no one, apart from the sender and intended recipient, suspects the existence of the message. The word steganography is of Greek origin and means "concealed writing" (steganos (στεγανός) - covered or protected", and graphei (γραφή) - "writing"). Contrarily to cryptography, that protects the contents of a message, steganography can be said to protect both messages and communicating parties.

ancient societies, including those living in Mesopotamia, Egypt, India and China[5,6].

What makes cryptology[h] different from steganography is the presence of a *code*: a (complex) mathematical tool that allows hiding the message in such a way that only authorised parties (that are aware of the code) can decrypt it. So that, the two main goals for cryptographist (sender and intended receiver) are to be able to communicate in a form unintelligible to unauthorised third parties, and to authenticate the communication channel, so to prove that the message was not intercepted and that they are really talking to the expected counterpart.

Up to the development of immaterial communication (*e.g.* radio and telegraph), crypto-messages have been conveyed on paper so that two parties had to physically share the code, either in the form of a machine or an algorithm, at a former time[i]. In such a paper–world interception was easy to detect, apart from the case of untrusted messengers. Since then, cryptography was restricted to military and diplomatic affairs.

After that, the moving to tele–communication has first meant the search for more complex coding then, after the Second World War, the slow discard of specific crypto-machines, replaced by specific computational algorithms.

Telecommunication is intrinsically conveyed over "open" channels so that avoiding interceptions is impossible and physical security was no longer guaranteed by the trust of messengers. Thus, it was necessary to code the message in a way that would have been "absolutely" impossible for any intruder to find the code key. History tells us that this has not been the case. The famous breaking of the German ENIGMA chipers at Bletchey Park, England, during the last World War, was a key military advantage for the allied forces in defeating the nazists.

The idea behind ENIGMA and others crypto mechanical machines, was to make the decoding so complex that no man would have been able to solve it. The solution came from Alan Turing that applied sophisticated mathematical tools for gaining inside the code. It is worth

---

[h] Cryptology embodies *cryptography* (the art of code making) and *cryptoanalysis* (the art of code breaking).

[i] Brief histories of crypto techniques and machineries can be found on the web (see for example http://en.wikipedia.org/wiki/History_of_cryptography).

noticing that to accomplish this goal, the first digital computer was built: crypto-analysis was "the midwife of computer science"[j].

The security of a cipher alphabet was, by that time, demanded to complex electro-mechanical machines able to shuffle, at the sender station, the original message into a pseudo-random sequence of symbols and to revert the process at the receiver. The quest for complexity took cryptology into a completely novel scenario.

Modern cryptology is based on the use of *crypto–keys* that coding and decoding algorithms appropriately manipulate to "encrypt" and "decrypt" the message. In the ideal case the crypto–messages should result in random sequences of symbols so that deciphering can happen only statistically with the attacker testing every possible key[k].

The increasing in complexity triggered extensive research in the field.

The 1949 famous Shannon paper[7] "Communication Theory of Secrecy Systems"[l] is, with no doubts, the foundational theory for modern, classical and quantum, cryptology. In it, Shannon, addresses separately secrecy and authenticity as two different aspects of communication security. While his focus was on secrecy, only thirty-five years later, G. J. Simmons gave a complete theory for authenticity[8].

Shannon defined *unbreakability* as the proof of perfect secrecy showing that only codes that make use of secret keys longer than or equal to the number of bits contained in the message can be unbreakable.

This proof was successfully applied, under some constrains[m], to the *one-time-pad* cryptosystem[n]. In this case the coding algorithm consists in adding, modulo the number of needed symbols, the digitalised form of the message with a random string of the some length (the one time key).

---

[j] Attributed to R. L. Rivest, one of the co-discoverers of the famous RSA public key cryptosystem (*see below*).

[k] The length of the *key* is the critical ingredient for the security of the cipher. The longer the key the higher is the number of possible keys an intruder has to test.

[l] The paper represents a *declassified* version of confidential report "A Mathematical Theory of Cryptography" Memorandum MM 45-110-02, Sept. 1, 1945, Bell Laboratories.

[m] The *one-time-pad* cryptosystem is unbreakable if and only if the key: (a) has, at least, the same length of the message, (b) is truly random, and (c) is never re-used.

[n] The *one-time-pad* cryptosystem, published in 1926, was actually invented in 1917 at AT&T by Gilbert Vernam[9]. Although it is provably secure, it is not of easy implementation and was never used extensively.

At the receiver, the reverse operation (subtraction of the some key modulo the number of symbols) guarantees the unveiling of the message.

In such a system, the security issue is transferred from the message to the secrecy of the one-time key making this approach impossible for conveying secret information over public communication channels as it requires an unbreakable channel for sharing the key. This leads to the *key distribution* problem. Moreover, it can be proven that, in a classical physics scenario, any communication channel can always be passively monitored, without the legitimate users being aware of the ongoing interception. Then, the challenge was to find a, possibly very simple and cheap, coding/decoding algorithm whose security was not "by principle" but "practical". The computational complexity required for obtaining the key from the interception of the message has to be so high that the probability of getting it in a reasonable time is very low.

## 2.1. *Public Key Cryptosystem*

In more recent times, let's say in the internet era, communication security is much more than an only military and diplomatic issue. Every time we exchange private information over the web, such an algorithm prevents any attempt for an intruder to catch the real message. So, before jumping into the quantum world and there make use of the above theoretical concepts, it is necessary to make a last step on the classical staircase to better focus on the *key distribution* problem that QC mainly addresses.

The most sophisticated classical cryptosystems actually in use are known as public–key cryptosystems[o]. In this case, the algorithm makes use of two types of key, public and private. Such an algorithm requires one key pair per user. One key of the pair must be a private one (always secret), while the other can be public (often widely available), so that no secure channel is needed for *key distribution*. If the private key stays

---

[o] Public–key cryptosystems were known to British security agencies few years earlier than they were "officially" discovered by cryptologists in late '70s of the last century. The milestone was an internal report at CESG, a state UK agency for communication security, by J.H. Ellis, "The possibility of Secure Non-secret Digital Encryption" in January 1970.

secret, it is possible to use the same pair indefinitely. The key distribution issue is solved by making the key known to everybody. The concept is similar to a safe (the sender private key) in which the sender closes the message with a padlock (the public key) sent there by the receiver. At this point, the message is safely closed and the only one that can access it is the receiver that has in hands the (private) key to open the padlock. Asymmetric algorithms rely, for their effectiveness, on a class of mathematical problems, called one-way functions. These functions require relatively little computational power to execute, but vast amounts of power to reverse, if reversal is possible at all[p].

Public key systems are based on asymmetric key encryption[10]. The key used for coding at the sender is different but mathematically related to the key used for decrypting at the receiver.

The most famous *public-key-system* is, by far, the RSA system[11].

Once again, being not possible to realise an unbreakable cryptosystem, security is demanded to complexity.


## 3. The Birth of QC

As often happens in science, while RSA and asymmetric algorithms have satisfied the "commercial" quest for a "practically" secure system, scientists have continued the search for provably secure ones. Moreover, security of RSA, resides on the not disproven fact that no efficient factorization algorithm is known for classical computers able to break it in reasonable times.

Earlier than Feynman urged for a quantum computer, there were some attempts to break the quantum wall for computation theory. In particular, in 1973 Holevo found a fundamental limit on the amount of information that can be gained from a quantum state[12]. While two years later, Poplavskii proved that it would have been impossible for a classical computer to simulate a dynamical quantum system[13]. Furthermore, we now know that if quantum computers will be ever available, RSA could

---

[p] The simplest example of one-way function is multiplication of (large) prime numbers. It is easy and computationally not expensive to multiply prime numbers while, given the result of such a multiplication, it is very difficult, and computationally expensive, to find the original prime factors.

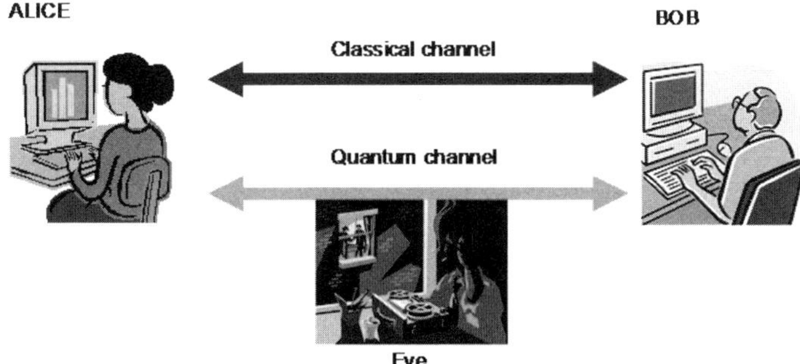


Fig. 1. The setting of a generic QC link: Alice and Bob are connected by a quantum channel, into which Eve can tap without any restriction other than the laws of physics; and by an authenticated classical channel, into which Eve can only listen to.

be, then, easily broken by Shor's quantum algorithm[14], able to factorize large numbers in a very efficient manner exploiting entangled states[q].

The quantum door was open.

Before giving a look inside, we need to get familiar with some terminology used in QC[r]. A generic setting for a QC protocol is given in Fig. 1. **Alice** will always be the sender, **Bob**, the receiver and **Eve** will always represent the *eavesdropper*. A **quantum channel**[s] is a transmission channel open to the public and on which anyone can manipulate information, contrarily to a **classical channel** where classical information is conveyed in public but only *listening* is allow for third parties. A **quantum cryptographic link** will, then, require Alice and Bob to talk over a classical channel to authenticate their selves and send classical information forth and back. Then, a quantum channel will convey quantum information between the two.

As we have seen in the latest days of classical cryptography, the open problem was essentially the *key distribution* process. In general, a QC protocol will provide Alice and Bob with nearly identical shared keys.

---

[q] The Shor algorithm has been the first demonstration that future quantum computer can perform some tasks more efficiently than classical ones. It has been already experimentally realized in different context[15].

[r] The names Alice, Bob and Eve are borrowed from classical communication theory.

[s] Formally, a quantum channel is represented through an Hamiltonian describing the quantum evolution of the information carrier system. In theoretical QC, it is used to model, under a QM approach, any possible operation that Eve can perform.

Then, the two parties, have to compare their strings in order to estimate the error level between the respective keys. These errors can be caused by Eve interceptions, channel imperfections (as losses), and detectors' inefficiencies and/or dark counts. It is physically impossible[t] to distinguish among these types of errors. This implies that to theoretically evaluate the level of security of a specific protocol, one has to assume all the errors due to eavesdropping. At the end of the quantum transmission, the raw keys are not completely equal. In order to distil a common trusted key, Alice and Bob perform some classical operations[u] on raw data.

The main question QC tries to answer is: "Is really possible to generate and distribute, in a provably secure way, a sequence of truly random numbers to form a shared trusted key?"

QM provides the answer. It tells us that any quantum measurement destroys the information the quantum system is carrying[v]. Moreover, the so called *no-cloning* theorem[19] states that: "it not possible to *clone* the state of a single *quantum*". So that, if we will be able to prepare single quanta systems and encode the information in their quantum state, then, only one user can faithfully receive that information. Then, in an ideal quantum world, we can surely answer affirmatively to the above – indeed simple – question. The conceptual scheme is: a) Alice prepares single quanta systems; b) in well definite (and known to her) quantum state; c) faithfully transmits them to Bob site; d) Bob measures their quantum state.

It is this simple *proof-of-principle* scheme that has triggered several proposals for QC protocols[w]. The first one was illustrated in 1984 by

---

[t] They are formally equivalent and represented by the same quantum mechanical model.

[u] In particular, Alice and Bob will perform *information reconciliation* and *privacy amplification*[16]. These two, barely classical protocols, are intended for removing the unavoidable erroneous bits and reducing the knowledge on the key Eve could have gained. It has been proven that if, and only if, the error rate between the keys is lower than a certain threshold[17], Eve's key knowledge can be reduced to an arbitrary small value.

[v] The *measurement problem* is a long debated open question in QM. General overviews on the problems, connected to the measurement problem in QM, can be found in Ref. 18.

[w] The number of proposed QC protocols is quite high and novel schemes seldom appear. A list of review papers on the argument is given in Refs. 20.

Charles Bennett and Gilles Brassard, and is, nowadays, worldwide known as the BB84 protocol[21]. It translates into a realistic environment the truly quantum concept of *Conjugate Coding*[22], earlier introduced by Wiesner[x]. In this seminal paper he recognised that, thanks to the possibility of realizing coherent superposition of quantum states, it would have been possible to convey over a single channel two distinct types of coded onto a single quantum system. At the same, thanks to the Heisenberg indeterminacy principle, only one of this information would be accessible at the end of the transmission.

Later, it has been found that also systems, more complex than single quanta ones, can convey quantum information provided that they can follow the spirit of Wiesner: a) they can be prepared as coherent superposition; b) there is an easy access, on them, to a pair of quantum non-commuting observables obeying to the Heisenberg principle.

These basic ideas are easily translated into language more familiar to telecommunication people. Single quanta system will be single photons while bright, laser-like, beams will play as complex systems.


## 4. Single Photons Systems

The Wiesner *Conjugate-Coding* idea was rather soon translated into the language of single photons. As a matter of fact, they represent the natural candidates for realizing single quanta states and the fastest and most reliable way of transmitting information.

Conjugate coding relies on two-level systems[y] and onto the possibility of choosing over at least two representations.

---

[x] In his abstract Wiesner wrote: "This paper treats a class of codes made possible by restrictions on measurement related to the uncertainty principle. Two concrete examples and some general results are given." This sentence marks, historically, the birth of quantum codes with no classical analogue. The paper was written in 1973 but was not published until 1983 as a contributed paper in volume promoted by the Association for Computing Machinery Special Interest Group on Algorithms and Computation Theory (ACM SIGACT). It is worth noting that in the same volume, Gilles Brassard published a provocative contribution on the claimed security of public-key-cryptosystems[23].

[y] The mathematics associated with two-level systems is very simple. The simplest example of two-level systems, for a physicist, is a spin ½ particle. All the formal instruments used for describing these systems are valid for any kind of two-level system.

Since BB84, a number of different schemes has been proposed. They can be divided into two main categories[z]: single photons (or discrete variables - DV) and continuous variable (CV) systems. It would be rather complicated to give reasons for this plethora of proposals, some of them indeed very similar. The most critical point seats on the fact the in real world application it is very difficult to translate QM. It is a theory fitting very well to simple and closed systems. As soon as one lets these systems interact with the external world they loose, at least partially, their quantum character. In a quantum language, QC is a wonderful tool in the pure state realm while it become less practical for mixed "decohered" states[25]. In addition, it is possible to distinguish, in each of the two general classes, two subclasses: entanglement based and not entanglement based protocols. They are defined by the use, or not, of entangled states as a quantum resource for enhancing protocol security.

After a long debate it is now widely accepted that entanglement is useful but not necessary to mark with the "quantum" certificate a QC protocol[26].

Looking closer to the original Wiesner idea, it becomes clear how entangled states entered the QC game. One of the *conjugate coding* pillars is the superposition principle; entanglement is not other than this principle applied to multi-partite systems. A state of a multi-partite system is entangled if and only if it is not possible to express it as the direct product of states describing a single subsystem[27].

Going back to single photons systems, we list in Ref. 28 some of the most relevant DV, entangled or not, protocols that have reached, up to now, a good level of reliability.

Here we describe, in a possibly simple language, the BB84 protocol in order to highlight features that are common to most of the DV protocols: a) the use of different measurements basis; b) the introduction of (truly) random elements; c) the requirement for classical communication at the end of quantum transmission.

---

A single photon is a two-level system if its polarization degrees of freedom are taken into account.

[z] In more recent years, since 2002, a third class of protocols, usually named "Distributed-phase-reference protocols" [24], has appeared. This last class will not be considered in this contribution.

In the spin ½ particles language, BB84 uses four quantum states (*e.g.* up, down, right and left $|\uparrow\rangle,|\downarrow\rangle,|\rightarrow\rangle$, and $|\leftarrow\rangle$)[aa], in pairs forming two distinct bases[bb]. One assigns the bit[cc] value 0 (1) to the states up $|\uparrow\rangle$ and left $|\leftarrow\rangle$, and the value 1 (0) to down $|\downarrow\rangle$ and right $|\rightarrow\rangle$. The protocol is conceptually very simple: a) Alice chooses randomly to send out a single photon in one of the four states[dd]; b) upon arrival, Bob measures the polarization of the incoming photon in one of the two bases, chosen randomly and independently from the random number used by Alice; c) for each measured photons Bob announces, on a public channel, the measurement base he had used (but not the measurement results); d) Alice reveals whether Bob's base matches her original choice or not; e) both, Alice and Bob, discharge data belonging to not-matched measurements.

At this point they have in hand a string of perfectly correlated bits.

Unfortunately, in this ideal scenario, we have not accounted for any possible photon losses, detection failures, eavesdropping etc. etc. ...

The cleverest strategy for Eve would be the *catch and resend* one.

She can take one photon out of the quantum channel; measure its polarization (in one of the two basis at her choice); send a different photon to Bob in the polarization state she has measured. If, by 50% chance, she has chosen the same base Alice has, then she got the same information Bob would have in his hand. On the contrary, if she has chosen the other base, Bob will receive a photon prepared in a state no more correlated to the one sent by Alice. In this way, Eve is introducing random errors on the key. To detect such a scheme of intrusion Alice and

---

[aa] In the photons polarization language it amounts to consider, horizontal, vertical and $\pm 45°$ polarizations.

[bb] These basis have to be maximally conjugated in the sense that any scalar product of the states belonging each to a different base should give ½; *i.e.* $|\uparrow\rangle\cdot|\rightarrow\rangle = 1/2$

[cc] In QC bits are replaced by qubits (or q-bits). It is of common sense that a bit can be either 0 or 1. On the contrary, thanks to the superposition principle, a qubit can be 0, 1, or any superposition of them. Thus, a generic qubit can be written as $|\psi\rangle = \alpha|0\rangle + \beta|1\rangle$ provided that $|\alpha|^2 + |\beta|^2 = 1$.

[dd] How she accomplishes this, in principle, simple task, is a delicate subject out of the scope of the present contribution. However, a good starting point would be Ref. 29 where a debate on true random numbers obtained from random physical processes is illustrated. The first optical implementation was, however, suggested in 1994[30].

Bob will have to publicity compare part of their final key. In any case, even if they decide not to compare their result, it is easy to prove, by simple statistical arguments, that the maximum knowledge Eve can gain amounts to the 25% of the key.

DV QC protocols have so far reached a good degree of maturity and several QC links over standard telecom infrastructures[ee] have been realised[31]. Moreover, up to 144 km links in air have been demonstrated with different protocols in order to prove the feasibility of satellite based QC distributed systems[32].

There are a few companies around the world commercialising plug-and-play QC systems. The first one was MagiQ technologies in the US[ff].

## 5. Continuous Variable Quantum Cryptography

Fifteen years later than BB84, the first proposal for DV-QC, in 1999, it was recognised that it would have been possible to encode quantum information onto bright beams[33]. In this particular realization, the key bits would have been coded as phase and amplitude modulation of entangled carrier beams[gg].

Contrarily to the DV case, where BB84 doesn't exploit entanglement, the latter was firstly considered as a prerequisite for obtaining high levels of security for CV protocols. The first CV proposals borrowed from DV protocols the three key elements: a) the use of different measurements; b) the introduction of random elements both at the transmitter and at the

---

[ee] It has to be noticed that, contrarily to the CV case (*see below*) the integration of DV QC in existing network cannot be immediate. As a matter of fact, DV protocols require single photons sources and single photon (low noise) detection that are not at all a standard in existing telecom communication protocols.

[ff] Quantum key distribution systems based on DV are on sale at MagiQ Tech. (USA) and id-Quantique (Switzerland). Very recently, Quintessence Lab, an Australian based company, has started to commercialize CV systems. The (not completed) list of, even very big, companies having their own QC research program includes: HP, NEC, Mitsubishi, IBM, NTT, Selex. It is indeed quite difficult to find exhaustive information on these industrial programs with QC being considered a strategic research of sure military interest.

[gg] In the quantum language phase and amplitude of optical field are described by a pair of non-commuting observables: the field quadratures. This pair obeys to an uncertainty principle like the one cast for position and momentum of a particle. The label continuous variable comes from the fact that these operators can assume a continuum of values.

receiver; c) the requirement for classical communication at the end of transmission in order to distil the key from a longer data set. The technologic challenge, in the CV context, is, on the contrary, rather different and simpler relying, more than in the DV case, on techniques and methods already integrated in standard telecom network as, *e.g.*, the use of modulation coding/decoding and homodyne and/or heterodyne detection.

However, these first proposals were still conceived with binary alphabets[34]. Moreover, the technology of single photons was, that time, more reliable than the one used for realizing bright quantum states. In addition, the promised performances were not so different from the DV case, already close to industrialization[hh]. So that, even if their security was proven[35] to be very close to the one attainable in the DV case, they were not considered as a real alternative to DV systems for being taken out of the labs for everyday usage. Furthermore, all these protocols, based on squeezed and/or EPR beams, required significant quantum resources and were significantly susceptible to decoherence.

The turning point for CV QC was the finding that more sophisticated modulation techniques would have strongly enhanced the system performances. In particular, the use of Gaussian modulation[36] makes it possible to use multi-digit alphabets strongly enhancing the attainable bit rate, thus exceeding the one of the DV case. This intuition, together with the study of properly designed classical protocols for obtaining the binary string of the cryptographic key from Gaussian modulated signals, paved the way to a rapid development of different systems of increasing performances. The idea, originally applied to an entangled based protocol, was soon extended to coherent states[37]. The use of coherent states, states laying at the classical-quantum boundary, allows to overcome decoherence limitations.

Shortly afterward, no-switching protocols[38] have further simplified the CV scheme demonstrating that, at the receiver, simultaneous measurement of two non-commuting observables can be made with no

---

[hh] MagiQ was founded in 1999 with the precise aim to develop a commercial quantum key distribution system.

random switching of the measurement apparatus, thus allowing to greatly simplifying the setup and enhancing the allowed bit rates[ii].

It is important to note that these last two protocols are, today, the most promising approaches for realizing commercial system[jj]. In particular, the coherent-state encoding, introduced in Ref. 37, has been implemented and repeatedly improved by using standard telecom components[40]. In the following, we will shortly describe this protocol that can be considered the progenitor of all the CV ones. In particular, we will try to enlighten the characteristics of a general CV QC scheme.

As in the DV case, the conceptual scheme of a CV protocol is very simple: a) Alice prepares a random sequence of signals, mapping a random variable $a$ onto a quantum system; b) she sends, over a quantum channel, these signals to Bob; c) Bob measures the signals extracting a random variable $b$ (strongly) correlated to Alice's $a$; d) Alice and Bob, over a public channel, perform classical post-processing for obtaining the shared secret key from their own raw data ($a$ and $b$).

The Gaussian modulated coherent state protocol, initially proposed by Grosshans and Grangier[37], consists of a quantum part followed by a properly designed classical post-processing. The former starts with the generation[kk], at Alice site, of the coherent state $|X_A + iY_A\rangle$ where $X_A$ and $Y_A$ are randomly Gaussian distributed variables with 0 mean value and a given variance (chosen by Alice). This state is, then, sent over a quantum channel. Upon arrival Bob measures either $X_B$ or $Y_B$ by homodyne detection[ll]. This procedure is repeated several times. At this point, quantum communication ends. Alice and Bob have each an ensemble of

---

[ii] In the last decade a number of different proposals all aiming at improving the CV performance, in terms of security and reliable transmission range has appeared. A surely not complete list is given in Ref. 39.

[jj] The only CV based commercial system actually on the market is produced by Quintessence Lab, a spin-off company of the Australian National University in Canberra and it is based on a non-switching protocol. Possibly, also a French consortium among Thales, Telecom Paris Tech and the Institute d'Optique (CNRS) is close to commercialise a CV system but there are no official news on this.

[kk] A coherent state of this form can be easily generated by modulating a strong carrier beam[41].

[ll] As already said, non-switching protocols have demonstrated that this randomization is not necessary. On the contrary, Bob can freely measure both variables so doubling the number of data he and Alice can use for the classical part.

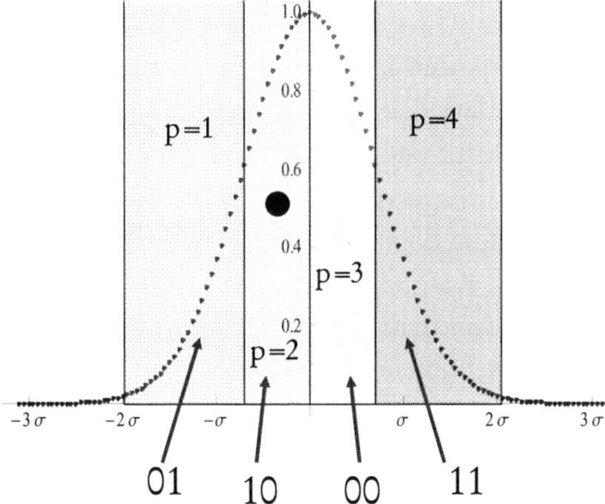



Fig. 2. Graphic representation of the sliced reconciliation protocol that allows distilling a multi-bit string of digitalized value from a Gaussian distributed random variable. The distribution is sliced into *n* (in Fig. *n=4*) regions each statistically containing the same number of data. Then, to each region a (multi-)bit value is assigned following an algorithm designed to increase data distinguish-ability. Details on the sliced protocol can be found in Refs. 42.

Gaussian distributed data. We note that the vacuum noise, entering the system through loss mechanisms and/or interception, makes the values obtained by Bob different but correlated to the values encoded by Alice (a scaling factor can be estimated by comparing a small set of data).

The classical post-processing comes straightaway. Bob tells, on a classical channel, which observable $X_B$ or $Y_B$ he has measured for each time slot. Then, useless data are discharged by Alice: at this point, they share two Gaussian distributed set of correlated variables. They need to transform continuous values[mm] into a binary string. This is accomplished by the *sliced* reconciliation protocol; also used to discharge data affected by transmission errors. Standard privacy amplification can be, at the end, applied for extracting the secret key.

In Fig. 2 a graphic representation of the sliced protocol[42] is given. The Gaussian distribution of the data is sliced into *n* regions of equal statistical weight. Then, to each region a string of *n/2* bit is assigned by an algorithm designed for increasing data distinguish-ability. Then, each

---

[mm] Data span a continuum of values in the limit of the detector resolution.

measurement gets its value from the statistical region it falls in. Note that such a protocol converts into a digital binary string not the single datum but its position in the data distribution. In this way, the only request for building a key is that Alice and Bob data are similarly distributed.

As in the DV case, we have omitted eavesdropping. A complete and comprehensive discussion on security issue goes beyond the scope of this contribution. However, Ref. 35 and the list given as Ref. 43 give a rather complete survey on the security issue for CV protocols.

We conclude this section by noting that both DV and CV protocols require sophisticate mathematical tools for the so call reconciliation and privacy amplification steps. QC seats in between different fields of science involving mathematics, physics, and information and communication technologies.


## 6.  Conclusions

As a summary of this contribution we wish to conclude by comparing, shortly, DV and CV protocols.

Single photons, DV, protocols require specially designed systems where single photon detection, actually showing very low efficiency (10%) in the telecom window, plays a crucial role for reaching high bit-rates. Moreover, key rates are limited also by the problem of realising single deterministic sources of single photon states. At the same time, the protocols are quite simple and security has been unconditionally proven.

On the contrary, CV systems employ weak coherent light readily attainable in standard lasers. Homodyne detection guarantees efficiencies as high as more than 90% with commercial components. The rate is strongly enhanced by multi-bit transmission Gaussian protocols up to Mbit per second. The integration of CV systems in standard existing telecom infrastructure is rather immediate.

While decoherence, actually, limits intrinsically the distance a CV link can reach, in the DV case it "only" reduces the attainable key rate.

For the above reason a definitive answer to the crucial question whether CV is better or not than DV is not possible. We believe that, actually, the two approaches will continue to be pursued.

# References


1. M. Planck, *Annalen der Physik*, **4**:553 (1901).
2. J.S. Bell, "Speakable and Unspeakable in Quantum Mechanics (Collected papers on quantum philosophy)", (Cambridge Univ. Press, 1987, ISBN 0521523389).
3. E. Schrödinger, *Naturwissenschaften*, **23**:807, 823, 844 (1935), *English transl. in* Proc. of the American Philosophical Society, **124**:323 (1980).
4. A. Einstein, B. Podolsky and N. Rosen, *Physical Review*, 1935, vol. 47, p. 777.
5. David Kahn, "The Codebreakers: The Story of Secret Writing", (Macmillan, New York, 1967, ISBN 0684831309).
6. Simon Singh, "The Code Book: The Science of Secrecy from Ancient Egypt to Quantum Cryptography", (Doubleday, New York, 1999, ISBN 0385495315).
7. C.E. Shannon, *Bell Syst. Tech. J.,* **28**:656 (1949).
8. G.J. Simmons "Authentication theory/coding theory", *in* Advances in Cryptology: Proceedings of CRYPTO 84, Lecture Notes in Computer Science **196**:411 (Springer, Berlin, 1985, ISBN 0387156585).
9. G. S. Vernam, *J. Am. Inst. Electr. Eng.* **45**, 109 (1926).
10. W. Diffie and M.E. Hellman, *IEEE Trans. Inf. Theory*, **IT-22**, 644 (1976).
11. R. Rivest, A. Shamir and L. Adleman, "On Digital Signatures and Public-Key Cryptosystems", MIT Lab. For Computer Science, Technical Report, MIT/LCS/TR-212 (January 1979).
12. A. S. Holevo, *Problems of Information Transmission*, **9**:177, (1973).
13. R. P. Poplavskii, Soviet Physics Uspekhi, **18**:222 (1975).
14. P.W. Shor, "Algorithms for quantum computation: discrete logarithms and factoring" *in* Proceedings of the 35th Annual Symposium on Foundations of Computer Science (ed. Goldwasser, S.) pp. 124–134 (IEEE Computer Society Press, 1994), *see also* J. Siam, *Comput.* **26**:1484 (1997).
15. L. M. K. Vandersypen, M. Steffen, G.Breyta, C. S. Yannoni, M. H. Sherwood and I. L. Chuang, *Nature* **414**, 883-887 (2001), C.-Yang Lu, D. E. Browne, T. Yang and J.-W. Pan, *Phys. Rev. Lett.* **99**, 250504 (2007), B. P. Lanyon, T. J. Weinhold, N. K. Langford, M. Barbieri, D. F. V. James, A. Gilchrist and A. G. White, *Phys. Rev. Lett.* **99**, 250505 (2007), E. Martín-López, A. Laing, T. Lawson, R. Alvarez, X.-Qi Zhou, J. L. O'Brien, *Nature Photonics* **6**, 773–776 (2012).
16. C. H. Bennett, F. Bessette, G. Brassard, L. Salvail and J. Smolin, *J. of Cryptology* **5**, 3 (1992).
17. H.-K. Lo and H. F. Chau, *Science* **283**, 2050 (1999); D. Meyers, *J. of the ACM,* **48**, 351 (2001); H. F. Chau, *Phys. Rev. A,* **66**, 060302(R) (2002).
18. C. W. Helstrom, "Quantum Detection and Estimation Theory", (Academic Press Inc., New York, 1976, ISBN 0123400503); J. A. Wheeler and W. H. Zurek, "Quantum Theory and Measurement", (Princeton Univ. Press, Princeton, 1984, ISBN 0691083169).
19. W. K. Wootters and W. H. Zurek, Nature, **299**:802 (1982).

20. R. J. Hughes, D. M. Alde, P. Dyer, G. G. Luther, G. L. Morgan and M. Schauer, *Contemporary Physics*, **36**, 149 (1995); S. J. D. Phoenix and P. D. Townsend, *Contemporary Physics*, **36**, 165 (1995); N. Gisin, G. Ribordy, W. Tittel and H. Zbinden, *Rev. Mod. Phys.* **74**, 145 (2002); V. Scarani, H. Bechmann-Pasquinucci, N. J. Cerf, M. Dušek, N. Lütkenhaus and M. Peev, *Rev. Mod. Phys*, **81**, 1301 (2009).

21. C.H. Bennett and G. Brassard, "Quantum Cryptography: Public Key Distribution and Coin Tossing", Proceedings of IEEE International Conference on Computers Systems and Signal Processing, Bangalore India, December 1984, pp.175-179 (1984).

22. S. Wiesner, "Conjugate Coding", *in* ACM SIGACT News - A special issue on cryptography, Vol. **15** Issue 1, pp.78-88 (1983).

23. G. Brassard, "An optimally secure relativized cryptosystem", *in* ACM SIGACT News - A special issue on *Cryptography*, Vol. **15** Issue 1, pp. 28-33 (1983).

24. K. Inoue, E. Waks and Y. Yamamoto, *Phys. Rev. Lett.*, **89**, 037902 (2002).

25. D. Buono, G. Nocerino, A. Porzio and S. Solimeno, *Phys. Rev. A*, **86**, 042308 (2012); G. Nocerino, D. Buono, A. Porzio and S. Solimeno, "Survival of continuous variable entanglement over long distances", *to appear in Physica Scripta* (2013).

26. C.H. Bennett, G. Brassard and A. K. Ekert, *Scientific American*, **267**, 50 (1992).

27. R. F. Werner, *Phys. Rev. A* **40**, 4277 (1989).

28. A. K. Ekert, *Phys. Rev. Lett.* **67**, 661 (1991); C. H. Bennett, G. Brassard and N. D. Mermin, *Phys. Rev. Lett.* **68**, 557 (1992); D. Bruß, D., *Phys. Rev. Lett.* **81**, 3018 (1998); J.-M. Mérolla, Y. Mazurenko, J.-P. Goedgebuer and W. T. Rhodes, 1999, *Phys. Rev. Lett.* **82**, 1656 (1999); H. Bechmann-Pasquinucci and N. Gisin, *Phys. Rev. A.* **59**, 4238 (1999); W. Tittel, W. J. Brendel, H. Zbinden and N. Gisin, *Phys. Rev. Lett.* **84**, 4737 (2000); V. Scarani, A. Acín, G. Ribordy and N. Gisin, *Phys. Rev. Lett.* **92**, 057901 (2004); W. Boucher and T. Debuisschert, *Phys. Rev. A* **72**, 062325 (2005); F. A. Bovino, P. Varisco, A. Martinoli, A. M. Colla, G. Castagnoli, G. Di Giuseppe and A. V. Sergienko, *Int. J. of Quantum Information* **3**, 1 (2005); K. Tamaki and H.-K. Lo, *Phys. Rev. A* **73**, 010302(R) (2006).

29. M. Isida and H. Ikeda, Ann. Inst. Stat. Math. **8**,119 (1956).

30. J. G. Rarity, P. C. M. Owens, and P. R. Tapster, J. Mod. Opt. **41**, 2435 (1994).

31. M. Bourennane, F. Gibson, Anders Karlsson, A. Hening, P. Jonsson, T. Tsegaye, D. Ljunggren and E. Sundberg, *Opt. Express* **4**, 383 (1999); D Stuck, N Gisin, O Guinnard, G Ribordy and H Zbinden, *New J. Phys.* **4**, 41 (2002); H. Kosaka, A. Tomita, Y. Nambu, T. Kimura and K. Nakamura, *Electron. Lett.* **39**, 1199 (2003); C. Gobby, Z. L. Yuan and A. J. Shields, *Appl. Phys. Lett.* **84**, 3762 (2004); X.-F. Mo, B. Zhu, Z.-F. Han, Y.-Z. Gui and G.-C. Guo, *Opt. Lett.* **30**, 2632 (2005); I. Choi, R. J. Young and P. D. Townsend, *Opt. Expr.*, **18**, 9600 (2010); S.-H. Sun, H.-Q. Ma, J.-J. Han, L.-M. Liang and C.-Z. Li, *Opt. Lett.*, **35**, 1203 (2010).

32. T. Schmitt-Manderbach, H.Weier, M. Fürst, R. Ursin, F. Tiefenbacher, T.Scheidl, J. Perdigues, Z. Sodnik, C. Kurtsiefer, J. G. Rarity, A. Zeilinger and H. Weinfurter,

*Phys. Rev. Lett.* **98**, 010504 (2007); C. -Z. Peng, J. Zhang, D. Yang, W.-Bo Gao, H.-Xin Ma, Hao Yin, He-Ping Zeng, Tao Yang, Xiang-Bin Wang, and Jian-Wei Pan, *Phys. Rev. Lett.* **98**, 010505 (2007);

33. T. C. Ralph, *Phys. Rev. A* **61**, 010303 (1999); **62**, 062306 (2000);

34. M. Hillery, Phys. Rev. A **61**, 022309 (2000); M. D. Reid, Phys. Rev. A **62**, 062308 (2000);

35. D. Gottesman and J. Preskill, Phys. Rev. A **63**, 022309 (2001);

36. N. J. Cerf, M. Lévy, and G. Van Assche, Phys. Rev. A, **63**, 052311, (2001);

37. F. Grosshans and P. Grangier, Phys. Rev. Lett. **88**, 057902 (2002); F. Grosshans, G. van Assche, J. Wenger, R. Tualle-Brouri, N. J. Cerf, and P. Grangier, Nature **421**, 238 (2003);

38. C. Weedbrook, A. M. Lance, W. P. Bowen, T. Symul, T. C. Ralph, and P. K. Lam, Phys. Rev. Lett. **93**, 170504 (2004); Phys. Rev. A **73**, 022316; A. M. Lance, T. Symul, V. Sharma, C. Weedbrook, T. C. Ralph, and P. K. Lam, Phys. Rev. Lett. **95**, 180503 (2005);

39. Ch. Silberhorn, T. C. Ralph, N. Lütkenhaus, and G. Leuchs, Phys. Rev. Lett. **89**, 167901 (2002); A. Porzio, V. D'Auriaa, P. Aniellob, M.G.A. Paris, and S. Solimeno, Opt. and Las. in Eng. **45**, 463 (2007); S. Pirandola, S. Mancini, S. Lloyd, and S. L. Braunstein, Nature Phys. **4**, 726 (2008); A. Leverrier, and P. Grangier, Phys. Rev. Lett. **102**, 180504 (2009); Phys. Rev. A 83, 042312 (2011); C. Weedbrook, S. Pirandola, S. Lloyd, and T. C. Ralph, Phys. Rev. Lett. **105**, 110501 (2010); P. Jouguet, S. Kunz-Jacques, and A. Leverrier, Phys. Rev. A **84**, 062317 (2011);

40. J. T. Lodewyck, J., T. Debuisschert, R. Tualle-Brouri, and P. Grangier, Phys. Rev. A **72**, 050303 (2005); J. Lodewyck, T. Debuisschert, R. Garcı´a-Patro´n, R. Tualle-Brouri, N. J. Cerf, and P. Grangier, Phys. Rev. A **76**, 042305 (2007); Phys. Rev. Lett. **98**, 030503 (2007); S. Fossier, E. Diamanti, T. Debuisschert, A. Villing, R. Tualle-Brouri, and P. Grangier, New J. Phys. **11**, 045023 (2009);

41. G Mauro D'Ariano, Martina De Laurentis, Matteo G A Paris, Alberto Porzio and Salvatore Solimeno, J. Opt. B: Quantum Semiclass. Opt., **4**, S127 (2002)

42. N.J. Cerf, S. Iblisdir, and G. Van Assche, Eur. Phys. J. D **18**, 211 (2002); Gilles Van Assche, Jean Cardinal, and Nicolas J. Cerf, IEEE Trans. Inf. Th., **50**, 394 (2004);

43. F .Grosshans, and N. J. Cerf, Phys. Rev. Lett. **92**, 047905 (2004); R. García-Patrón and N. J. Cerf, Phys. Rev. Lett. **97**, 190503 (2006); M. Navascués, F. Grosshans, and A. Acín, Phys. Rev. Lett. **97**, 190502 (2006); R. Renner, Nature Phys. **3**, 645 (2007); S. Pirandola, S. L. Braunstein, and S. Lloyd, Phys. Rev. Lett. 101, 200504 (2008); R. Renner and J. I. Cirac, Phys. Rev. Lett. **102**, 110504 (2009); A. Leverrier and P. Grangier, Phys. Rev. A **81**, 062314 (2010). A. Leverrier, F. Grosshans, and P. Grangier, Phys. Rev. A **81**, 062343 (2010).

# METAMATERIALS AND THE MATHEMATICAL SCIENCE OF INVISIBILITY

André Diatta, Sébastien Guenneau*, André Nicolet and Fréderic Zolla
*Institut Fresnel (UMR CNRS 6133)*
*Aix-Marseille Université 13397 Marseille cedex 20, France*
**E-mail : sebastien.guenneau@fresnel.fr*

In this chapter, we review some recent developments in the field of photonics: cloaking, whereby an object becomes invisible to an observer, and mirages, whereby an object looks like another one (say, of a different shape). Such optical illusions are made possible thanks to the advent of Metamaterials, which are new kinds of composites designed using the concept of transformational optics. Theoretical concepts introduced here are illustrated by finite element computations.

## 1. Introduction

In the past six years, there has been a growing interest in electromagnetic Metamaterials[1], which are composites structured on a sub-wavelength scale modeled using homogenization theories[2]. Metamaterials have important practical applications as they enable a markedly enhanced control of electromagnetic waves through coordinate transformations which bring anisotropic and heterogeneous[3] material parameters into their governing equations, except in the ray diffraction limit whereby material parameters remain isotropic[4]. Transformation[3] and conformal[4] optics, as they are now known, open an unprecedented avenue towards the design of such metamaterials, with the paradigms of invisibility cloaks.

In this review chapter, after a brief introduction to cloaking (section 2), we would like to present a comprehensive mathematical model of metamaterials introducing some basic knowledge of differential calculus.

The touchstone of our presentation is that Maxwell's equations, the governing equations for electromagnetic waves, retain their form under coordinate changes. A modern formalism of differential calculus is that of differential forms (section 3), among which Withney forms offer a natural framework for finite elements in curvilinear coordinates. Illustrative numerical simulations are provided in section 4 (cloaking via a singular transform) and section 5 (cloaking via a non-singular transform). Section 6 summarizes the results presented in this chapter.

## 2. General Aim of Cloaking

In 2006, Pendry *et al.*[3] and Leonhardt[4] independently showed the possibility of designing a cloak that renders any object inside it invisible to electromagnetic radiation. This coating consists of a metamaterial whose physical properties, the electric permittivity and the magnetic permeability, are deduced from a coordinate transformation in the Maxwell system. The anisotropy and the heterogeneity of the parameters of the coat work as a deformation of the space around the object we want to hide by bending the wavefront around it and enabling waves to emerge on the other side in the original propagation direction without any perturbation, see Fig. 1 for the principle of cloaking. The experimental validation of these theoretical considerations was given, a few months later, by an international team involving the former authors who used a cylindrical cloak consisting of concentric arrays of split ring resonators. This cloak makes a copper cylinder invisible to an incident plane wave at 8.5 GHz[5] as predicted by the numerical simulations[6].

However, there are other types of cloaking, such as an invisibility cloak with a negative refractive index theorized by Milton and Nicorovici in 2006[7] and one with a low refractive index proposed by Alu and Engheta in 2005[8]. Such plasmonic routes to cloaking do not detour the wave trajectories, but rather counteract the scattering of an obstacle, e.g. the circular cylinder, with the resonance of a coating.

We shall not discuss these alternatives to cloaking via geometric transforms, although they exist in their own right, and are equally interesting from the physical viewpoint.

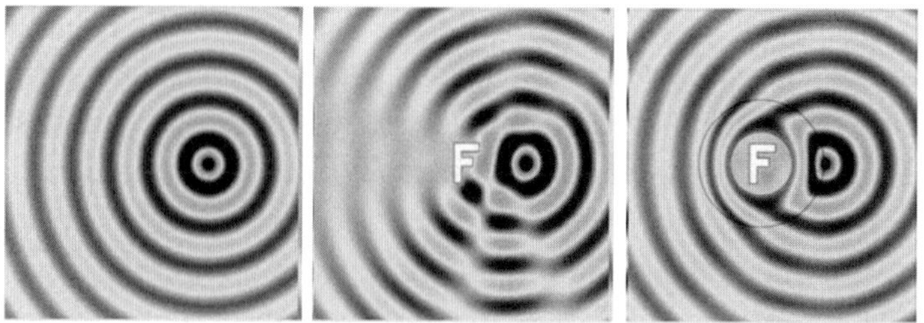

Fig.1. Principle of cloaking : (left) A point source radiating in an homogeneous isotropic medium; (middle) A point source radiating in an homogeneous isotropic medium in the presence of an infinite conducting F-shaped scatterer; (right) A point source radiating in an homogeneous isotropic medium in the presence of an infinite conducting F-shaped scatterer surrounded by an invisibility cloak (an anisotropic heterogeneous ring).

## 3.  Formalism of Differential Forms

One of the most appropriate frameworks for dealing with Maxwell equations is the formalism of exterior calculus on differential forms. In such a setting, the magnetic field **H** and the electric field **E** are differential 1-forms, whereas the current density **j** the magnetic flux density **B** and electric flux density **D** (also termed electric displacement) are differential 2-forms. The charge density $\rho$ is a 3-form. Maxwell equations in the time-harmonic domain, encapsulate their relationship as follows

$$\mathbf{dH} = \mathbf{j} - i\omega\mathbf{D}, \quad \mathbf{dE} = i\omega\mathbf{B}, \quad \mathbf{dD} = \rho, \quad \mathbf{dB} = 0. \tag{1}$$

where **d** is the exterior derivative of differential forms[15].

Note that, in the first two and the last two relations of Eq. (1), the exterior derivative **d** plays the roles of **curl** and **div**, respectively. Note also that the fundamental property **dd**= 0 always applies[15]. Remark that the relation $\mathbf{dE}= \iota\omega\mathbf{B}$ means that **B** coincides with the exact 2-form $\mathbf{B} = \mathbf{d}(- \frac{i}{\omega} \mathbf{E})$ and hence implies, in particular, the relation **dB=0**.

Furthermore, the equation $\mathbf{dH} = \mathbf{j} - i\omega\mathbf{D}$ also implies that $\mathbf{dj} = i\omega\mathbf{dD} + \mathbf{d^2H} = i\omega\mathbf{dD} = i\omega\rho.$

Now consider a metric $\mathbf{g}$ on, say $R^n$, that is a non-degenerate symmetric bilinear map from the space of vector fields on $R^n$ to the space of smooth scalar valued functions. More precisely, two vectors $\mathbf{V_p}$ and $\mathbf{V_p}'$ originating from a point p are mapped into a scalar $g\,(\mathbf{V_p}, \mathbf{V_p}')$.

The bilinearity means that this correspondence depends linearly on both arguments $\mathbf{V_p}$ and $\mathbf{V_p}'$. Making the point p vary over $\mathbf{R^n}$, or equivalently, using vector fields, $\mathbf{V}$, $\mathbf{V}'$ gives rise to a smooth function $\mathbf{g(V,V')}$ on $R^n$.

The Hodge star operator $*$ associated to the metric $\mathbf{g}$, is a linear operator on differential forms, bijectively mapping the space of p-forms to that of (n-p)-forms on $R^n$, for every fixed p which could be 0, 1,2,... or n. It is a very convenient tool, allowing to define a scalar product

$$(\alpha, \beta) := \int_{\mathbf{R^n}} \alpha \wedge * \beta \tag{2}$$

on the space of differential p-forms, on $R^n$. In the above equations (2), the symbol $\wedge$ stands for the (skew-symmetric) exterior (wedge) product on differential forms. It could be shown that, endowed with the scalar product (2), the space of differential p-forms on $R^n$ becomes a Hilbert space. This property is then used here, in a weak formulation of Maxwell's equations, to derive the magnetic permeability and the electric permittivity coefficients of a transformed medium. To keep things simple, we will also restrict ourselves to the case where is the Euclidean metric. Written in Cartesian coordinates $(x_1,......x_n)$ with corresponding local basis $(\frac{\partial}{\partial x_1}, ... \frac{\partial}{\partial x_n})$ for vector fields, the metric has a symmetric matrix with (i,j) entry $g_{i,j} = \mathbf{g}\left(\frac{\partial}{\partial x_i}, \frac{\partial}{\partial x_j}\right) = \delta_{i,j}$, where $\delta_{i,j}$ are the Kronecker symbols.

In our case of interest here, n=3 and p=0, 1, 2 or 3. The global usual basis of $R^3$ are denoted by $(x_1, x_2, x_3) = (x, y, z)$, endowed with the Euclidean metric. In this case we have

$$*1 = dx \wedge dy \wedge dz, \quad *dx \wedge dy \wedge dz = 1, \quad *dx = dy \wedge dz, \quad *dy \wedge dz = dx,$$

$$*dy = dz \wedge dx, \quad *dz \wedge dx = dy, \quad *dz = dx \wedge dy, \quad *dx \wedge dy = dz.$$

$$\tag{3}$$

Let us recall that the differential exact 1-forms $\mathbf{d_x}, \mathbf{d_y}, \mathbf{d_z}$ form the (local) basis for covectors, dual to $(\frac{\partial}{\partial x}, \frac{\partial}{\partial y}\frac{\partial}{\partial z})$. Using equation (3), we can derive the following interesting property of the Hodge star operator.

That is, if $\alpha = \alpha_x \mathbf{dx} + \alpha_y \mathbf{dy} + \alpha_z \mathbf{dz}$ and $\beta = \beta_x \mathbf{dx} + \beta_y \mathbf{dy} + \beta_z \mathbf{dz}$ are two differential 1-forms on $\mathbf{R}^3$, then the following holds

$$\alpha \wedge *\beta = (\alpha_x \beta_x + \alpha_y \beta_y + \alpha_z \beta_z)\, \mathbf{dx} \wedge \mathbf{dy} \wedge \mathbf{dz}$$
$$= (\alpha_x, \alpha_y, \alpha_z)(\beta_x, \beta_y, \beta_z)^T \, \mathbf{dx} \wedge \mathbf{dy} \wedge \mathbf{dz}. \tag{4}$$

Next, we use the following, which we may call the change of volume rule under change of coordinates.

Suppose we have another coordinate system $(u,v,w)$, with the following change of coordinates $(x, y, z) = (x (u, v,w)), y(u,v,w)), z(,v,w))$. The Jacobian matrix of this change of coordinates is

$$\mathbf{J}(u, v, w) = \begin{pmatrix} \dfrac{\partial x}{\partial u} & \dfrac{\partial x}{\partial v} & \dfrac{\partial x}{\partial w} \\[2mm] \dfrac{\partial y}{\partial u} & \dfrac{\partial y}{\partial v} & \dfrac{\partial y}{\partial w} \\[2mm] \dfrac{\partial z}{\partial u} & \dfrac{\partial z}{\partial v} & \dfrac{\partial z}{\partial w} \end{pmatrix}$$

Using the determinant of the above matrix, the formula below indicates how the volume changes as we switch from a system of coordinates to another:

$$\mathbf{dx} \wedge \mathbf{dy} \wedge \mathbf{dz} = \det(\mathbf{J})\, \mathbf{du} \wedge \mathbf{dv} \wedge \mathbf{dw}. \tag{5}$$

The relations

$$\mathbf{dx} = \frac{\partial x}{\partial u}\mathbf{du} + \frac{\partial x}{\partial v}\mathbf{dv} + \frac{\partial x}{\partial w}\mathbf{dw}, \quad \mathbf{dy} = \frac{\partial y}{\partial u}\mathbf{du} + \frac{\partial y}{\partial v}\mathbf{dv} + \frac{\partial y}{\partial w}\mathbf{dw},$$
$$\mathbf{dz} = \frac{\partial z}{\partial u}\mathbf{du} + \frac{\partial z}{\partial v}\mathbf{dv} + \frac{\partial z}{\partial w}\mathbf{dw},$$

could be written in a compact way as

$$\begin{pmatrix} \mathbf{dx} \\ \mathbf{dy} \\ \mathbf{dz} \end{pmatrix} = \mathbf{J} \begin{pmatrix} \mathbf{du} \\ \mathbf{dv} \\ \mathbf{dw} \end{pmatrix}.$$

Now we can also deduce, the relation between the expressions of a given 1-form $\alpha$ in the two coordinate systems, as follows

$$\alpha = \alpha_x \mathbf{dx} + \alpha_x \mathbf{dy} + \alpha_x \mathbf{dz}$$

$$= (\alpha_x, \alpha_y, \alpha_z) \begin{pmatrix} \mathbf{dx} \\ \mathbf{dy} \\ \mathbf{dz} \end{pmatrix} = (\alpha_x, \alpha_y, \alpha_z) \mathbf{J} \begin{pmatrix} \mathbf{du} \\ \mathbf{dv} \\ \mathbf{dw} \end{pmatrix} \tag{6}$$

On the other hand, we could also write directly

$$\alpha = \alpha_u \mathbf{du} + \alpha_v \mathbf{dv} + \alpha_w \mathbf{dw} = (\alpha_u, \alpha_v, \alpha_w) \begin{pmatrix} \mathbf{du} \\ \mathbf{dv} \\ \mathbf{dw} \end{pmatrix}. \tag{7}$$

Thus we get the equality

$$(\alpha_u, \alpha_v, \alpha_w) = (\alpha_x, \alpha_y, \alpha_z) \, \mathbf{J} \tag{8}$$

From Eqs. (5) and (7), we can rewrite (4) and see how the 3-form $\alpha \wedge *\beta$ is transformed under the change of coordinates:

$$\alpha \wedge *\beta = (\alpha_u, \alpha_v, \alpha_w) \, \mathbf{J}^{-1} \, ((\beta_u, \beta_v, \beta_w)\mathbf{J}^{-1})^T \, \det(\mathbf{J}) \, \mathbf{du} \wedge \mathbf{dv} \wedge \mathbf{dw}$$

That is,

$$\alpha \wedge *\beta = (\alpha_u, \alpha_v, \alpha_w) \, \mathbf{J}^{-1}\mathbf{J}^{-T} \, \det(\mathbf{J}) \, (\beta_u, \beta_v, \beta_w)^T \, \mathbf{du} \wedge \mathbf{dv} \wedge \mathbf{dw} \tag{9}$$

This is easily generalized to more general piecewise differentiable maps between two domains.

Now, for simplicity, we place ourselves in a region where there are no charges or currents. Under these considerations, Eq. (1) now reads as follows

$$\mathbf{dH} = i\omega\mathbf{D}, \quad \mathbf{dE} = i\omega\mathbf{B}, \quad \mathbf{dD} = 0, \quad \mathbf{dB} = 0.$$

For example, we can eliminate $\mathbf{H}$ in the equation $\mathbf{dH}=i\omega\mathbf{D}$ and get the wave equation of the electric field $\mathbf{E}$ in homogeneous media, and using the Hodge operator, we get

$$\mathbf{d}(\mu^{-1} * \mathbf{dE}) - \omega^2 \varepsilon * \mathbf{D} = \mathbf{0}.$$

Here, the electric field $\mathbf{E}$ is seen as an unknown. The above allows us to use Eq. (2) to write the weak formulation as follows

$$\int_\Omega \mu^{-1} * \mathbf{dE} \wedge \mathbf{d\bar{E}'dx} - \omega^2 \int_\Omega \varepsilon * \mathbf{E} \wedge \mathbf{\bar{E}'dx} = 0, \tag{10}$$

for every element $\mathbf{E}'$ of $H_0(\mathbf{curl}, \Omega)$.

In an isotropic medium, the electric permittivity, $\varepsilon$ and the electromagnetic permeability $\mu$ are scalar valued functions (indeed of the form of a scalar function times the identity matrix). So starting our transformation from an isotropic medium occupying a domain D, the above equalities allow us to extract the following relations between the expressions of the electric permittivity and the magnetic permeability in the different coordinate systems[18]

$$\varepsilon' = \varepsilon \mathbf{T}^{-1} \text{ and } \mu' = \mu \mathbf{T}^{-1}, \text{ where } \mathbf{T} = \frac{\mathbf{J}^T\mathbf{J}}{\det(\mathbf{J})}. \tag{11}$$

Remark that, we have started from an isotropic and homogeneous medium, so that $\varepsilon$ and $\mu$ are constant numbers. And yet, because of the structure of the tensor $\mathbf{T}$ in the transformed medium, we end up having tensors $\varepsilon'$ and $\mu'$ depending on the space variable. Hence, the transformed medium is always highly anisotropic and inhomogeneous.

It could be checked, using exactly the same process, that if we start from anisotropic and inhomogeneous medium with tensors $\varepsilon$ and $\mu$ then Eq. (10) takes the form

$$\varepsilon' = \mathbf{J}^{-1}\varepsilon\mathbf{J}^{-T}\det(\mathbf{J}) \text{ and } \mu' = \mathbf{J}^{-1}\mu\mathbf{J}^{-T}\det(\mathbf{J}), \tag{12}$$

which defines an anisotropic and inhomogeneous transformed medium.

Note that, one can write the transformation in any coordinate systems of their choice, and to get back to Cartesian ones in order to use formulas (11) or (12), one just needs to perform a composition of the elementary Jacobians involved, as in Eq. (14) of Section 4.2.

## 4.  Cloaking with a Singular Transformation

In this section, we discuss the physics of cloaking via a singular transform blowing up a point onto a disc of radius $R_1$, while mapping a disc of radius $R_2$ onto itself. We note that this transform was introduced in 2003 by Greenleaf et al.[9] in the context of inverse problems for the conductivity equation.

### 4.1. *Invisibility*

Using the same notations as in Section 3, we show in Figure 2 the effect of the previous transform on the metric of space. The distorted metric is associated with an anisotropic heterogeneous medium defined through the Jacobian $\mathbf{J}$ of the transform via the formula (11). Such an invisibility cloak is defined by tensors of permittivity and permeability with extreme

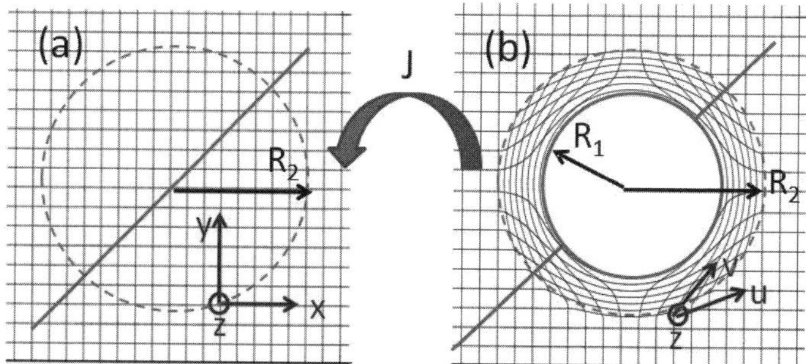


Fig. 2.   Effect of singular transform on space metric: (a) Metric of space with coordinate system (x,y,z) for a homogeneous isotropic medium; (b) Metric of space with coordinate system (u,v,z) for a locally heterogeneous anisotropic medium inside the invisibility cloak which is an annulus of inner radius $R_1$ and outer radius $R_2$.

coefficients at the inner radius of the cloak (vanishing and infinite eigenvalues). It leads to perfect cloaking, as numerically illustrated in Figure 1 (right). However, one can also give a twist to such a cloak and use it to create a mirage effect[6,10], as we discuss in the next paragraph.

## 4.2. *Mirage Effect*

Transformation Optics/Electromagnetics heavily relies on the equivalence principle for Electromagnetics stating that, any electromagnetic medium creates geometry and vice versa. The electromagnetic properties of such a medium entirely characterize the metric (including non-Riemannian metrics, such as spacetime metrics, depending on the cases) of the geometry. The converse also holds true.

The trick in generalizing Pendry's cloak and using it as a means to produce optical illusions and mimesis also falls along these lines. The aim of this section is expose a proposal initiated by two of us[10] to show that not only the whole (circular) region bounded by $\|\mathbf{x}'\| = R_1$ is equivalent to an optical void, but any defect hidden inside the cloak $R_1 < \|\mathbf{x}'\| < R_2$ will automatically acquire the same optical/electromagnetic properties as its inverse image via the transformation used to design the cloak, see Figure 1.

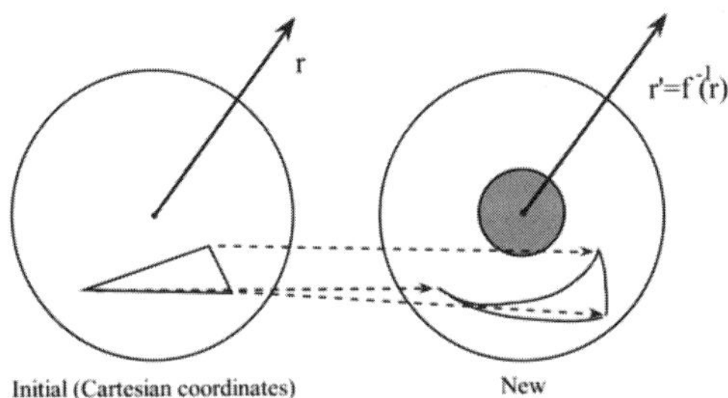


Fig. 3. Construction of the transformation $f : \mathbf{r} = (x, y, z) \mapsto \mathbf{r}' = (x', y', z')$ giving rise to the generalized cloak. When the material properties are piecewise defined, a pushforward of the geometry involving the inverse transformation is useful.

Any incoming wave will see such an object as its counterpart (inverse) placed in vacuum, and will bear this resemblance to an observer. The transformation used maps $\mathbf{x} = (x, y, z)$ to $\mathbf{x}' = (x', y', z')$ is given by

$$\mathbf{x}' = \left( \frac{R_2 - R_1}{R_2} + \frac{R_1}{\| \mathbf{x} \|} \right) \mathbf{x}, \tag{13}$$

where $\| \mathbf{x} \|$ is the norm (modulus) of $\mathbf{x}$.

Now we can choose to use cylindrical coordinates $(r, \theta, z)$ which are related Cartesian coordinates $(x, y, z)$ by

$$x = r\cos(\theta),\ y = r\sin(\theta),\quad z = z$$

with corresponding Jacobian matrix

$$\mathbf{J}_{xr}(r, \theta, z) = \begin{pmatrix} \cos(\theta) & -r\sin(\theta) & 0 \\ \sin(\theta) & r\cos(\theta) & 0 \\ 0 & 0 & 1 \end{pmatrix} = \begin{pmatrix} \cos(\theta) & -\sin(\theta) & 0 \\ \sin(\theta) & \cos(\theta) & 0 \\ 0 & 0 & 1 \end{pmatrix} \begin{pmatrix} 1 & 0 & 0 \\ 0 & r & 0 \\ 0 & 0 & 1 \end{pmatrix}.$$

Note the appearance of the matrix of rotation through an angle $\theta$ whose inverse is the rotation through an angle $-\theta$. The inverse of this Jacobian is then easy to compute

$$\mathbf{J}_{xr}^{-1} = \mathbf{J}_{rx} = \begin{pmatrix} 1 & 0 & 0 \\ 0 & \dfrac{1}{r} & 0 \\ 0 & 0 & 1 \end{pmatrix} \begin{pmatrix} \cos(\theta) & \sin(\theta) & 0 \\ -\sin(\theta) & \cos(\theta) & 0 \\ 0 & 0 & 1 \end{pmatrix} = \begin{pmatrix} \cos(\theta) & \sin(\theta) & 0 \\ -\dfrac{1}{r}\sin(\theta) & \dfrac{1}{r}\cos(\theta) & 0 \\ 0 & 0 & 1 \end{pmatrix}$$

We may then rewrite the transformation (13) as mapping $(r, \theta, z)$ to new coordinates $(r', \theta', z')$. It maps $\mathbb{R}^3$ to $\mathbb{R}^3$ minus a cylinder over a disk $D_1$ of radius $R_1$, in such a way as it is the identity outside a cylinder over a disk $D_2$ of radius $R_2 > R_1$ and inside the cylindrical hollow region over the annulus $R_1 < r \le R_2$, it reads

$$r' = \frac{R_2 - R_1}{R_2} r + R_1, \quad \theta' = \theta, \quad z'=z, \tag{14}$$

so that the origin is blown up to the circle of radius $R_1$. All discs and circles referred to here, are contained in the xy-plane and centered at the origin. Setting $\alpha = \frac{R_2 - R_1}{R_2}$, the Jacobian of this transformation is

$$\mathbf{J}_{r'r}(r,\theta,z) = \begin{pmatrix} \alpha & 0 & 0 \\ 0 & 1 & 0 \\ 0 & 1 & 1 \end{pmatrix} \tag{15}$$

To deduce the Jacobian in Cartesian coordinates, we compose the three elementary Jacobians as follows

$$\mathbf{J} := \mathbf{J}_{x'x} = \mathbf{J}_{x'r'}\mathbf{J}_{r'r}\mathbf{J}_{rx} \tag{16}$$

so that, we have

$$\mathbf{J}^{-1} = (\mathbf{J}_{rx})^{-1}(\mathbf{J}_{r'r})^{-1}(\mathbf{J}_{x'r'})^{-1} = \mathbf{J}_{xr}\mathbf{J}_{rr'}\mathbf{J}_{r'x'}$$

$$= \begin{pmatrix} \cos(\theta) & -r\sin(\theta) & 0 \\ \sin(\theta) & r\cos(\theta) & 0 \\ 0 & 0 & 1 \end{pmatrix} \begin{pmatrix} \dfrac{1}{\alpha} & 0 & 0 \\ 0 & 1 & 0 \\ 0 & 0 & 1 \end{pmatrix} \begin{pmatrix} \cos(\theta) & \sin(\theta) & 0 \\ -\dfrac{1}{r'}\sin(\theta) & \dfrac{1}{r'}\cos(\theta) & 0 \\ 0 & 0 & 1 \end{pmatrix}$$

We are now ready to use formula (11) in order to derive the electromagnetic properties of the cloak. We get

$$\mathbf{T}^{-1} = \begin{pmatrix} (\mathbf{T}^{-1})_{11} & (\mathbf{T}^{-1})_{12} & 0 \\ (\mathbf{T}^{-1})_{21} & (\mathbf{T}^{-1})_{22} & 0 \\ 0 & 0 & \dfrac{r'-R_1}{\alpha^2 r'} \end{pmatrix} \tag{17}$$

where the coefficients are given as follows

$$(\mathbf{T}^{-1})_{11} = 1 - \frac{R_1 \sin^2(\theta')}{r'} + \frac{R_1 \cos^2(\theta')}{r' - R_1},$$

$$(\mathbf{T}^{-1})_{22} = 1 - \frac{R_1 \cos^2(\theta')}{r'} + \frac{R_1 \sin^2(\theta')}{r' - R_1}, \tag{18}$$

$$(\mathbf{T}^{-1})_{12} = (\mathbf{T}^{-1})_{21} = \frac{R_1 \cos(\theta') \sin(\theta')(R_1 - 2r')}{(R_1 - r')r'}.$$

We can write these expressions only in terms of $(x', y', z')$, using

$$r' = \sqrt{(x')^2 + (y')^2}, \qquad \theta' = 2 \arctan\left(\frac{y'}{x' + \sqrt{(x')^2 + (y')^2}}\right), \qquad z' = z.$$

The electromagnetic permeability $\mu$ and electric permittivity $\varepsilon$ of the cloak are then ready to be implemented, applying (11).

Now following Nicolet et al.[10] we can apply the transformation (14), but this time, the disc $D_2$ contains an object of arbitrary electromagnetic properties. The cloaks now mimics the object originally contained in $D_2$. A numerical simulation with the Getdp freeware is shown in Figure 4.

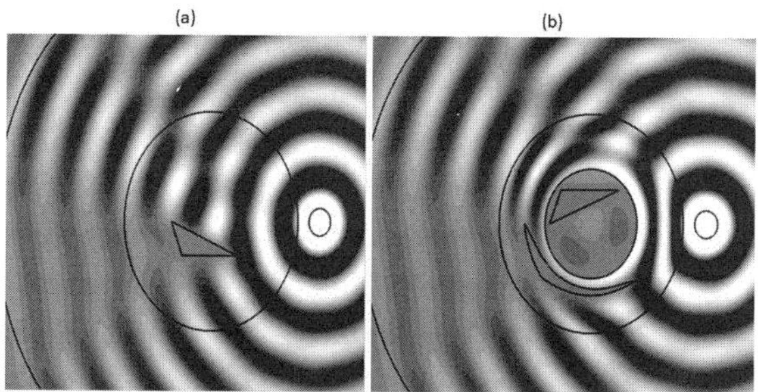


Fig. 4. (a) Conducting triangular cylinder scattering cylindrical waves radiated by an electric line source (white disc with a black perimeter on the right-hand side) located nearby. (b) Distorted conducting cylinder placed inside a cloak $R_1 < \|\mathbf{x}'\| < R_2$, producing the same scattering pattern as in panel (a). Note that the triangular cylinder placed inside the invisibility region $\|\mathbf{x}'\| \leq R_1$ does not affect the scattering, and could have in principle any desired shape and constituting material.

## 5. Cloaking and Mirage Effect with a Non-Singular Transformation

The mathematical model given below (and proposed by two of us[12]) provides a general setting for the design of a cloak aimed at concealing a region having the property that all of its points are within line-of-sight from a fixed vantage point. We call a star domain, every domain with such a property. Discs and Squares in 2D, ellipsoids in 3D ... are examples of star domains. The designed cloaks play the double role of displaying any desired optical effect, in particular acquiring the same optical properties as other different objects on the one hand, as well as allowing to hide any defects inside it, just as an ordinary cloak would do, on the other hand. The mathematical model underlying this design is given by a nonsingular transformation mapping the two domains (objects to be made equivalent) on each other. Such a transformation is made to preserves every line passing through the chosen vantage point. It is a more general piecewise smooth and nonsingular transformation sending a shape to another one of a different geometry. Depending on whether objects to be mimicked live in isotropic or anisotropic media, formulae (11) and (12) will be used, accordingly. However, as noted in Section 3, the obtained cloak itself will always be a heterogeneous and anisotropic

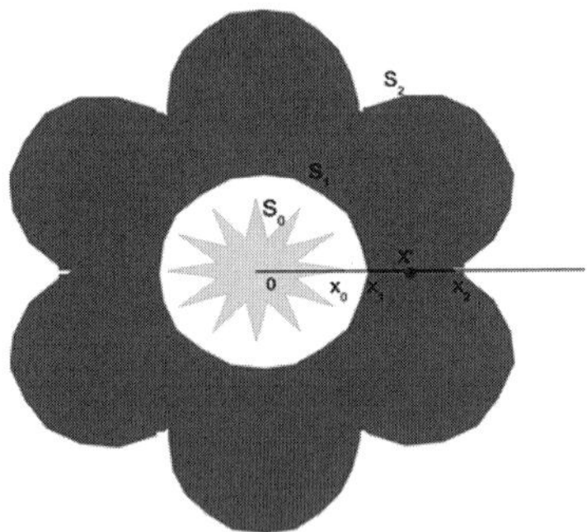


Fig. 5. Construction of a generalized non-singular cloak for mirage effect. The transformation with inverse (19) shrinks the region bounded by the two surfaces $S_0$ and $S_2$ into the region bounded by $S_1$ and $S_2$. The curvilinear metric inside the carpet (here, a dark gray flower) is described by the transformation matrix T, as in Eq. (11). This is designed to play the double role of mimesis and cloaking: any types of objects located within the region $D_1$ bounded by the surface $S_1$ will be invisible to an outer observer while the region itself still scatters waves like an object $D_0$ bounded by $S_0$ (here, a star). In the limit of vanishing light gray region, the transformation matrix T becomes singular on S1 (ordinary invisibility cloak).

medium and yet will look to an external observer just as the object we wish to mimic. For the sake of simplicity in applications, we will consider here that, our imaginary objects to be mimicked are surrounded by an isotropic region, so that the permeability $\mu'$ and permittivity $\varepsilon'$ can be implemented using Eq. (11), to make the two objects acquire equivalent electromagnetic properties.

Here is a short mathematical description of the transformation and the derivation of the material properties of the designed cloak. Suppose three bounded star domains $D_0$, $D_1$, $D_2$ in $R^n$ with, say, n = 2,3,..., have piecewise smooth arbitrary boundaries $\partial D_j$, j=0,1,2. We chose them nested, that is, $D_2$ contains $D_1$ which in turn, contains $D_0$ and they share the same vantage point, say $\underline{0}$.

The domain $D_0$ is our reference, the one to be mimicked by $D_1$ and thereby by the cloak itself. The hollow region $D_2/D_1$ will be endowed with Neumann conditions on its inner boundary $\partial D_1$. It is meant to be the model for the cloak and in which any type of defect could be hidden, but will still have the same electromagnetic response as the region $D_0$ with an infinitely conducting boundary $\partial D_0$.

The transformation coincides with the identity map outside $D_2$, that is, in $R^n \backslash D_2$, but bijectively (indeed, at least, piece-wise diffeomorphically) maps the hollow region $D_2 \backslash D_0$ to the hollow region $D_2 \backslash D_1$, so that $\partial D_2$ stays point-wise fixed, while $\partial D_0$ is mapped to $\partial D_1$. We may divide the domains Dj into subdomains, such that each part of the boundaries lying inside $\partial D_j$ is an arbitrary smooth hypersurface.

Now for the construction of the transformation, let $\underline{x} = (x^1, x^2,..... x^n)$ be a point in D1, $D_0$, whose coordinates are given in a system of coordinates centered at the chosen vantage point $\underline{0}$. A line segment issued from $\underline{0}$ and passing through $\underline{x}$, meets the boundaries $\partial D_0$, $\partial D_1$, $\partial D_2$ at the unique points $\underline{x}_0 = (x^1_0, x^2_0,..... x^n_0)$, $\underline{x}_1 = (x^1_1, x^2_1,..... x^n_1)$, $\underline{x}_2 = (x^1_2, x^2_2,.... x^n_2)$, respectively. For numerical implementations, we directly use the inverse $\underline{x}' \mapsto \underline{x}$ of the transformation, in the coordinates system. It is given by

$$x^i = x^i_0 + \alpha_i \, (x'^i - x^i_1), \text{ where } \alpha_i = \frac{x^i_2 - x^i_0}{x^i_2 - x^i_1}. \qquad (19)$$

In the 3-space with coordinates $(x^1, x^2, x^3) = (x, y, z)$, we can write this transformation as

$$
\begin{cases}
x = x_0 + \alpha \ (x' - x_1), \text{ with } \alpha = \dfrac{x_2 - x_0}{x_2 - x_1} \\[2em]
y = y_0 + \beta \ (y' - y_1) \text{ with } \beta = \dfrac{y_2 - y_0}{y_2 - y_1} \\[2em]
z = z_0 + \gamma \ (z' - z_1), \text{ with } \gamma = \dfrac{z_2 - z_0}{z_2 - z_1}
\end{cases}
\tag{20}
$$

The cylindrical cases (cylinders over plane curves, such as triangles, squares, ellipses, sun flower-like cylinders, and so forth) are given by the following transformation mapping the region enclosed between the cylinders $S_0$ and $S_2$ into the space between $S_1$ and $S_2$. Its inverse reads

$$
\begin{cases}
x = x_0 + \alpha \ (x' - x_1), \text{ with } \alpha = \dfrac{x_2 - x_0}{x_2 - x_1} \\[2em]
y = y_0 + \beta \ (y' - y_1) \text{ with } \beta = \dfrac{y_2 - y_0}{y_2 - y_1} \\[2em]
\qquad\qquad z = z'.
\end{cases}
\tag{21}
$$

Applying formula (11), material properties can be derived via the matrix representation of the tensor $\mathbf{T}^{-1}$, which now has the following simpler form[12]

$$
\mathbf{T}^{-1} = \begin{pmatrix} T_{11}^{-1} & T_{12}^{-1} & 0 \\ T_{12}^{-1} & T_{22}^{-1} & 0 \\ 0 & 0 & T_{33}^{-1} \end{pmatrix}
\tag{22}
$$

with coefficients

$$
T_{11}^{-1} = \frac{a_{11}}{a_{33}}, \ T_{12}^{-1} = \frac{a_{12}}{a_{33}}, \ T_{22}^{-1} = \frac{a_{22}}{a_{33}}, \ T_{33}^{-1} = a_{33}
\tag{23}
$$

where the functions $a_{ij}$ are given by

$$a_{11} = \left(\frac{\partial x}{\partial y'}\right)^2 + \left(\frac{\partial y}{\partial y'}\right)^2 , \quad a_{12} = -\left(\frac{\partial x}{\partial x'}\frac{\partial x}{\partial y'} + \frac{\partial y}{\partial x'}\frac{\partial y}{\partial y'}\right),$$

$$a_{22} = \left(\frac{\partial x}{\partial x'}\right)^2 + \left(\frac{\partial y}{\partial x'}\right)^2 , \quad a_{33} = \frac{\partial x}{\partial x'}\frac{\partial y}{\partial y'} - \frac{\partial x}{\partial y'}\frac{\partial y}{\partial x'}.$$

$$(24)$$

All we have to do now, is to plug in the specific values of all those $a_{ij}$ $a_{ij}$ and then obtain $\mu'$ and $\varepsilon'$, via formula (11). The form of the functions $x_i$, i=0,1,2 depend on the shapes of the regions $D_i$ involved in the cloaking process. So are the $a_{ij}$.

### 5.1. *Squaring the Circle*

Following the above, we may want to make a circle acquire the same (electromagnetic) signature as a virtual small square sitting inside its enclosed region and sharing the same centre used as a vantage point[12].

By doing so, such a circle will have the same appearance to an observer just as a small square. Consider the two diagonals of the square, they part the region enclosed between the small square and the outer circle into four sectors. In each sector, following segments issued from the origin (common centre of both the square and circle), the transformation shrinks diffeomorphically the region between the small square and the outer circle into the circular annulus bounded by the inner and outer circles, in such a way as the sides of the square are mapped to the inner circle and the outer circle stays fixed point-wise.

First, let us recall[12] that in 2D, if a piece of the boundary of a star domain $D_i$ is part of a line of the form $y = a_i x + b_i$ then clearly the line through the origin and a point $(x', y')$ intersects this piece of boundary at

$$(x_i, y_i) = (\frac{b_i x'}{y' - a_i x'}, \frac{b_i y'}{y' - a_i x'}) \qquad (25)$$

or, if this piece of boundary is a vertical segment with equation $x = c_i$ at

$$(c_i, c_i \frac{y'}{x'}). \qquad (26)$$

In the case where this piece of boundary is part of an ellipse with equation

$$\frac{(x-a)^2}{c_i^2} + \frac{(y-b)^2}{d_i^2} = 1$$

then[10], a segment issued from (a,b) passing through $(x',y')$ will intersect such a piece of boundary at $(x_i',y_i')$ with[12]

$$x_i = a + \frac{x'-a}{\sqrt{(x'-a)^2/c_i^2 + (y'-a)^2/d_i^2}},$$

$$y_i = a + \frac{y'-a}{\sqrt{(x'-a)^2/c_i^2 + (y'-a)^2/d_i^2}} \tag{27}$$

For the construction of our cloak, we use for $D_0$ a region bounded by a square with sides of length $L_0 = 0.2$ centered at the vantage point (a,b) = (0,0) so that in both the leftmost and rightmost parts, formula (26) applies to give

$$(x_0, y_0) = (L_0, L_0 \frac{y'}{x'}) \tag{28}$$

whereas in both the uppermost and lowermost we will invoke formula (7) to obtain

$$(x_0, y_0) = (L_0 \frac{x'}{y'}, L_0) \cdot \tag{29}$$

For both regions $D_1$ and $D_2$, we use two discs, bounded by circles with radii $R_1 = c_1 = d_1 = 0.2$, $R_2 = c_2 = d_2 = 0.4$, respectively and all centered at (a,b) = (0,0). So here formulae (27) apply in all sectors to give

$$(x_i, y_i) = (R_i \frac{x'}{\sqrt{x^2+y^2}}, R_i \frac{y'}{\sqrt{x^2+y^2}}), \ i=1,2. \tag{30}$$

Plugging Eqs. (28)-(30) in Eq. (21) then in Eqs. 23 and 24, we explicitly obtain $\mathbf{T}^{-1}$ using (27) and readily deduce the material properties of the cloak (annulus of radii $R_1$ and $R_2$ and) via formula (11) as shown in Figure 6.

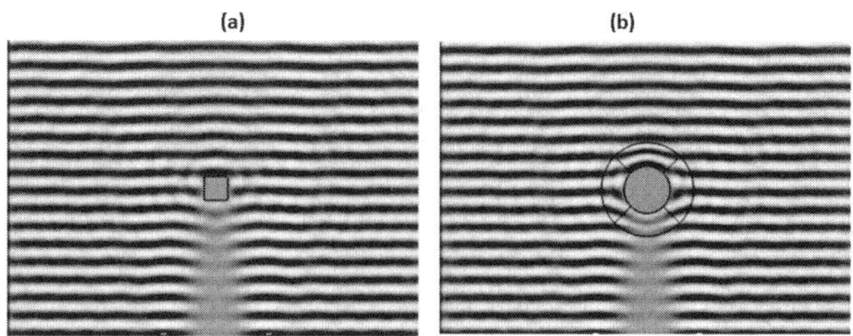


Fig. 6.  A plane wave incident from above on (a) a small perfectly conducting square cylinder and (b) a circular nonsingular cloak is scattered in exactly the same way.

## 5.2. *Optical Illusions with Squares*

Here we apply exactly the same procedure as in section 5.1 in order to design square-like cylindrical cloaks that acquire the same optical signature as an as smaller square-like cylinder as we want. Hence such cloaks will in particular appear to external observer as a very small cylinder. Here, all the domains involved have linear boundaries, so that we may chose formulas (25) and (26) as shown in Figure 7. The same algorithm can be used to design any nonsingular generalized cloak of this kind, where the domains involved have a mixture of elliptic and/or linear boundaries. Such a cloak will then make cross-like, sunflower-like, polygram-like, ellipse-like, … objects have the same electromagnetic signature as one another and in particular  look like one another to an external observer.

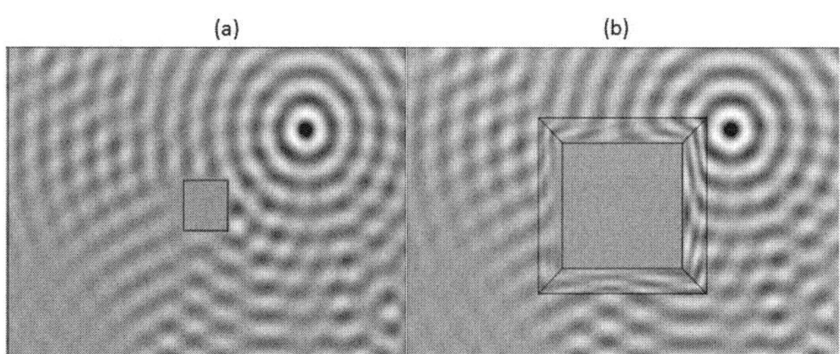


Fig. 7. (a) Scattering by a point source in the presence of a small infinite conducting square obstacle. (b) Scattering by a point source in the presence of a large infinite conducting square obstacle. The scattered fields outside the region occupied by the large square obstacle are identical in (a) and (b).

## 6. Conclusion

In this chapter, some recent developments in the theory of cloaking, using metamaterials, are presented. In particular, in sections 4.2 and 5 we discussed two different approaches to mirage effects in invisibility cloaking.

The design of mimesis cloaks we discussed in section 5 (see Diatta and Guenneau[12]), only involves nonsingular transformations which are determined by the optical properties we would like to confer to our cloaked objects. It maps the two objects to be made equivalent, in a one-to-one smooth way. In this way, any object can acquire the same optical properties as any chosen other one. However, such a design requires Neumann conditions on the inner boundaries.

In contrast, the generalized cloaking technique as in section 4, generally involves a singular transformation that gives a given object the appearance of any desired one without requiring Perfect Conducting Boundary (Neumann) conditions. It is an extension of the mirage effect discussed in Nicolet et al.[10] whereby a point source located in the transformed medium seems to radiate from a shifted location, to finite size bodies, which undergo the geometric transformation (13). As a result, an infinite conducting object placed inside the heterogeneous anisotropic coating of the cloak, appears to an external observer as another infinite conducting one.

## Acknowledgments

A.D. and S.G. thank the European Research Council for its support through ERC Starting Grant ANAMORPHISM.

## References

1. J. B. Pendry, A. J. Holden, D. J. Robbins and W. J. Stewart, *IEEE Trans. Microw. Theory Tech.* **47**, pp. 2075–84 (1999).
2. S. Guenneau and F. Zolla, *Prog. Electromagn. Res.* **27**, 91–127 (2000).
3. J. B. Pendry, D. Schurig, and D. R. Smith, *Science*, **312**, 1780 (2006).
4. U. Leonhardt, *Science* **312**, 1777 (2006).

5.　D. Schurig, J. J. Mock, B. J. Justice, S. A. Cummer, J. B. Pendry, A. F. Starr and D. R. Smith, *Science* **314**, 977 (2006).

6.　F. Zolla, S. Guenneau, A. Nicolet and J.B. Pendry, *Opt. Lett.* **32,** no. 9, pp. 1069–1071 (2007).

7.　G. W. Milton and N. A. Nicorovici, *Proc. R. Soc.* A, **462**, pp. 3027–59 (2006).

8.　A. Alu and N. Engheta, *Phys. Rev.* E **95,** 016623 (2005).

9.　A. Greenleaf, M. Lassas and G. Uhlmann, *Math. Res. Lett.* **10**, pp. 685–93 (2003).

10.　A. Nicolet, A. Zolla, F. and C. Geuzaine*, IEEE Trans. Magn*, **46,** no. 8, pp. 2975 – 2981 (2010).

11.　J. Li and J. B. Pendry, *Phys. Rev. Lett.* **101** 203901 (2008).

12.　A. Diatta and S. Guenneau, *J. Opt.*, **13**, no. 2, pp. 024012–024022 (2011).

13.　G. Dupont, M. Farhat, A. Diatta, S. Guenneau and S. Enoch, *Wave Motion*, **48**, issue 6, pp. 483-496 (2011).

14.　A. Diatta, G. Dupont, S. Guenneau and S. Enoch, *Opt. Express*, **18**, no. 11, pp. 11537-11551 (2010).

15.　J. Berenger , *J. Computational Phys.*, **114**, no. 2, pp. 185–200 (1994).

16.　A. Bossavit, *PIER* **32**, pp. 45–64, (2001).

17.　A. Nicolet, J. F. Remacle, B. Meys, A. Genon and W. Legros, *J. Applied Physics*, **75**, no. 10, pp. 6036–6038 (1994).

18.　F. Zolla, G. Renversez, A. Nicolet, B. Kuhlmey, S. Guenneau and D. Felbacq, Imperial College Press (2005).

# LED ILLUMINATION: ILLUMINOTECHNICAL, OPTICAL, METROLOGICAL AND SAFETY ISSUES


Franco Docchio[1,2*], Luca Fumagalli[1], Giorgio Libretti[1] and Paolo Tomassini[1]

*[1]T-LED srl, and Q-Tech srl, Via G. Matteotti 161, 25087 Rezzato BS, Italy*
*[2]Laboratory of Optoelectronics, University of Brescia,*
*Via Branze 38, 25123 Brescia Italy*
**E-mail: franco.docchio@ing.unibs.it*


The chapter focuses on various aspects related to LED illumination, i.e., one of the most challenging aspects of everyday life in terms of energy saving, environmental sustainability. Illuminotechnical aspects of LED illumination are highlighted. The R&D of our company towards a new generation of high efficiency lenses is presented. The instrumentation and the metrological references for street illumination measurements are shown; finally, safety and security issues related to the use of LED luminaires are discussed.

## 1. Introduction

Solid state lighting is considered to be the unique solution for street- and building illumination, as far as energy saving and quality enhancement of our environment are concerned[1]. Indeed, LED technology has grown to a maturity level that allows LED-based illumination to cover almost all needs in a number of outdoor and indoor installations. This maturity is demonstrated by the very high efficiencies reached by the latest generation of LED chips (in excess of 140 lm/W for high color temperature LEDs, even at high operating currents), the quality of phosphors used to convert blue light into white light, the efficiency of LED drivers (which are indeed a critical component to guarantee a LED long lifetime), and new materials and heat exchanger designs for heat removal from the LED chips.

Outdoor illumination is conceived to be one of the key applications of LEDs, for the benefit of the community and the quality of life, with a higher attention to energy saving issues and environmental respect.

Public illumination (for roads, crossings, roundabouts, parkings) is the LED illumination application with broadest development. At the same time, also private outdoor illumination is increasing, as many industrial premises are progressively converting to LED illumination for energy saving purposes and environmental sustainability.

The design of efficient and uniform, LED-based, illumination systems is a mix of different competences and skills, involving aesthetic design, optical technology, electronic/mechanical design, as well as metrologic skills. In particular, optical design is a key aspect of efficient LED-based illumination projects. In fact, even if single, high-power LEDs are supposed to be inherently high-efficiency components, the light path from the chip to the road may be composed of a number of interfaces, such as secondary lenses and plates, which considerably decrease the amount of extracted light, and hence the overall efficiency.

Thus, assembling a battery of LEDs into street- or park luminaires is more than a matter of arranging the LEDs in lines or matrices: an accurate choice of the lenses that shape the output light while maintaining a high extraction efficiency is mandatory.

This optimization task requires, as mentioned, a noticeable ensemble of metrologic skills and instrumentation to measure, for a given lamp assembly geometry, the angular distribution of the radiation, and to use this information to produce the illumination layout by means of conventional illuminotechnical software tools.

Metrological skills are also extremely helpful to perform accurate luminous emittance measurements. These, accompanied by the knowledge of the road reflectance data, help in deriving the luminance information, which is essential in determining the comfort of the travelers (drivers, bikers, as well as pedestrians).

Comfort is also closely related to safety and security. Blue light hazard has become more and more important these days[2], and LED emission spectra, characterized by a reasonably high blue-edge component, deserve some attention. In addition, the peculiar spectral distribution of LEDs, poses some problems related to the visibility of LED-based luminaires for some classes of the population.

T-LED, a newborn high-tech company specialized in the design of high-efficiency radiation-coupling optics for LED-based illumination, and in the design of LED lamps for outdoor illumination, has gathered from the parent Laboratory of Optoelectronics of the University of Brescia the necessary optical and metrologic skills to become one of the most experienced designers of high-efficiency and safe components and systems for LED illumination, both public and private.

This chapter will focus on the most important aspects in LED illumination, together with the achievements of the company in the field.

Section 2 will focus on the photometry of LED radiation, Section 3 is mainly devoted to the illuminotechnical aspect, Section 4 deals with optical design achievement of the company for the production of tailored lenses for high-power LEDs, Section 5 focuses on the metrologic side, Section 6 on the safety and security aspects of LEDs and of LED illumination. Finally Section 7 outlines the issue of a "Dark sky" approach of the 21$^{st}$ century illumination.

## 2.  Basic Radiometry and Photometry of LEDs

LEDs are incoherent light sources, and their photometric features are regulated by the norms applicable to all incoherent sources, although the rather unique primary beam characteristics (mainly a narrow divergence) have initially placed them under the laser safety requirements. LEDs are composed of a semiconductor chip of Silicon Carbide (SiC) or Gallium Nitride (GaN) from where blue light is emitted. This light is converted into visible light by means of suitable phosphors contained within lens-shaped plastic matrices (mainly Polymethylmethacrylate, PMMA), which are placed in front of the chip and form its primary lens. The lifetime characteristics of LEDs are determined by (i) the intrinsic lifetime of the chip, and the lifetime of the converting phosphors, which can undergo degradation with time and with use.

LEDs most common radiometric and photometric quantities and units of measure are, like for other incoherent sources[3,4]:

- The overall radiant flux W [Watt] (Radiometric quantity);
- The overall emitted Photon Flux Φ [lumen] (Photometric Quantity);

The two quantities are related by the relation that involved the CIE eye sensitivity curve:

$$\Phi = K_m \int_{380}^{760} W_\lambda \, V(\lambda) d\lambda \qquad (1)$$

where $W_\lambda$ is the spectral flux, $V(\lambda)$ is the CIE spectral distribution for photopic (daylight) vision (normalized to 1 at the wavelength of 550 nm), and $K_m$ is the conversion constant, whose value is 683 lm/W. From Eq. (1), a single – wavelength light source at the wavelength of 550 nm (peak of visual sensitivity) and of 1 Watt power, has a luminous flux of 683 lumens.

Related photometric quantities, used routinely in illuminotechnics, are:

- The luminous intensity I, expressed in *candelas*, (symbol *cd, 1 cd=1 lm/sr*), equivalent to:

$$I = \frac{\Delta\phi}{\Delta\Omega} \qquad (2)$$

where $\Delta\Omega$ is the solid angle of the emission,

- The Illuminance E, expressed in *lux* (symbol *lx, 1 lx=1 lm/m$^2$*). It is the ratio of the luminous power received and the area of the surface:

$$E = \frac{\Delta\phi}{\Delta A} \qquad (3)$$

- The Luminance L, expressed in cd/m$^2$. It is the ratio between the luminous intensity of a surface and the apparent area of the surface itself (i.e., the Area of the surface multiplied by the cosine of the angle formed between the normal to the surface and the observation direction):

$$L = \frac{\Delta I}{\Delta A \cos\theta} = \frac{\Delta\phi}{\Delta A \Delta\Omega \cos\theta} \qquad (4)$$

## 3. Illuminotechnics: Road Illumination Calculation

In Europe, the EN 13201:2003 Standard[5] is the consolidated reference for the evaluation of the luminaires to be used in road illumination. The Norm is divided in three parts: (i) Part 2 – Performance requirements, (ii)

Part 3 – Performance Calculations, (iii) Part 4 – Methods of measuring lighting performance. Italian Standard UNI 11248:2007[6], published in 2008, is the general Standard for road illumination, which refers to the above mentioned EN 13201:2003, and which has replaced the older 10439:2001 Standard. The Standard UNI 11095:2011[7] is a specific Standard related to the illumination of tunnels, which has to be used in conjunction with 11248:2007 for the calculation and measurement of luminance at the interface street-gallery. According to these Standards, each road is assigned a conventional illuminotechnical reference category (e.g., ME1 for motorways, with a minimum average Luminance requirement of 2 cd/m$^2$, ME3c for urban roads, with a minimum average Luminance requirement of 1 cd/m$^2$)[5]. Each category can be influenced by suitable "influence parameters" depending on environment, social contexts (e.g., likelihood of criminal events), geographic contexts (e.g. slopes). This correction, and the subsequent risk analysis, lead to the so called "illuminotechnical project category".

The 13201-2:2003 Standard defines, for each category, the performance requirements for a correct illumination of the road. For this purpose, the above set of categories can be grouped into three macroareas: (i) the macroarea where it is possible to perform luminance calculations based on parameters such as luminous flux, asphalt reflectivity, height of luminaires, etc.; (ii) the macroarea where this is not possible (e.g., crossings, commercial roads, etc.), and (iii) the macroarea mainly devoted to pedestrians and bikers. The 13201-3:2003 Standard contains guidelines for the calculation of luminance where applicable, given the geometrical characteristics of the luminaires and of the road. Finally, the fourth part of the guide contains guidelines on the correct performance of measurements of the Luminous Intensity from a metrological standpoint.

The evolution of LED illumination in public transportation systems has evolved quite slowly, so far, for a number of reasons. Firstly, LEDs are characterized by a narrow emission aperture. Assembling individual LEDs in matrices insides luminaires that were originally intended for traditional lamps (Mercury or Sodium) led to a first generation of luminaires that seldom fulfilled the Standard's requirements, yielding unsatisfactory luminance maps on the ground.

Attempts at correcting the Intensity pattern by introducing additional lenses to the luminaires somewhat improved the luminance characteristics, however at the expense of a reduction in the efficiency of the luminaire, which was much less competitive with respect to traditional ones. Secondly, the non-ideal heat dissipation housings also reduced the lifetime of the components, decreasing by far the strategic advantage of LEDs. Finally, from the social standpoint, the purported advantage of LEDs to be individually dimmable arose complaints in relation to unsatisfactory illuminations of roads/parks and a consequently increased possibility of crimes.

## 4. The Optics of LEDs: High Performance Aspherical Lenses for LED chips

The need to overcome the limitations in the fulfillment of the requirements imposed by the EN 13201:2003 Standard, and at the same time to increase device efficiency in LED-based luminaires, motivated T-LED srl to entirely re-design optical elements, which could be tailor-made for different types of luminaires for all the above mentioned illuminotechnical categories.

In general, available illuminotechnical CADs allow the placement of the luminaires (luminous flux, height, inter-element distance) starting from the intensity angular distribution of the elements to be mounted on the luminaire. Our approach was somewhat different from the traditional one: while the operator usually relies on existing angular distributions, we chose a "Reverse Engineering" approach. In general, for each illumination application, we simulate an intensity distribution which can yield the ideal luminance distribution according to the Standard, and immediately validate it using the CAD (in our case we used the Oxytech CAD tool[8]). This optimum angular distibution then becomes our input parameter for the design and production of the luminaire, in terms of (i) the number and distribution of the LEDs in the luminaire, (ii) the LED type and power, (iii) the angle of each LED with respect to the basement of luminaire support, and, especially, (iv) the lens.

Concerning the LED angle, in particular, we found that the common attitude to manufacture convex basements with the expectation of an overall broader illumination angle should be corrected. In fact, in many cases the winning solution can be represented by flat basements, provided that suitable lenses are coupled to the individual components. The use of a flat basements, moreover, optimizes the design of high efficiency heat sinks to remove the heat dissipated by the LED.

In turn, the ideal lenses, for LED illumination proved to be free-form lenses (i.e., non spherical lenses). The design of the lens can be derived from the desired angular distribution of the luminaire, taking into account the number of components to be mounted on the luminaire.

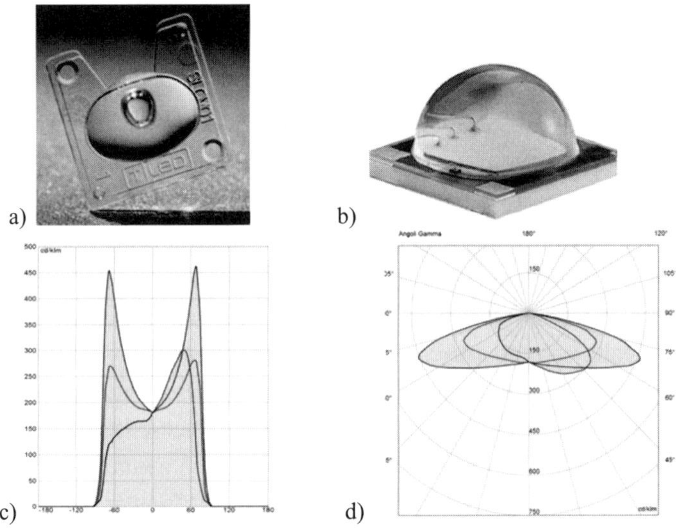

Fig. 1. Details of T-LED lens SL-CM01 for street illumination. (a) Photo of the lens; (b) Sketch of the Cree XM-L LED; (c) Cartesian plot of relative intensity, with red curve: angular distribution along street direction, blue curve: angular distribution perpendicular to street direction, black curve: angular distribution at 18° with respect to street direction, vertical Units: Candelas/Kilolumen; (d) Polar plot of relative intensity, same as for (c).

As a result, Figure 1 shows our free form lens Model SL-CM01 which was designed specifically for LED "Cree XM-L". Equipping a luminaire with a set of 48 LEDs with free-form lenses resulted in a LED luminaire in excess of the UNI-EN 13201:2003 illuminotechnical category M3a, (Luminance better than 1.0 cd/m$^2$), allowing luminaire placement with ratios of pole distance to height better than 4,4:1.

Figure 1.a) shows a photograph of the plastic lens, suitably designed for the CREE XM-L LED sketched in Figure 1.b). Cartesian and polar intensity distributions are plotted in Figure 1.c and 1.d respectively. Red, blue and black lines refer to the longitudinal, transverse and oblique (18°) directions. In the latter figures, intensity values are expressed in candelas over kilolumens (cd/klm).

The typical bat-wing distribution in the longitudinal direction, ideal for road illumination, is evident, along with a distribution in the transverse direction that extends beyond the road border, as required by the specifications. Lenses like the one in the figure are available also in square modules of 2x2, such as in Figure 2, and in linear modules.

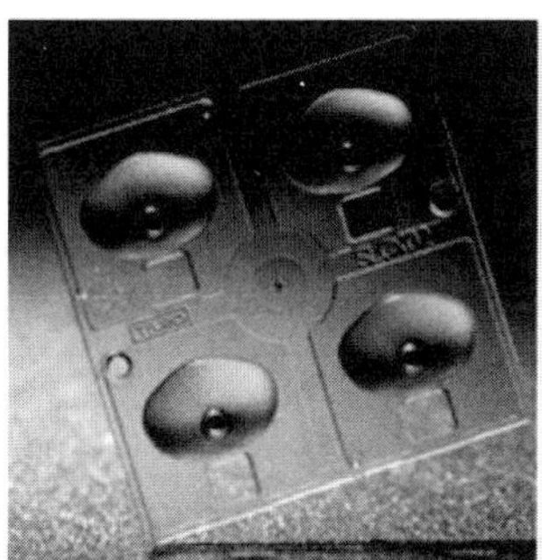

Fig. 2.　2 x 2 lens module for LED illumination.

Figure 3 shows, for the same lens/LED assembly in Figure 1, the false-color (3.a) photometric solid rendering of the intensity distribution, together with the software-calculated iso-illuminance (isolux, 3.b) map at street level. We note the uniform illuminance over a very wide portion of the illuminated street, favoring an increased distance-to height ratio between adjacent luminaires.

Finally, Figure 4 shows a software-calculated luminance map of the street, highlighting the excellent uniformity of luminance all through the carriageway. The performance of the illumination project obtained with this LED/lens selection is in accordance to the 13201:2003 Standard ME3a class, with 4,4:1 distance to height ratio.

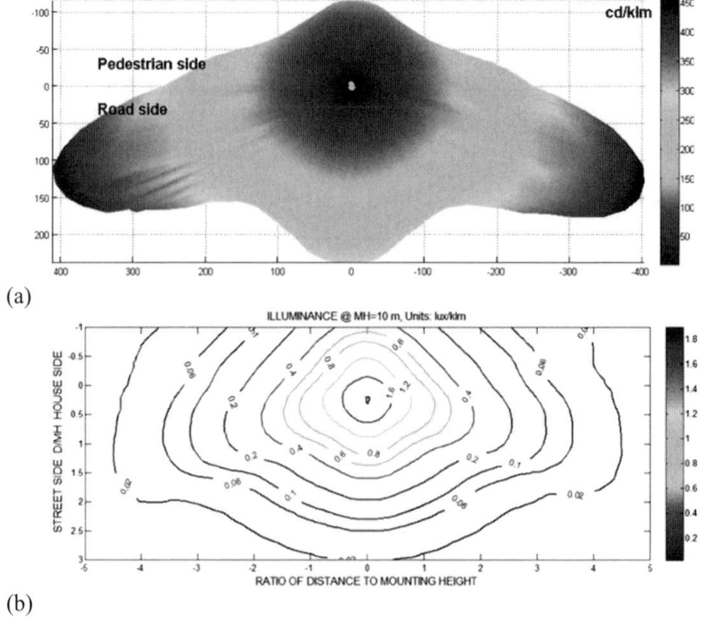



(a)

(b)

Fig. 3. Solid photometric plot, and illuminance plot at street level, of LED-Lens assembly of Fig. 1. (a) Solid photometric plot, distances are in cm; (b) isolux map at street level, abscissa is expressed in ratio of distance to height of luminaire, ordinate is expressed as distance in meters from edge of carriageway (negative values correspond to pedestrian pathway).

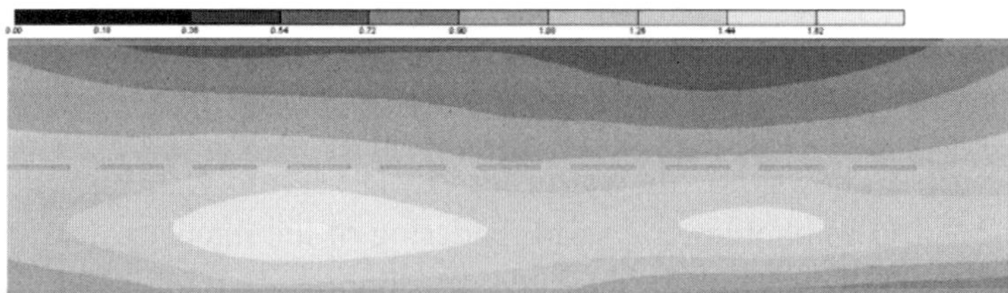

Fig. 4. Luminance plot at street level.

## 5. Metrology: Measurement Techniques and Instrumentation

In addition to optical design, photometry, illuminotechnical and vision skills, metrological skills are crucial for a correct approach to LED illumination. Vision science and measurement is a discipline where a correct metrological approach is generally operator-dependent, as the observer's perception is important in the measurement process. This is why in LED illumination the performance of light illumination system is often a subject for the so-called "soft metrology".

From strict methodological viewpoints, Illuminance and Luminance measurements and calculations are, by definition, linked to photopic (daylight) vision, since the use of the photopic sensitivity curve leads to the definition of the lumen and of its derived quantities candela and lux.

However this may be not completely correct, due to the fact that LED illumination is commonly measured in a dark context where scotopic vision responses, or the intermediate mesopic vision response, may be more appropriate, leading to correction factors for the results.

Measurement devices and techniques are based on laboratory, and in-field measurement instruments. The former allow the measurement of radiation angular patterns of individual LED devices or of complete luminaires, to derive the files which feed the illuminotechnical CADs or are used to verify the performance of special lenses. The most popular is the gonioradiophotometer, whereby the light source is mounted on a set of rotating units, one for each axis of rotation, and the illuminance is measured at a fixed location at a distance with respect to the luminaire. An example of this category of instruments, produced by the partner company Q-Tech srl, is depicted in Figure 5.

In contrast to Sodium or Mercury luminaires, whose emission spectrum is well-defined, LED emission depends on the type and concentration of the phosphors used in the primary lens. A whole region of the CIE colorimetric curve can therefore be covered, with "colder" (i.e., more blue) or "warmer" (i.e., more red) color effects: color ranges can be used for different applications (exteriors, interiors, food, art, etc.). Spectral information is therefore very useful in the characterization of luminaires for various types of applications, and integrate photometric information. This may be even more important in the near future, as the technological evolution of the main LED producers is to manufacture "all-in-one", bulb-like, LED lamps, operated either AC or DC, with a number of blue LEDs and the external bulb covered by light converting phosphors to produce white light.

*F. Docchio et al.*

Fig. 5. (a) Goniophotoradiometer Mod. QT-R1 by Q-Tech srl; (b) detail of the rotator unit.

Luxmeters are the most popular photometric instruments to measure illuminance. They measure the ratio of the total photometric flux impinging on the sensing surface, to the area of the surface itself. As such, luxmeters are not sensible to the direction of arrival of the light.

In contrast, luminance meters are portable instruments suitable for luminance measurement: a single sensor or a CCD camera are coupled to an objective whose aperture defines the acceptance solid angle, and therefore the apparent surface $\Delta A\cos\theta$. Luminance meters can function as colorimeters.

The need for accurate luminance measurements in all situations requires the use of precision mobile instrumentation. In fact, if in the illuminotechnical macroarea 1 (see Section 3) luminance information may be directly calculated from illuminance data (here the luxmeter is sufficient), this is not true for the other macroareas where no calculation is possible. In this case mobile luminance meters should scan the region of interest to provide luminance maps which characterize the road tract.

An example is given by the Tiresia mobile Laboratory of the National Metrologic Institute of Italy (I.N.Ri.M.), depicted in Figure 6.a), especially designed for the monitoring of motorway tunnels in Italy.

Figure 6.b) shows a luminance map obtained with the equipment, and demonstrates the superiority of CCD-based luminance meters with respect to conventional ones[9,10].

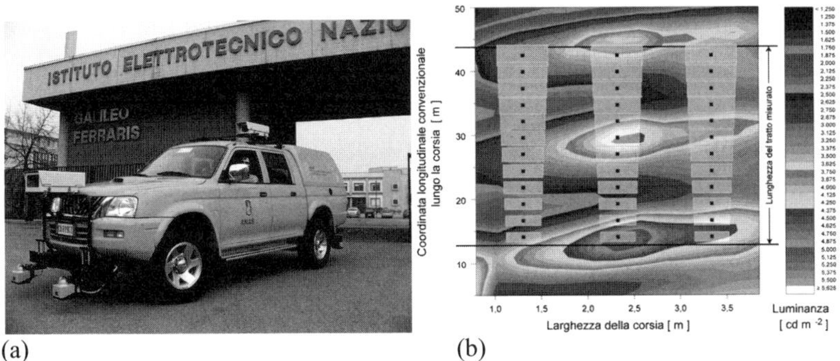

(a)                                            (b)

Fig. 6. (a) The mobile Laboratory "Tiresia" of I.N.Ri.M., which measures luminance on roads and tunnels; (b) A typical map of luminance obtained by a CCD luminance meter mounted on Tiresia and performing measurements in accordance to the 13201-4:2003 Standard. Courtesy of P. Iacomussi, I.N.Ri.M.

## 6. Safety and Security Issues related to LED Technology

As previously stated, LEDs were initially considered under the category of coherent light sources, and submitted to the safety Standard IEC/EN 80625-1[11]. This was due to the fact that, until recently, LEDs were low-power, highly collimated sources, with red-to IR spectral emission, and their use was mainly confined to pointing, display and remote control uses. Only recently the radiant flux of LEDs skyrocketed, the blue LED technology got to full maturity, and the use of phosphors gave birth to the present generation of white light LED sources. As a consequence, the use of LEDs as everyday illumination light sources became more and more relevant upon the above applications. LEDs are to-date common illumination sources, replacing incandescent bulbs and even fluorescent lamps for exterior and interior illumination (and presently for TV and

Computer displays). It is obvious that the attention towards LED safety became a matter of Standards related to everyday non-coherent light exposure.

LED safety is presently included in the European Standard CEI EN 62471:2009 - Photobiological Safety of lamps and lamp systems[12], which is being adopted by all UE States as National Standards. This Standards sets Exposure Limits (Els) for all light sources of everyday use, with respect to the photobiological effects of these sources on eye and skin of the exposed subjects.

Three main photobiological mechanisms determine the limits: (i) the UV (Actinic and near-UV) Action Spectum, mainly restricted to the UV region of the Spectrum, (ii) the blue light hazard spectrum (from 380 to 500 nm), and (iii) the thermal hazard spectrum (380 to 1,400 nm). Of these, only the two latter have relevance in LED illumination, since LEDs for everyday use do not exhibit UV radiation.

Blue light hazard is a relatively recent aspect of safety, not included in previous hazard studies where photobiological effects were mainly confined into the UV region. This effect is anyway relevant, since long exposure to blue light (which easily penetrates into the eye and reaches the cornea, still having a relevant photochemical activity) can damage the retina and the macula, favoured also by the low sensitivity of the eye to blue radiation and, hence, by the limited capability of the eye to protect itself against it by closing the pupil or blinking.

Moreover, the LED has unique properties with respect to conventional extended sources, being classified as a "point-like source". This may pose some problems in the correct classification of LEDs in relation to the exposure risk.

Concerning LED-based luminaires, risk classification is somehow questionable. It is evident that street luminaires cannot be classified as "no risk" (Level 0), except for distances of at least a couple of meters from the source. This is, in general, safe for the users, as they travel at a considerable distance from the sources (in excess of 5 m), but may pose some problems for the maintenance staff, who may get closer to the luminaire.

Another intriguing aspect is related to the spectral content of LED emission. All white LEDs are characterized by an intense blue peak (the original LED emission) centered around 460 nm, and a red-shifted wideband emission characteristic of the phosphors. Between the two emission bands there is an emission gap centered around 490 nm. The appearance of the LED color depends on the ratio of the two bands. In Figure 7.a two spectra of (i) a "daylight" (i.e., blue-shifted) and "warm" (i.e., red-shifted) LEDs are shown together with the CIE spectral lines.

According to their spectra, LEDs are classified in terms of Color Temperature (which is the well-known parameter related to the peak emission wavelength of a blackbody radiator satisfying Wien's law[3]). Therefore each LED is characterized by a color temperatures: Figure 7.b shows, as an example, the range of emission spectra of Cree LEDs in the CIE space.

Unfortunately, commercial LED catalogues (see, for instance, Ref. 13) specify, for a certain product, a broad range of color temperatures; this means that their installation in street luminaires may result in non-uniform street illumination color. Binning, i.e., the purchase of selected color temperature LEDs, is always possible, although at increased component costs.

Color temperature is a significant aspect in LED illumination, as compared to Sodium and Iodides illumination. Blue light, in addition to an increased risk of photochemical effects at the retinal level, presents other disadvantages. The first is related to the age-related transmission of ocular media to the blue region of the spectrum: wavelengths below 500 nm are strongly attenuated[14] by the ageing eye with respect to the wavelengths above 500 nm: as a consequence, elderly travelers may perceive the street as much darker than young ones. In addition, blue light illumination has been reported to hinder the production of melatonin. Finally, it is well known that light scattering in a medium is proportional to the inverse of the fourth power of the wavelength: therefore, in aged patients with even moderately scattering crystalline lenses and vitreal bodies, blue light can create a greater amount of discomfort.

For all these reasons there seems to be a great attention, in the most recent regional directives, to limit the installation of illumination luminaires to those equipped with LEDs with color temperatures below 3,500 K. Unfortunately, these "warm light" LEDs exhibit a conversion efficiency in the range 70-90 lm/W, well below their blue counterpart (about 140 lm/W), and this may limit the effective energy saving advantage.

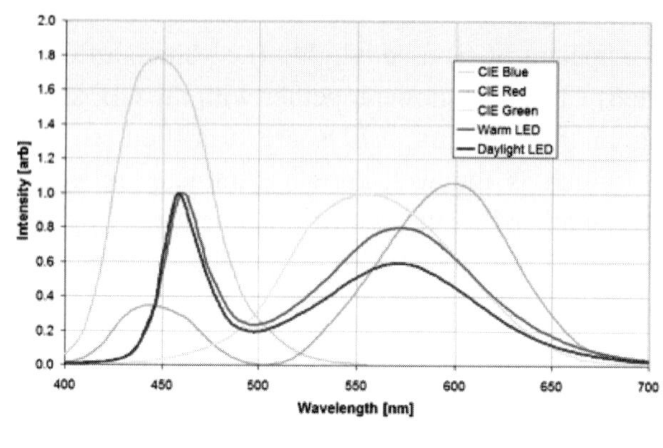

(a)

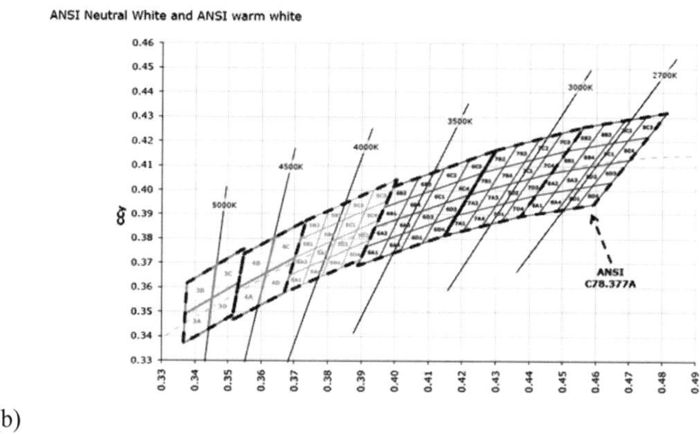

(b)

Fig. 7. Spectral distribution of LED emission. (a) Typical emission spectra of two LEDs with different color temperatures. Note the blue peak of primary LED emission, and the red components of phosphors. LEDs present a typical emission hole around 480 nm; (b) CREE LED emission in CIE chart: grey letters: neutral white, red letters: warm white. Solid lines define color temperatures.

## 6. Towards a Dark Sky: LED Illumination against Light Pollution

Public illumination Plans, according to Regional and National Laws (such as the Law of Regione Lombardia 17:2000) are specific projects, and technical specifications, aimed at governing all new illumination installations on public roads, crossings, parks, roundabouts, commercial centers. Major constraints on the installations, in addition to public safety and energy saving, are related to fight against light pollution, as a way to protect the environment. This "dark sky issue" is important, since a dark and starry sky at night is one of the natural beauties of nature, and it is important to preserve it against all kinds of polluting agents, including artificial light.

LED-based illumination is expected to play a key role in this context, since the natural directionality of the LED sources is likely to minimize the amount of light scattered upwards, even in the case of building, bridges, and monuments illumination.


## 7.  Conclusions

This work outlined some of the major issues related to LED illumination, i.e., technology advances, efficiency, safety, energy saving. Although at least in Italy Public Administrations are to some extent reluctant to massively adhere to LED illumination systems, probably because of the high luminaire costs and the uncertainty about the effective lifetimes of the LEDs, the situation is improving, as Administrations increasingly request illumination plans to cover Underground stations or newly built parks, and private Industries are increasingly installing LED luminaires inside their premises. The impact of free-form lenses to dramatically improve the angular distribution of the luminaire has been presented, and using this new approach a new generation of high efficiency, long lifetime illumination systems can replace the existing ones, favoring economical and environmental sustainability.


## References


1. *White paper: Solving the system-level thermal management challenges of LEDs*, Mentor Graphics, www.mentor.com, 1-8.
2. W. T. Ham, H. A. Mueller and D. H. Sliney, *Nature*, **260**, pp. 153-155 (1976).
3. D. H. Sliney and M. Wolbahrst: *Safety with lasers and other optical sources*, Plenum Press (1980).
4. CIE Publication n. 17, IEC 50845, International Light Vocabulary.

5. European Standard 13201:2003, Light and lighting. Part 2 – Performance requirements; Part 3 – Calculation of performance; Part 4: Methods of measuring lighting performance.

6. Italian Standard UNI 11248:2007, Road lighting, Selection of lighting classes.

7. Italian Standard UNI 11095:2011, Luce e illuminazione - Illuminazione delle gallerie stradali.

8. www.oxytech.it.

9. P. Iacomussi, "Metrological applications for lighting engineering", *Tutto_Misure 1*, 65 (2012).

10. P. Fiorentin and A. Scroccaro, "Illuminance measurements from a multi-luminance meter", Tutto_Misure 2, 111 (2012).

11. European Standard EN 80625:2004, Safety of lasers.

12. European Standard EN 62741:2009, Photobiological safety of lamps and lamp systems.

13. www.cree.com.

14. E. A. Boettner and J. R. Wolter, *Invest. Ophthalmol.* **1**, 776 (1962).

# FIBER OPTIC SENSOR TECHNOLOGY FOR OIL AND GAS APPLICATIONS

Morten Eriksrud* and Jon Thomas Kringlebotn

*Optoplan AS*
*P.O. Box 1963, N-7448 Trondheim, Norway*
**E-mail: morten.eriksrud@optoplan.com*

This chapter provides an overview of the fiber optic sensing technologies currently in use within the oil and gas business. Brief descriptions of the various sensing principles as well as field experiences are given.

## 1. Introduction

In the late 1980's work started to explore the use of fiber optic sensing technology within the oil and gas business[1]. This was motivated by an increasing demand from oil companies for permanent sensors in hydrocarbon producing offshore wells to facilitate production from these costly wells. At the same time the telecommunication industry was about to be revolutionized by the fiber optic technology and this created a base for advanced optical sensing technology through adoption of components developed for telecommunication applications.

The main incentives for introducing optical sensing system to the oil and gas industry was the high reliability potential in the downhole and subsea environment and the potential for providing a suit of sensors for different measurements in this environment. The first development of optical sensors for permanent downhole installations was reported in 1991[2]. It was a pressure and temperature gauge system based on optical excitation and interrogation of micro-machined resonant sensors. The sensing principle is illustrated in Figure 1. Key system components included sensor and sensor housing, downhole optical cable and topside optoelectronic equipment for system operation. This fiber optic well monitoring system was then installed in an onshore gas well in 1993 at the Sleen field in the Netherlands followed by an offshore installation in 1994 at the Gyda field in the North Sea[3]. These installations were successful and clearly documented continuous operation of fiber optic sensing technology in harsh downhole environment.

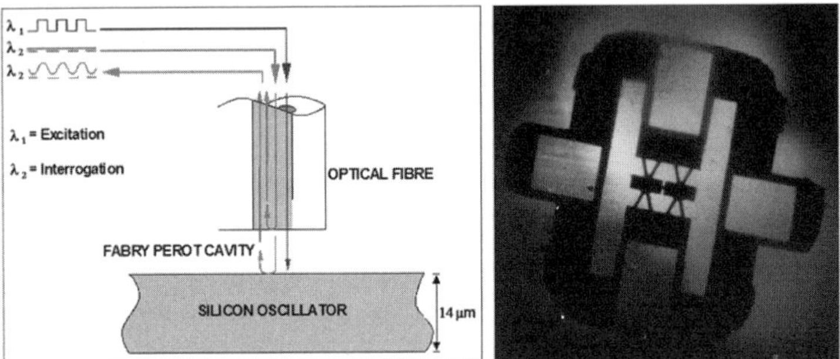


Fig.1. The silicon oscillator is excited by wavelength $\lambda_1$ modulated at a sweeping frequency. When this modulation frequency matches the resonant frequency of the oscillator, the length of the Fabry Perot cavity is modulated accordingly and this frequency is monitored by the second wavelength $\lambda_2$ .The resonant frequency depends on the stress level in the resonator and when packaged properly in a silicon tube, any external pressure impacts directly the oscillator stress level and thereby the optical phase modulation in the Fabry Perot cavity. A picture of the micro-machined resonator is also included (Courtesy Optoplan AS).

These pioneer installations were followed by significant development work and several more practical optical sensing principles have been investigated. Over the last two decades oil companies have been heavily involved in this development. The reason is an increasing demand for reliable permanent reservoir monitoring systems. In offshore fields only 35-50% of the hydrocarbon resources are typically recovered from the reservoirs. Enhanced recovery can be achieved by monitoring the reservoir characteristics and using the monitored data to optimize the production conditions. However, very reliable sensor systems of high accuracy and stability are required due to the high installation and maintenance cost of offshore wells of high complexity. Currently the measurements of most interest for permanent reservoir monitoring are:[4-7]

- pressure and temperature;[3, 8, 9]
- temperature profile (distributed);[5, 10, 11]
- phase fraction and corresponding flow rates;[12, 13]
- seismic (reservoir imaging).[14-17]

Basically, many different optical sensing principles can be utilized for these types of measurements. Currently, within the commercial oil and gas market the two dominating principles are related to Fiber Bragg Grating (FBG) technology and Raman backscattering technology.

The FBG technology either uses the FBG as the sensing element or the FBG as an optical reflector in an interferometric design where the fiber between two FBGs is the sensing element. The reflected wavelength from an FBG or the optical phase through a fiber between two FBGs is dependent on strain and temperature and the mesurand is normally converted into strain through a transduction mechanism.

Temperature compensation techniques might be required. The FBG technology provides single point measurements, but multiple sensor points along a single fiber are implemented by multiplexing. The Raman backscattering technology utilizes a light scattering process inherent in the optical fiber glass material and the whole fiber itself represents a continuous temperature probing element. The Raman backscatter technology provides distributed temperature profile along the fiber and is known as Distributed Temperature System (DTS).

Downhole permanent reservoir monitoring is based on sensors installed in producing wells or injector wells. Typical length of these wells is 2-8 km with a typical number of individual sensing points being less than 20. Distributed measurements, as DTS, can be installed in combination with individual sensing points.

More recently, seabed permanent reservoir monitoring has attracted significant attention. In this case a large seafloor area is covered by a cable network containing individual sensing points. The seafloor area can be $30 - 150$ km$^2$ with a typical number of individual sensing points in the range of $2000 - 10000$. Each sensing point can include four sensor channels (3 accelerometers and one hydrophone).

Another application of optical sensors within the oil and gas field is integrity or conditional monitoring. This relates mainly to pipelines along the seabed and riser cables from the platform down to the seabed[18, 19].

Such systems are based on strain monitoring and/or temperature monitoring[20-23], but are still in an early phase for commercial implementation in the oil and gas industry.

## 2.  Application Requirements

The downhole environment is characterized by high pressure (typical 100-1000 bar) and temperature (typical 80 - 200 $^{\circ}$C). The most challenging requirement for a sensor system is the temperature requirement. A key advantage of fiber optic sensors compared to electronic sensors is the ability to withstand higher temperatures for reliable long term operations. Fiber optic sensors are completely passive devices with no active components in contrast to electronic sensors. They require no downhole or subsea electrical power with the interrogation unit remotely positioned topside.

The main value of having a downhole permanent pressure and temperature gauge system is to optimize production, improve reservoir drainage and save cost on well surveillance intervention. Well intervention, i.e. re-entering a well to carry out a maintenance operation, is complex and expensive. Therefore reliable and robust sensing systems are required for permanent installation. Continuous and real time high quality data information about the behavior of the well will provide increased production from the well[9]. Typical figures can be in the range of 10 – 20% increased production.

High stability in the measurements is required and extreme resolutions are targeted by the oil companies for reservoir and well management purposes. Pressure resolution better than 0.1 psi and temperature resolution better than 0.1 $^{\circ}$C are requested.

A key design issue for fiber optic sensors is the packaging required to withstand the long term environmental load experienced in oil and gas producing wells. The package design will basically determine the mechanical and optical integrity of the sensing system. Another challenge is the actual installation of the sensors in the well, particularly in subsea field developments where wellheads are located on the seafloor. This requires "wet" optical connections underwater, in contrast to the "dry" connections that can be made on a platform. An additional

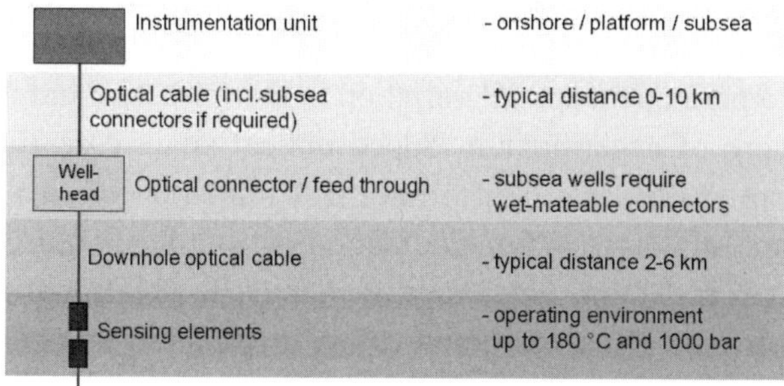


Fig. 2. Key components of a downhole optical sensing system.

key component is the optical downhole cable (see Figure 2). This will be exposed to corrosive environment and needs to provide sufficient protection against hydrogen induced losses at the elevated well temperatures. The optical fibers are typically contained in a metal tube free to move with some excess length to avoid tension on the fiber during installation and operation. The cable is normally clamped on the tubing in a conventional way.

## 3. Fiber Optic Sensing Technology

This section describes the basic principles of the main fiber optic sensing technologies that currently are in use for reservoir monitoring. This includes FBG technology for multiplexed single point measurements and Raman backscattering technology for distributed measurements. Other backscattering technologies under evaluation for additional distributed measurements are also briefly discussed. All these technologies utilize the optical fiber as the sensing element and the mesurands can modulate (i) the wavelength and/or the amplitude of reflected or scattered light from the fiber core, and/or (ii) the optical phase of light propagating in the fiber core. These characteristics of an optical fiber are inherent in the glass material under exposure of temperature, strain and hydrostatic pressure. In addition, glass material of optical fibers is an excellent mechanical material. It behaves elastically under tension until it breaks – in contrast to polymers and metals. Sensor designs where the mesurand converts into fiber strain (or optical phase changes) are ideal to achieve highly linear sensitivity responses with negligible hysteresis.

### 3.1. *Fiber Bragg Grating Sensing Technology*

An FBG is a permanent spatial variation in the refractive index imprinted along the core of a germanium doped optical silica core fiber through side-exposure of UV laser light fringes[24]. A standard FBG has a length of ~1 cm with a refractive index modulation period of ~0.5μm and reflects light with a reflection spectrum centered at a Bragg wavelength determined by the actual refractive index modulation profile in the core of the optical fiber. The sensor potential of FBGs was early realised[25] due to the fiber wavelength encoded sensor signals through the temperature and strain dependence of the Bragg wavelength and the corresponding wavelength multiplexing potential, and several sensing configurations and demodulation principles have been explored[26].

The Bragg wavelength, $\lambda_B = 2n_{eff}\Lambda$, of an FBG with uniform period $\Lambda$ will shift linearly with fiber strain $\varepsilon$ and change in temperature $\Delta T$, with a relative wavelength shift given by:

$$\Delta\lambda_B/\lambda_B = (\alpha + \xi)\Delta T + (1 - p_e)\,\varepsilon \tag{1}$$

where $\alpha$ is the thermal expansion coefficient of the fiber, $\xi$ is the thermo-optic coefficient, $\xi = (1/n_{eff})\, dn_{eff}/dT$, of the fiber core, where $n_{eff}$ is the effective refractive index seen by the guided light mode, and $p_e$ is the photoelastic constant accounting for the change in refractive index with strain. For a typical germanium doped silica core fiber; $\alpha = 0.55*10^{-6}\,°C^{-1}$, $\zeta = 5.9*10^{-6}\,°C^{-1}$, and $p_e = 0.22$. For an FBG with a Bragg wavelength at 1550 nm in the telecommunication C-band the resulting temperature and strain sensitivity will be 10 pm/°C and 1.2 pm/μstrain, respectively.

The Bragg wavelength of an FBG can be detected by several interrogation techniques. A common technique is to use a broadband source (semiconductor light emitting diode or fiber superfluorescent light source) and a tunable optical filter, in combination with a stabilized wavelength reference and passive optical routing components, as illustrated in Figure 3. Wavelength resolutions below 0.1 pm (Figure 3) can be achieved by using advanced peak-detection algorithms.

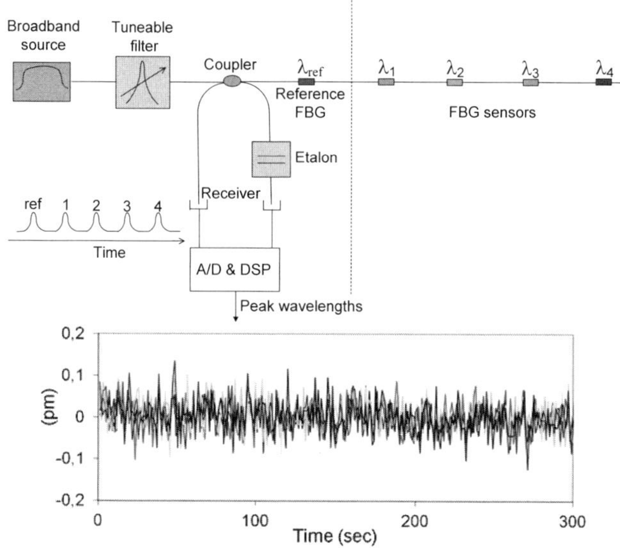


Fig. 3. FBG wavelength readout design for interrogation of wavelength multiplexed FBG sensors. A reference FBG and an etalon filter are used to achieve high wavelength reproducibility. Lower curve shows repeatability in peak wavelength detection from 6 stabilized FBGs (1535–1560 nm). The rms deviation is 0.04 pm for 1 sec measurement time. (Courtesy Optoplan AS).

Monitoring changes in the Bragg wavelength is the sensing principle used for P/T gauges, containing two FBGs which respond equally to temperature, but differently to pressure. The sensor packaging is converting hydrostatic pressure to strain on the FBG. At high temperatures it is difficult to achieve sufficient stable performance due to mechanical creep with conventional optical fiber attachment techniques and bonding materials. A highly stable solution has been realized with a compressive design (increased pressure introduces increased compression) using a large diameter glass rod waveguide with a standard size core where the FBGs are imprinted[27]. Typical diameter of this glass rod structure is 1-4 mm with a length in the order of a cm, and requires advanced glass machining technology to be fabricated.

The glass rod wavelength sensor design with imprinted FBGs has also been used to realize a temperature point-sensor array system[28].

Several waveguide sensors are spliced in series along a single optical fiber, where the glass rod structure isolates the FBG from any strain exposed to the optical cable between the sensors. This array temperature system with its high temporal resolution at specific locations can be combined with the complete spatial information provided with DTS.

Another sensor approach, commonly used for dynamic measurements such as flow measurements and seismic measurements, is to establish Fabry-Perot interferometers based on pairs of FBG reflectors with an optical fiber coil in-between. The interferometer resonance modes will shift with temperature and strain induced changes in optical phase delay of the interferometers. Weakly reflecting FBGs separated by fiber coils (of typical 1-100m length) are used to form two-beam interferometers. Multiplexing along the same fiber of such interferometric sensors can be performed using (i) Time-Division Multiplexing (TDM), where a serial array of equal wavelength FBGs having equidistant spacing are used, and/or (ii) Wavelength Division Multiplexing (WDM) through the wavelength encoded reflection of the FBGs. In a reflective TDM based sensor array, with equidistant FBGs, multiple reflections are inevitably to occur, where the resulting cross-talk depends strongly on the reflectivity of the FBGs.

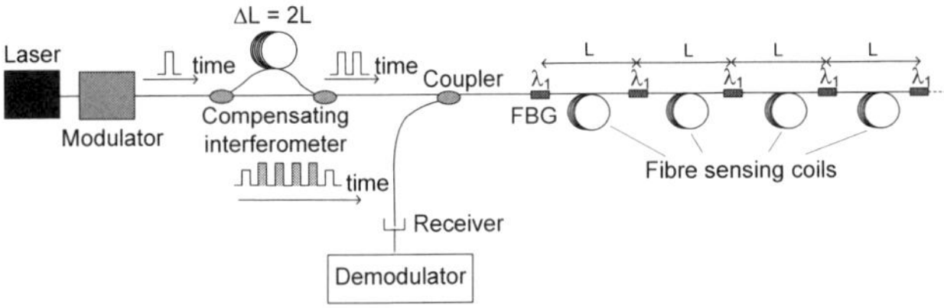


Fig. 4. FBG based interferometric design with TDM multiplexing, where the fiber sensing coils are located between FBGs operating at the same wavelength. WDM multiplexing is added by including additional groups of sensor coils between FBGs operating at other wavelengths.

However, the impact of these multiple reflections can be efficiently removed through use of an advanced layer-peeling algorithm[29], allowing higher reflectivities and improved power budget.

A reflective array TDM network can also be realized with a coupler-based Michelson interferometer design, where couplers and mirrors are used instead of FBGs, and with WDM implemented by means of optical add/drop multiplexer components[30]. Another multiplexing approach with a coupler-based Michelson interferometer system is to combine WDM, using add/drop multiplexers, with Frequency Division Multiplexing (FDM), where the optical carrier frequencies are modulated at different frequencies[31].

The FBG-based design has the advantage of dramatically reducing the number of optical components (couplers and add/drop multiplexers) and the corresponding splices required to build a large-scale sensor system. Continuous FBG arrays, which can be coiled directly onto the sensor elements, are now made with high yield using fully automated manufacturing systems[32], reducing the number of splices.

For a two-beam interferometer, i.e. a coupler-based Michelson design or FBG-based Fabry-Perot design with low reflectivity FBGs, the measured reflected interference signal can be written as

$$P = P_0 + P_1 \cos (\Phi) \tag{2}$$

where $\Phi = 2\pi v\tau$ is the two-way phase change between the two FBGs, where $v$ ($= c/\lambda$) is the interrogating laser frequency and $\tau = 2Ln/c$ is the two-way time delay between the two FBG, where $L$ is the fiber length between the two FBGs, $n$ is the refractive group index, and $c$ is the speed of light. The relative change in phase, $\Delta\Phi/\Phi$, with changes in fiber strain and temperature is given by Eq. (1), replacing $\Delta\lambda_B/\lambda_B$ by $\Delta\Phi/\Phi$. The resolution of an interferometer with $L = 10m$ can be as low as $1\text{-}10\mu\text{rad}/\sqrt{\text{Hz}}$, which corresponds to a strain resolution of $0.01\text{-}0.1$ pstrain/$\sqrt{\text{Hz}}$. The minimum detectable phase change ($\delta\Phi \sim 1/\sqrt{\text{SNR}}$) is determined by the Signal-To-Noise Ratio (SNR) at the receiver. Several noise reduction schemes can be implemented in the interrogating instrument to reduce noise for instance induced by Rayleigh backscattering and laser frequency fluctuations[33]. Currently DFB fiber lasers are normally used due to the very low frequency noise characteristics. Since the recorded interference signal has a periodic non-linear dependence on the phase $\Phi$ (see Eq. 2), there is a need to linearize the response to unambiguously measure/follow the induced phase changes. This can be done by means of the Phase-Generated Carrier (PGC) technique[31].

The FBG-based Fabry-Perot interferometric design is used for in-well flow measurements[12], using a TDM network with low reflectivity of the FBGs, and for in-well seismic measurements[14], using WDM as the multiplexing scheme with higher reflectivity of the FBGs. Large-scale seabed seismic sensing network has been realized by combining TDM and WDM in an FBG-based interferometric design[34]. Other optical solutions for permanent seabed seismic systems utilize coupler-based Michelson interferometeric design with combined TDM and WDM[35] or WDM and FDM[36].

A key issue for these ultra-high resolution interferometric sensing systems is the mechanical packaging of the actual sensor stations. Various mechanical transducer designs are used to convert pressure and acceleration to fiber strain. Air-backed mandrels to accommodate the coiled sensing fiber are used with hydrophones, while accelerometers are using a moving-mass principle to convert acceleration to fiber strain.

### 3.2. *Raman Scattering Sensing Technology*

The dominating light scattering mechanism in silica glass is the Rayleigh scattering which determines the minimum loss in optical fused silica core fibers. Raman scattering is another scattering mechanism in the glass material, which results in much lower optical scattering efficiency (typical three order of magnitudes less), but provides a shift (both upwards and downwards) in optical frequency of the scattered light. The shift is caused by interaction of incoming light (photons) with thermally induced molecular vibrations (phonons) in the fused silica core and has a typical value of 13.5 THz. The Raman spectral band at decreased frequency is called the Stokes component while the component with increased frequency is denoted Anti-Stokes. The Stokes band is caused by photon interaction with molecules in the ground vibrational state, while the Anti-Stokes band is a result of interaction with molecules in a thermal excited state. Therefore the optical intensity in the Anti-Stokes band is strongly influenced by temperature, while the optical intensity in the Stokes band is only weakly dependent on temperature. The ratio of optical intensity of scattered light in the two spectral bands, R(T), at the scattering point can be expressed as[37]

$$R(T) = A\ (\lambda_S/\lambda_A)^4 \exp(-(h\Delta v/k\ T)) \tag{3}$$

where $\lambda_S$ and $\lambda_A$ is the wavelength of the Stokes and Anti-Stokes scattered light, h is Planck's constant, $\Delta v$ is the Raman frequency shift, k is the Bolzmann constant and T is the absolute temperature of the glass. This ratio can be measured along an optical fiber by using an optical time-domain reflectometry (OTDR) set-up as illustrated in Figure 5. This will provide a temperature profile along the length of the optical fiber.

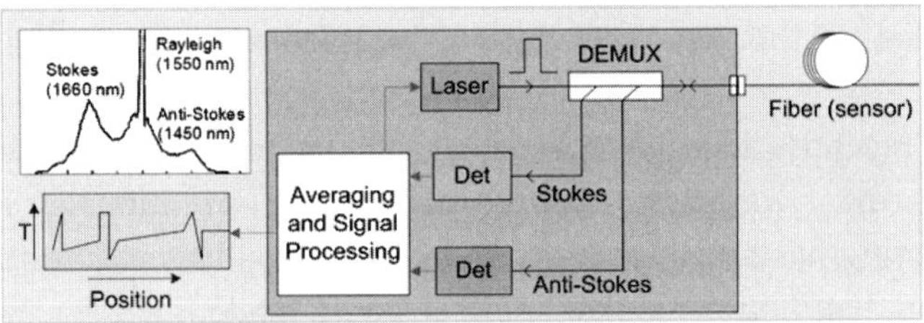


Fig. 5. Basic Raman backscatter measurement design.

However, as long as the ratio between the Anti-Stokes and Stokes intensity, defined in Eq. 3 at the scattering point, is not measured at the point of scattering in the fiber, both the Stokes and Anti-Stokes signal will be attenuated when propagating back to the detectors in the instrument. Since the two signals are at different wavelengths, they will generally suffer different loss. This problem is enhanced in multimode fibers where modal differential losses also might occur[38]. The advantage of multimode fiber, however, is a higher backscattering coefficient than experienced in a single mode fiber. Different methods are in use for compensation of the propagation losses and today temperature changes in the range of 0.1 - 1°C can be accurately monitored by this Raman backscattering technique that has become a well established method for distributed downhole temperature monitoring.

### 3.3. *Other Scattering Sensing Technologies*

Other scattering techniques that can provide distributed measurements are the Rayleigh scattering and the Brillouin scattering mechanism.

Brillouin scattering results from an interaction between propagating light and low-frequency phonons (lattice vibrations) and provides a frequency shift (~ 11 GHz at 1550 nm) much less than the Raman scattering. This frequency shift is proportional with the longitudinal strain in the optical fiber (~ 0.5 GHz/%strain)[39]. By using an optical time-domain reflectometry type of instrument to detect local variations in the Brillouin shift, the local strain distribution along the fiber can be monitored. The measurements are complicated by the fact that the Brillouin shift also is temperature dependent. The Brillouin backscattering technique targets strain monitoring in a variety of field applications[40, 41], but have so far found limited use within the oil and gas sector.

More recently distributed acoustic sensing has been developed by utilizing the Rayleigh backscattering. Conventional Rayleigh backscattering has been used for the last 30 years to measure the actual attenuation distribution along an optical single mode fiber. The technique is known as Optical Time Domain Reflectometry (OTDR). Launching a coherent laser pulse into a single-mode fiber in an OTDR arrangement, external phase perturbations can be located with some indication of the level of perturbation[42]. If acoustic signals provide these phase perturbations, one should then be able to monitor the distribution of the acoustic signals through interference effects along the fiber length. The technique, named as Distributed Acoustic Sensing (DAS), is now being introduced into the oil and gas industry[43]. A major problem, however, of such simple disturbance sensors is that they basically are incapable of determining the full acoustic field, but work is ongoing on various array configurations and processing approaches to overcome this limitation.


## 4.  Field Experiences

Technology for permanent reservoir monitoring requires extensive field testing before its acceptance by the oil companies. Successful pilot installations in real environment are normally necessary before one is allowed to make permanent installations. Pilot testing requires strong co-operation with the field owners (oil companies).

Field testing with fiber optic sensing for downhole applications started in the 1990's with strong support by BP and Shell[7] and successful field trials resulted in a growing market for optical reservoir monitoring. Field testing of fiber optic solutions for seabed seismic application started in 2006 with support from Statoil[16] and encouraging results led to increasing interest for fiber optic sensing network solutions for this application.

## 4.1. *Down hole Permanent Monitoring*

Initially, optical downhole sensing was limited to reservoir pressure monitoring and Distributed Temperature measurements (DTS). In 2005 were nearly 100 pressure/temperature gauges permanently installed and hundreds of distributed temperature monitoring applications was in use.

The first pressure gauge based on FBG wavelength sensing technology was installed in year 2000[5] and this sensing technology is today dominating for use in optical pressure gauges. Close to 500 wells worldwide has now been equipped with optical pressure gauges for permanent reservoir monitoring. These gauges are now qualified for continuous operation at 175 °C and have life expectancy of more than 15 years. Pressure resolution is better than 0.03 psi[44].

More recently advanced multifunction sensing systems are being installed combining P/T gauges with multiphase flow measurements and Distributed Temperature measurements (DTS)[45]. The multiphase flow meter is utilizing turbulent flow for primary measurements of flow velocity and speed of sound. These measurements are made by an FBG interferometric design where several sensor coils are wrapped around a pipe containing the flow. From induced strain in the fiber coils, caused by acoustic and vortical pressure disturbances, the velocity and speed of sound can be independently determined by using an advanced array processing algorithm[46]. In a 2-phase flow configuration the flow meter is deployed with at least one P/T gauge, while a 3-phase configuration requires two P/T gauges separated vertically in the well to determine the density of the fluid mixture. The achieved flow accuracy is +/- 1% for single phase (liquid and gas) and +/- 5% for multiphase[44].

It is expected that downhole multiphase flow measurements will play a major role in monitoring and optimizing well performance; especially for wells equipped with advanced well completions. Although several successful installations (mainly single and 2-phase flow)[44-46] have been made, demonstrating the ability of the flow meter to measure both phase fraction and volumetric flow rate, the use is still more limited than the simpler P/T gauge systems and Distributed Temperature Systems (DTS).

DTS requires only an optical fiber along the well to provide the temperature profile. A separate or dedicated optical multimode fiber is normally used with DTS, while the FBG sensing systems are interrogated through single mode fibers. Initially, DTS was introduced in the late 1990's for monitoring steam wells, but has now been used in a variety of applications both in onshore and offshore wells for flow assurance purposes and zonal contributions[47-51]. The use of DTS is based on the understanding of how fluids flowing in the wells are gaining or loosing heat as a result of geothermal and fluid effects and the DTS data is imported into various thermal models for further interpretation. Current DTS systems have temperature resolution close to 0.1 °C and usually a measurement interval of 1 m is used, which allows measurements up to 12 km length of fiber[52].

In- well seismic monitoring provides acoustic reservoir imaging utilizing an array of accelerometers installed down in the well. The first installation of a fiber optic multi-component system was made in a land gas well of TotalFinaElf in the southern part of France in 2002[53]. The installation successfully confirmed conventional vertical seismic profiling (VSP) where accelerometers recorded acoustic signals from a moving surface source (vibrator) as well as microseismic monitoring where high frequency events such as sound from cracking of rock are continuously recorded (passive monitoring) [15]. Time-lapse or 4D VSP imaging was shown as a powerful technique for gas water contact monitoring in a shallow formation[54]. A key issue for an in-well seismic system is the coupling to the formation and decoupling to the production tubing and both the sensor packaging and the coupling device need to be of slim design[55] due to limited space between casing and tubing. Sufficient decoupling has been demonstrated for seismic monitoring in actual flowing wells[56].

Only a handful in-well seismic monitoring systems have so far been installed with typical 5-8 three orthogonal accelerometers (3C stations)[57]. The most profiled installation is the one at the Valhall field in the North Sea in 2006[58] where both passive noise and active surface sources have been used. The recording with the active source was conducted simultaneously with the acquisition of a permanent seabed system and high-resolution images within a few hundred meters from the borehole complemented the seabed seismic images[58]. This might point on a future direction in combining borehole and seabed seismic systems.

Reliable performance requires proper installation of the sensors in complex well structures. Both the downhole equipment and the installation techniques[59, 60] have to meet strict requirements from the oil companies responsible for the well operation. In addition, a key issue has been to develop detailed installation procedures that ensure that the high reliability potential of optical sensing technology is maintained. Strong development focus has been on optical cable characteristics, optical splicing detailed procedures and optical penetrators and feedthroughs design.

The track record of optical sensors installed downhole in wells is now exceeding ten years. Initially there were some failure incidents, mainly related to fiber breakages located at splice or connection points. Further improvements in installation procedures have significantly reduced the occurrence of these failure incidents. It has recently been published that the in-well seismic systems installed in 2002 in the southern France[53] and in 2006 in the North Sea[58] is still in operation after 10 and 6 years, respectively[57]. This illustrates the potential of high reliability or long life time of in-well optical sensing technology.

Finally, some initial field testing of Distributed Acoustic Sensing (DAS) has recently been performed by utilizing already installed optical fibers for interrogation of PT gauges[61,62]. The most interesting applications seem to be monitoring of hydraulic fracturing operations[63, 64] and Vertical Seismic Profiling (VSP)[65-67], but more work has to be done to understand the qualitative character of the recorded data.

## 4.2. *Seabed Permanent Monitoring*

The seabed permanent monitoring is a much more complex and costly application than downhole monitoring and has developed from time-lapsed streamer surveys where hydrophone cables are towed in the sea above the reservoir while sound waves from towed air guns are reflected by the various sediment and rock layers and recorded by the hydrophones. Nearly thirty years on from its beginning, time-laps or 4D seismic monitoring has established itself as a valid and valuable reservoir monitoring tool[68].

Another approach to achieve improved illumination for reservoir monitoring was initiated nearly twenty years ago by laying retrievable receiver cables on the seabed surface[69]. The next step was to consider permanent installation of Ocean Bottom Cables (OBC) providing improved repeatability in time-lapsed seismic imaging. The first full-scale permanent installation was realized at the BP Valhall field in the North Sea in 2003 by using electrical cables containing conventional geophones and hydrophones[70]. At the same time some work started to evaluate how fiber optic sonar systems developed for military applications could be applied for large scale multiplexed sensor arrays for geophysical applications[71]. Another development route was to extend multiplexing capacity of existing optical downhole seismic sensing technology[55].

The first field testing of four component optical sensing solutions for permanent seabed reservoir monitoring (with three orthogonal accelerometers and one hydrophone in each sensor station) was conducted in 2006 as an outcome from two different development programs supported by Statoil[16, 17, 34] and Chevron[36], respectively.

The first pilot installation of a four component (4C) fiber optic system was conducted by Statoil at the Snorre field in the North Sea in 2008[72]. 10 km of fiber optic seismic cable was trenched into the seabed at a depth of around 1m and the sensors were interrogated through optical fibers (in a riser cable) connected to an instrument located at the platform. Seismic acquisition was performed and demonstrated real time data transfer and quality control and delivery of time-lapsed images within a few days[73]. The pilot system was delivered by Optoplan AS and contained sensors based on an FBG interferometric design[74, 75].

The breakthrough of this fiber optic sensing technology came later in 2008 when Optoplan AS was awarded a contract to supply a complete full-scale seabed permanent reservoir monitoring (PRM) system for installation at the Ekofisk field in the North Sea[76]. The system included a complete seabed network with 200 km seismic cable containing 4000 4C sensor stations, riser cable and topside instrumentation (in a fully equipped container). Installation of the system was successfully completed end of 2010[77]. Dedicated installation equipment had been developed where a rotating basket was used for the backbone cable and containers were used for the seismic cables[78]; in contrast to cable drums normally used for seabed cable installation. The cables were trenched at a depth of about 1.5 m and they covered a seabed area of 60 km$^2$.

This sensor network must be largest ever installed with fiber optic sensors. Nearly 16 000 sensor channels are multiplexed by combining wavelength, time and spatial multiplexing. During a seismic acquisition survey all sensor channels are recording data simultaneously and all recorded data are continuously transferred (SEGD format) onshore through a dedicated optical fiber telecommunication link. One shooting survey is generating a total of 350 Tbit of data for onshore processing.

By mid 2012 three repeated shooting surveys have been performed and all sensor channels are still recording data, illustrating the high reliability potential of optical fiber sensing systems for full-scale deployment. Processing of the recorded data has proven high-resolution time-lapse seismic images of the reservoir.

## 5. Conclusions and Outlook

Over the last decade fiber optic sensing technology has entered the commercial market within the oil and gas industry and the trend points to a significant volume increase for the next decade.

For down hole permanent reservoir monitoring fiber optic P/T gauges are already well established in competition with electrical gauges. DTS provides unique in-well measurements and is now widely used.

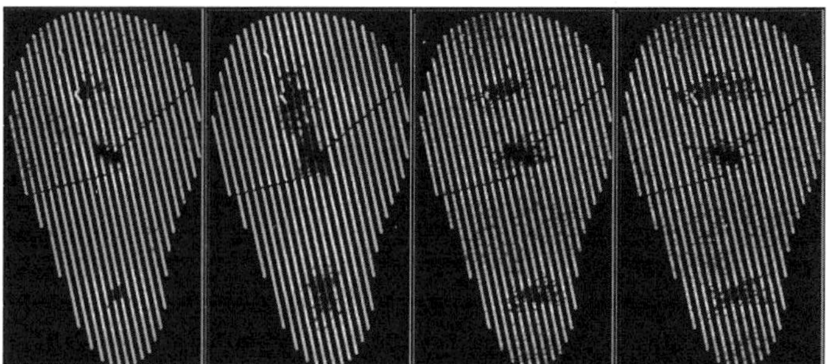

Fig. 6. Area display of seismic cables deployed with a total of 15 864 optical sensors when recording industrial noise at the field. The displays show from left to right the hydrophone and the accelerometer components x, y and z (not-rotated). The red regions are areas with high energy levels (extensive noise from field activity and supply vessels) and the blue regions are areas with less noise[79].

Currently, we are also witnessing novel fiber-optic technology solutions promising a wider range of down hole measurements[80]. In addition, we see a trend towards integrated solutions for subsea wells including wet-mate connectors for wellhead feedthroughs and marinization of interrogating units[81]. However, qualification of new downhole sensing technologies is a tedious and costly process due to the high mechanical complexity of advanced well structures and the risks involved during installation and operation of the sensing equipment. Another challenge is integrated interpretation of recorded data as input for production decisions.

Seabed permanent reservoir monitoring differentiates significantly from the downhole monitoring as it requires a cable network with extremely large number of sensing points (typically 4000–40000 individual sensor channels). It involves multi-discipline activities including costly marine installation in fields with complex subsea infrastructures. However, the increasing interest for improved imaging using wide azimuth acquisition techniques favors permanent seabed systems as these provide full azimuth and long offset data. In addition, the high reliability potential of fiber optic systems should results in no subsea maintenance, which is costly, over the life of the field (typical 15 - 25 years). The main question, however, raised regarding introduction of permanently installed seismic cables has been the uncertainty in defining the value creation or business value of such systems. The value creation will depend on remaining hydrocarbon reserves and remaining drilling expenditure and will vary from case to case. However, the Valhall system installed in the North Sea nearly ten years ago have been a commercial success and future extension of the system is being planned[82].

Several more systems are being planned - both in the North Sea and in deep water fields outside Brazil and in the Gulf of Mexico. The next fiber optic system to be installed (most likely second half of 2012) is at 1250 m water depth at the Jubarte field of Petrobras[83]. This system contains 700 4C stations with sensors using a coupler based interferometric design and will be supplied by PGS[84].

Key features of optical seismic sensing systems are (i) no electronic components or electrical power requirements at the sensing point, (ii) remote interrogation, (iii) long distance, low noise analogue transmission of multiplexed sensor data, (iv) excellent sensor performance and (v) high reliability[85]. These characteristics represent the main technical motivation for introducing fiber optic sensing technology for permanent reservoir monitoring.

# References


1. G. D. Pitt, J.A. Barnett, and G.K. Thornton, "Offshore Monitoring and Control using Optical Fibers", *Proceedings of Offshore Technology Conference,* OTC 4978 (1985).
2. B. Bjørnstad, T. Kvisterøy and M. Eriksrud, "Fiber Optic Well Monitoring System", *Proceedings of Offshore Europe Conference,* SPE 23147 (1991).
3. B. Bjørnstad and M. Eriksrud, "Fiber Optic Well Monitoring System", *Proceedings of Offshore Technology Conference,* OTC 7748 (1995).
4. A. D. Kersey, *"IEICE Trans. Electron.,* **E83-C**, No. 3, 400 (2000).
5. T. Kragas, B. Williams and G. A. Myers, "The Optic Oil Field: deployment and Application of permanent In-well Fiber Optic Sensing Systems for Production and Reservoir Monitoring", *Proceedings of SPE Annual Technical Conference,* SPE 71529 (2001).
6. M. Eriksrud, "Fiber Optic Instrumentation Systems in the Oil and Gas Industries", *14th International Conference on Optical Fiber Sensors,* Conference Technical Digest, We 3-1 (2000).

7.  E. J .Zisk, "Optical In-Well Permanent Monitoring - Initial Promise Now a Reality?, *Proceedings of Offshore Technology Conference,* OTC 17529, (2005).

8.  M. Eriksrud and H. Nakstad, "An Optical Fiber Instrumentation System for Pressure and Temperature Measurement in Oil and Gas Wells – From Design to Installation and Operation", *Conference on Lasers and Electro-Optics Europe 1998,* IEEE Conference Publications, paper CWP1 (1998).

9.  H. M. Frota and W. Destro, "Reliability Evolution of Permanent Downhole Gauges for Campos Basin Sub Sea Wells; A 10-Year Case Study", *Proceedings of SPE Annual Technical Conference, SPE 102700* (2006).

10. B. D. Carnahan, R. R.. W. Clanton, K. D. Kochler, G. O Harkins and G. Williams, "Fiber Optic Temperature Monitoring Technology", *Proceedings of SPE Western Regional Meeting,* SPE 54599 (1999).

11. G. Williams, G. Brown, W. Hawthorne, A. Hartog and P. Waite, "Distributed temperature sensing (DTS) to characterize the performance of producing oil wells", *Proceedings SPIE,* Vol. **4202**, 39 (2000).

12. T. Kragas, F.X. Bostick, C. Mayeu, D. Gysling and A. van der Speck, "Downhole Fiber-Optic Multiphase Flowmeter: Design, Operating Principle and Testing" , *Proceedings of SPE Annual Technical Conference,* SPE 77655 (2002).

13. T. Kragas, G. Myers, G. E. Pudy, E. D. Dennis and F.H. Rambow, "Downhole Fiber-Optic Multiphase Flowmeter: Field Installation", *Proceedings of SPE Annual Technical Conference,* SPE 77654 (2002).

14. S. Knudsen, G. B. Havsgård, A. Berg, H. Nakstad, T. Bostick, "Permanently installed high-resolution fiber optic 3C/4D seismic sensor system for in-well imaging and monitoring applications", *6$^{th}$ Pacific Northwest Fiber Optic Sensor Workshop, Proceedings SPIE,* Vol. **5278**, 51 (2003).

15. F. Bostick, S. Knudsen, H. Nakstad, J. Blanco and E. Mastin, "Permanently Installed Fiber Optic Multi-station 3C In-well Seismic Trial at Izaute Field", *EAGE 65$^{th}$ Conference, Extended Abstract,* F45 (2003).

16. M. Thompson, L. Amundsen, P. I. Karstad, J. Langhammer, H. Nakstad and M. Eriksrud, "Field trial of fiber-optic multi-component sensor system for application in ocean bottom seismic", *SEG 76$^{th}$ Annual Meeting, Expanded Abstracts,* 1148-1152 (2006).

17. M. Thompson, L. Amundsen, H. Nakstad, J. Langhammer, and M. Eriksrud, "Field Trial of a Fiber Optic Ocean Bottom Seismic System", *EAGE 69$^{th}$ Conference, Extended Abstract,* P192 (2007).

18. J. Frings, "Enhanced Pipeline Monitoring with Fiber Optic Sensors", *6$^{th}$ Pipeline Technology Conference,* Abstracts PTC (2011).

19. C. S. Dahl, M. Andersen, and M. Groenne, "Developments in managing Flexible Risers and Pipelines, A Suppliers Perspective", *Proceedings of Offshore Technology Conference,* OTC 21844 (2011).

20. M. Andersen, A. Berg and S. Saevik, "Development of an Optical Monitoring System for Flexible Risers", *Proceedings of Offshore Technology Conference*, OTC 13201 (2001).

21. S. R. K. Morikawa, C. S. Camerini, A. M. B. Braga and R. W. A. Llerena, "Real Time Continuous Structural Integrity Monitoring of Flexible Risers with Optical Fiber Sensors", *Proceedings of Offshore Technology Conference*, OTC 20863 (2010).

22. L. E. Hares, A. Strong and P. L. Stanc, "Midstream & Subsea Pipeline Condition Monitoring, *Proceedings of Offshore Technology Conference*, OTC 21767 (2001).

23. L. Zou and O. Sezerman, "Pipeline Corrosion Monitoring by Fiber Optic Distributed Strain and Temperature Sensors", *NACE Corrosion Conference,* Paper No 08146 (2008).

24. G. Meltz, W. W. Morey, and W. H. Glenn, *Opt. Lett.,* **14**, 823 (1989).

25. W. Morey, J. R. Dunphy, and G. Meltz, "Multiplexing fiber Bragg grating sensors," *Proceedings SPIE,* Vol. **1586**, 216 (1991).

26. A. D. Kersey, M. A. Davies, H. J. Patrick, M. LeBlanc, K. P. Koo, C. G. Askins M. A. Putnam and E. J. Friebele, *J. Lightwave Technol*, **15**, 1442 (1997).

27. T. W. MacDougall and P. E. Sanders, "Large-diameter waveguide Bragg grating components and their application in downhole oil and gas sensing," *Proceedings SPIE* Vol. 5589, 221 (2004).

28. D. Taverner, E. Dowd, J. Grunbeck, J. Dunphy, G. Daigle, R. Jones, D. Norton and T. MacDougall, "Optical temperature point-sensor array for oil and gas down-hole applications, *19$^{th}$ Int.Conf. on Optical Fiber Sensors*, Proceedings SPIE, Vol. 7004, 70040H-1 (2008).

29. O. H. Waagard, E. Rønnekleiv, S. Forbord and D. Thingbø, "Reduction of crosstalk in inline sensor arrays using inverse scattering," *19$^{th}$ Int. Conf. on Optical Fiber Sensors*, Proceedings SPIE, Vol. **7004**, 218 (2008).

30. G. A. Cranch and P. J. Nash, *J. Lightwave Technol.*, **19**, 687 (2001).

31. A. Dandridge, A. B. Tveten and A. D. Kersey,  *J. Lightwave Technol.*, **5**, 947 (1987).

32. J. T. Kringlebotn, H. Nakstad and M. Eriksrud, "Fiber optic ocean bottom seismic cable system - from innovation to commercial success", *International Conference on Optical Fiber Sensors*, Proceedings SPIE, Vol. **7503**, 277 (2009).

33. E. Rønnekleiv, O. H. Waagaard, D. Thingbø, and S. Forbord, "Suppression of Rayleigh scattering noise in a TDM multiplexed interferometric sensor system," *Proceedings of Conference on Optical Fiber Communications,* paper OMT4 (2008).

34. J. Langhammer, H. Nakstad, M. Eriksrud, A. Berg, E. Rønnekleiv, O. H. Waagaard and M. Thompson, "Fiber Optic Reservoir Monitoring: Field Trials and Results", *Proceedings of Offshore Technology Conference,* OTC 19391 (2008).

35. A. V. Strudley and P. J. Nash, "Efficient Optical Architectures for Permanent Seismic Reservoir Monitoring Arrays", *EAGE 69$^{th}$ Conference,* Extended Abstract, P191 (2007).

36. S. Maas and B. Bunn, "Fiber Optic 4C Seabed Cable for Permanent Reservoir Monitoring", *Proceedings of Offshore Technology Conference,* OTC 19445, (2008).

37. J. P. Dakin, D. J. Pratt, G. W. Bibby and J. N. Ross, *Electron. Lett.,* **21**, 569 (1985).

38. M. Eriksrud, A. R. Mickelson and S. Lauritzen, *J. Lightwave Technol,* **2**, 139 (1984).

39. H. Ohno, H. Naruse, M. Kihara and A. Shimada, *Optical Fiber Technology,* **7**, 45 (2001).

40. D. Inaudi and B. Glisic, "Reliability and field testing of distributed strain and temperature sensors", *SPIE Smart Structures and Materials Conference,* Proceedings SPIE Vol. **6166**, XX(2006).

41. D. Inaudi and B. Glisic, *J. of Pressure Vessel Technol.,* Vol. **132**, 011701-1(2010).

42. S. V. Shatalin, V. N. Treschikov and A. J. Rogers, *Applied Optics,* **37**, 5600 (1998).

43. T. Parker, S. V. Shatalin, M. Farhadiroushan, Y. I. Kamil, A. Gillies, D. Finfer and G. Estathopoulos, "Distributed Acoustic Sensing - A New Tool for Seismic Applications", *EAGE 74$^{th}$ Conference,* Extended Abstract, Y002 (2012).

44. C.G. Ferrais and L.E. Gonzales, "Phase Implementation to Real Time Well Testing Using Fiber Optical sensing Technology, San Alberto San Antonio Fields: Case Study – Part 1", *Proceedings of SPE Latin America & Caribbean Petroleum Engineering Conference,* SPE 139347 (2010).

45. B. Smith, M. Hall, A. Franklin, E. S. Johannesen and O.H. Unalmis, "Field-Wide deployment of In-Well optical Flowmeters and Pressure/Temperature Gauges at Buzzard Field", *Proceedings of SPE Intelligent Energy Conference,* SPE 112127 (2008).

46. O.H. Unalmis, E.S. Johannesen and L.W. Perry, "Evolution in Optical Downhole Multiphase Flow Measurement: Experience Translates into Enhanced Design", *Proceedings of SPE Intelligent Energy Conference,* SPE 126741 (2010).

47. O.S. Laurence and G.A. Brown, "Using Real-Time Fiber Optic Distributed Temperature Data for Optimising Reservoir Performance", *Proceedings of SPE/PS-CIM Int. Conference on Horizontal Well Technology,* SPE 65478 (2000).

48. M. Weaver, T. Kragas, J.Burman, D. Copeland, B. Phillips and R. Seagraves, "Installation and Application of Permanent Downhole Optical Pressure/ Temperature Gauges and Distributed Temperature Sensing in Producing Deepwater Wells at Marco Polo", *Proceedings of SPE Annual Technical Conference,* SPE 95798 (2005).

49. G. H. Lanier and G. Brown, "Brunei Field Trial of a Fiber Optic Distributed Temperature Sensor (DTS) System in a 100 m Open Hole Horizontal Oil Producer", *Proceedings of SPE Annual Technical Conference,* SPE 84324 (2003).

50. G. A. Brown, B. Kennedy and T. Meling "Using Fiber-Optic Distributed Temperature Measurements to Provide Real Time reservoir Surveillance Data on Wytch FaRM Field Horizontal Extended-Reach Wells", *Proceedings of SPE Annual Technical Conference,* SPE 62952(2000).

51. P. Wood, A. Simpson, B. Holland, G. Brown, G. Rogers, S. Anderson, E. Balster and M. Figueroa "Monitoring Flow and Completion Integrity of a North Sea Subsea HPHT Appraisal Well During an Extended Well Test Using Permanently Installed Fiber-Optic Temperature Sensors", *Proceedings of SPE Annual Technical Conference,* SPE 137120 (2010).

52. C. Costello, P. Sordyl, C. Hughes, M. Figueroa, E. Balster and G. Brown, "Permanent Distributed Temperature Sensing (DTS) Technology Applied in Mature Fields: A Forties Field Case Study", *Proceedings of SPE Intelligent Energy International Conference,* SPE 150197 (2012).

53. "Listening with light", *Offshore Engineer*, 26 - 30, April 2003.

54. J. Blanco, S. Knudsen, F. X. Bostick," Time-lapse VSP field test for gas reservoir monitoring using permanent fiber optic seismic system", *SEG 76$^{th}$ Annual Meeting,* Expanded Abstracts, 3447- 3451 (2006).

55. S. Knudsen, G.B. Havsgård, A. Berg, D. Thingbø, M. Eriksrud, T. Bostick, *Optical Fiber Sensors*, OSA Technical Digest, paper FB2 (2006).

56. S. Knudsen, G.B. Havsgård, A. Berg, T. Bostick, "Flow Induced Noise in Fiber-optic 3C Seismic Sensors for Permanent Tubing Conveyed Installations", *EAGE 68$^{th}$ Conference,* Extended Abstracts, D037 (2006).

57. G. Gaston and T. Bostick, "Permanent Downhole Optical Sensing Continues to Deliver value – Well Into the 4$^{th}$ Dimension", *EAGE 74$^{th}$ Conference,* Extended Abstracts, Y007 (2012).

58. B. E. Hornby, O. I. Barkved, O. J. Askim, F. X. Bostick and B. Williams, "Permanent Fiber-Optic Borehole Seismic Installation and Imaging at Valhall", *EAGE 69$^{th}$ Conference,* Extended Abstracts, P201 (2007).

59. E. Pruett, P. Abshire, B.M. Holt and J.A Sherrit, "Permanent Fiber Optic Monitoring at Northstar System Installation", *Proceedings of SPE Western Regional/AAPG Pacific Section Joint Meeting,* SPE 76748 (2002).

60. T. Kragas, E. Pruett and B. Williams, "Installation of In-well Fiber Optic Monitoring System", *Proceedings of SPE Annual Technical Conference,* SPE 77710 (2002).

61. K. Johannesen, B. Drakeley and M. Farhadiroushan, "Distributed Acoustic Sensing - A New Way of Listening to Your Well/Reservoir", *Proceedings of SPE Intelligent Energy Conference,* SPE 149602 (2012).

62. J. Mestayer, B. Cox, P. Wills, D. Kiyashchenko, J. Lopes, M. Costello, S. Bourne, G. Ugueto, R. Lupton, G. Solano, D. Hill and A. Lewis, "Field Trials of Distributed Acoustic Sensing for Geophysical Monitoring" *SEG 82$^{nd}$ Annual Meeting,* Expanded Abstracts, 4253-4257 (2011).

63. M.M. Molenaar, D.J. Hill, P. Webster, E. Fridan and B. Birch, "First Downhole Application of Distributed Acoustic Sensing (DAS) for Hydraulic Fracturing Monitoring and Diagnostics", *Proceedings of SPE Hydraulic Fracturing Technology Conference,* SPE 140561 (2011).

                    *M. Eriksrud and J.T. Kringlebotn*

64. M.M. Molenaar, E. Fridan and D.J. Hill, "Real-Time Downhole Monitoring of Hydraulic Fracturing Treatments Using Fiber-Optic Distributed Temperature and Acoustic Sensing", *Proceedings of SPE/EAGE European Unconventional Resources Technology*, SPE 152981 (2012).

65. K. N. Madsen, T. Parker and G. Gaston, "A VSP Field Trial Using Distributed Acoustic Sensing in a Producing Well in the North Sea", *EAGE 74th Conference, Extended Abstracts*, Y006 (2012).

66. C. Barberan, C. Allanic, D. Avila, J. Hy-Billiot, A. Hartog, B. Frignet and G. Lees, "Multi-offset Seismic Acqusition Using Optical Fiber Behind Tubing", *EAGE 74th Conference, Extended Abstracts*, Y003 (2012).

67. J. Mestayer, S. Grandi Karam, B. Cox, P. Wills, A. Mateeva, J. Lopez, D. Hill and A. Lewis, "Distributed Acoustic Sensing for Geophysical Monitoring", *EAGE 74th Conference, Extended Abstracts*, Y005 (2012).

68. D. G. Foster, "The BP 4-D Story: Experience Over The Last 10 Years And Current Trends", *Proceedings from the International Petroleum Technology Conference*, IPTC 11757 (2007).

69. M. Thompson and L. Amundsen, "Past and future trends. Two decades of Ocean Bottom Seismic experience in light of Moore's law", *SEG 79th Annual Meeting, Expanded Abstracts*, 3395- 3399 (2009).

70. O. I. Barkved, "Continuous seismic monitoring", *SEG 74th Annual Meeting*, Expanded Abstracts, 2537- 2540 (2004).

71. P.J. Nash, G.A. Cranch and D. J. Hill, "Large scale multiplexed fiber-optic arrays for geophysical applications", *Proceedings SPIE*, Vol. **4202**, 55 (2000).

72. A. Morton, M. Andersen and M. Thompson, "Field Trial of Focused Seismic Monitoring on the Snorre Field", *EAGE 71st Conference, Extended Abstracts*, X026 (2009).

73. A. Morton, M. Andersen, M. Thompson and T. Probert, "Focused Seismic Monitoring: From acquisition to interpretation in 48 hours", *SEG 79th Annual Meeting, Expanded Abstracts*, 3894- 3898 (2009).

74. H. Nakstad and J.T. Kringlebotn, "Realisation of a full- scale fiber optic ocean bottom seismic system", *14th Int. Conf.on Optical Fiber Sensors*, Proceedings SPIE, Vol. **7004**, 36 (2008).

75. M. Erikrud, J. Langhammer, H. Nakstad, "Towards the optical oil field", *SEG 79th Annual Meeting, Expanded Abstracts*, 3400- 3404 (2009).

76. P. G. Foldstad, "Fiber Optic Sensor Technology for Permanent Seismic Monitoring at Ekofisk", *Biennal Geophysical Seminar, Norwegian Petroleum Society (NPF)*, Expanded Abstracts, 100 (2010).

77. H. Nakstad and J. Langhammer, "Successful North Sea Fiber Optic PRM Installation", *EAGE Workshop on PRM – Using Seismic Data*, Abstracts, 10238 (2011).

78.  H. Nakstad, J. Langhammer and M. Eriksrud, "Fiber optic permanent reservoir monitoring breakthrough", *12$^{th}$ Int. Congress of the Brazilian Geophysical Society,* SO-24 (2011).

79.  H. Nakstad, J. Langhammer and M. Eriksrud, "Permanent Reservoir Monitoring Technology Breakthrough in the North Sea", *EAGE 73$^{rd}$ Conference,* Extended Abstracts, I044 (2011)

80.  J. M. V. A. Koelman, J. L. Lopez, J. H. H. M. Potters, "Optical Fibers: The Neurons For Future Intelligent Wells", *SPE Intelligent Energy Conference,* SPE 150203 (2012).

81.  B. K. Drakeley and S. Omdal, "Fiber Optics Sensing Systems for Subsea Applications – Sensing Capabilities, Applications, and the Challenges Being Faced in order to Provide reliable Transmission of Data for Online Reservoir Management", *SPE Intelligent Energy Conference, SPE 112205* (2008)

82.  O. I. Barkved, *Seismic Surveillance for Reservoir Delivery*, EAGE Publications, Chapter 10 (2012).

83.  E. A. Thedy, W. L. R. Filho and P. R. S. Johann and S. Seth, "Jubarte – PRM In Ultradeep Water and with an Innovative Optics System", *EAGE Workshop on PRM - Using Seismic Data,* Abstracts, 10251 (2011).

84.  J. B. Bunn, S. J. Maas and S. N. Seth, "The Underlying Technology Powering Two Game Changing Fiber-optic Seismic Sensing Systems- 1M Channel Land and Deepwater PRM", *EAGE 74$^{th}$ Conference,* Extended Abstracts, Y008 (2012).

85.  M. Eriksrud, "Towards the optical seismic era in reservoir monitoring", *First Break,* Vol. **28**,105 (2010).

# PHOTONIC SENSORS FOR
# FOOD QUALITY AND SAFETY ASSESSMENT


Anna Grazia Mignani[1]* and Raffaello Prugger[2]

*[1]CNR – Istituto di Fisica Applicata "Nello Carrara"*
*Via Madonna del Piano, 10 – 50019 Sesto Fiorentino (FI), Italy*
**E-mail: a.g.mignani@ifac.cnr.it*

*[2]Tecnoalimenti S.C.p.A.*
*Via G. Fara, 39 – 20124 Milano, Italy*



The most common spectroscopic techniques used for assessing quality and safety of intact food are reviewed. These make it possible for food to be checked in real time, without the need for intermediate chemistry, thus providing a "green" tool for analytics.


## 1. The Needs of the Food Industry

The food and drink industry is the largest industry in Europe with a turnover of 913 billion euros, followed by the automotive and chemical industries. With its 308,000 enterprises, this industry employs a share of 13.5% of the total workforce, corresponding to 4.3 million people; 63% are employed in small and medium industries. The said industry processes about 70% of the produce of primary production.

The food industry is only one step in the food supply chain, which is a complex sequence of various players linked by the production flow, from the field and ingredients, and extends to the finished product purchased, and then stored in the consumer's refrigerator. Food-chain players have a very different technological level, one that needs to be taken into account: e.g. the small farmer versus the large dairy industry. Therefore, any device to be distributed to food-chain players needs to be sufficiently user-friendly in order to be accepted by certain players.

The food industry is under pressure to assure compliance with food safety regulations, market requirements, and sustainability concerns. This implies a wide use of laboratory analyses and sensors. Recently, consumers have been increasingly expecting the food chain to demonstrate the quality and safety of its products. Globally speaking, fraud and counterfeiting of food origins and branded foods are increasingly economic and safety issues.

However, not only the amount of measurement data is important, but also the quality of the said data. First of all, the food industry does not require extreme precision in the measurements, but cares a lot about the robustness of the data. In fact, food regulations and consumers do not care about infinitesimal residues. Furthermore, much money would be saved if measurements could be provided at the right moment during the process without any delays and during the appropriate stage of the production process: for instance, rapid systems for avoiding the stocking of products until laboratory results are ready, which sometimes takes several hours or even days. Field contaminants can be better determined in raw materials at the very first stages of the food chain, rather than in the finished product, after much time and precious resources have already been spent during the manufacturing. Although it is of key importance, the generating of measurement data is not enough. They also need to be managed efficiently, in order to contribute to the competitiveness. Moreover, data need to be managed and accessible to all players throughout the supply chain.

## 2. Working Principles of Photonic Sensors for Food Applications

Photonics has been recognized by the European Commission as being a "Key-Enabling Technology", that is, a driving force capable of supporting, improving and boosting the quality of our lives. Photonic technologies that were initially developed for the telecommunications sector have subsequently generated a myriad of compact light sources, detectors, micro-spectrometers, fiber optics, and micro-photonic components that can be used effectively for the implementation of innovative sensors. The field of food control has exploited these opportunities, above all in order to respond to requests for compact, robust, and low-cost instrumentation for use in online practices by operators with modest technical skills.

Two approaches are possible for implementing photonic sensors:

- *Direct measurements*, when only light is used as a probe directly shone on the sample, and the information is given solely by the interaction between food and light[1].
- *Chemically-mediated measurements*, when an intermediate chemical transducer is used, and the information is indirectly obtained by optically interrogating the transducer [2,3].

Both approaches are suitable for food applications. Although the preference of food industry is for direct measurements, since these allow no-contact, rapid and online controls, chemically-mediated sensors are also attractive, especially when shaped as adhesive spots coupled with packaging.

This chapter presents examples of sensors that exploit spectroscopy for direct sensing in the case of the most common food commodities (milk, meat, oil, cereals, fruits and vegetables, etc), as well as sensor spots for the non-destructive testing of freshness in packed food or for indicating the correct functionality of the packaging itself.


## 3. Spectroscopy-based Sensors - Towards "Green Analytics"

Spectroscopy, which studies the interaction between light and matter in order to obtain information on chemical composition, is used for direct measurements. When light impinges on a sample, it can either be absorbed or scattered, resulting in an attenuation of the transmitted light. Both these effects provide information about the nature of the sample. When matter contains complex molecules or has a crystalline structure, quantized vibrational modes may be excited: in this case, a part of the absorbed and scattered light is re-emitted at wavelengths that are different from the original one. In the former case, we have fluorescence; in the latter, Raman scattering.

Optical spectroscopy exploits the entire 200-2500 nm wavelength range, including the ultraviolet (UV: 200-400 nm), visible (Vis: 400-780 nm), and near-infrared (NIR: 780-2500 nm) bands. It enables a rapid and non-destructive analysis without making use of reagents or chemical treatments, thus avoiding the problem of waste discharge. Indeed, spectroscopy is an ideal candidate for food analyses, especially as a part of automation in the food-processing industry [4]. Moreover, since it does not imply environmental side effects, spectroscopy is an ideal candidate for "green analytics"[5,6,7].

Indeed, food matrices are complex mixtures of many organic constituents, and, consequently, optical spectra appear as convoluted spectra, which may be difficult to interpret. In practice, optical spectra can be considered to be a fingerprint from which to extract multiple information regarding adulteration and contamination[8]. Raman spectra show sharp bands that identify the molecular composition, and can immediately lead to the detection of multiple components and their quantification, provided that a calibration is available. On the contrary, absorption and fluorescence spectra show broad peaks resulting from the convolution of the many overlapping bands, which are poorly resolved for the purpose of a multicomponent analysis. A straightforward multicomponent analysis from optical spectra can be achieved by using multivariate chemometric techniques[9,10,11] as summarized in Figure 1.

What is needed is a library of representative spectra and relative analytical data with which the spectrum of a test sample can be matched.

1) First, a data dimensionality reduction is carried out, which usually leads to a scatter plot in which samples are clustered according to the similarity of their spectra. This makes possible a preliminary inspection of data structure and the detection of which parameters are more likely to be correlated with spectroscopic data. Two of the most popular methods for data dimensionality reduction and classification are Principal Component Analysis (PCA), and Linear Discriminant Analysis (LDA)[12,13,14].

2) A more specific analytical tool is then chosen depending on the type of variable that has to be predicted, quantitatively or qualitatively. A "Calibration Matrix" is created, and the constituent of interest may be calculated from it by means of a linear combination of spectroscopic data. The calibration equation has associated statistics which define the closeness of the actual and predicted values. A scatter plot is usually created for the purpose of detecting any deviating data. Ideally, the

                    *A.G. Mignani and R. Prugger*

scatter plot should contain data points distributed evenly along a straight line with a narrow confidence limit.

3) A validation procedure is then applied in order to test the effectiveness of the calibration method. While the data dimensionality reduction is usually capable of identifying similarities among products, the correlation to quality indicators always requires the further steps of calibration and validation. A popular methods for calibration/validation models is the Partial Least Square (PLS) regression[15].

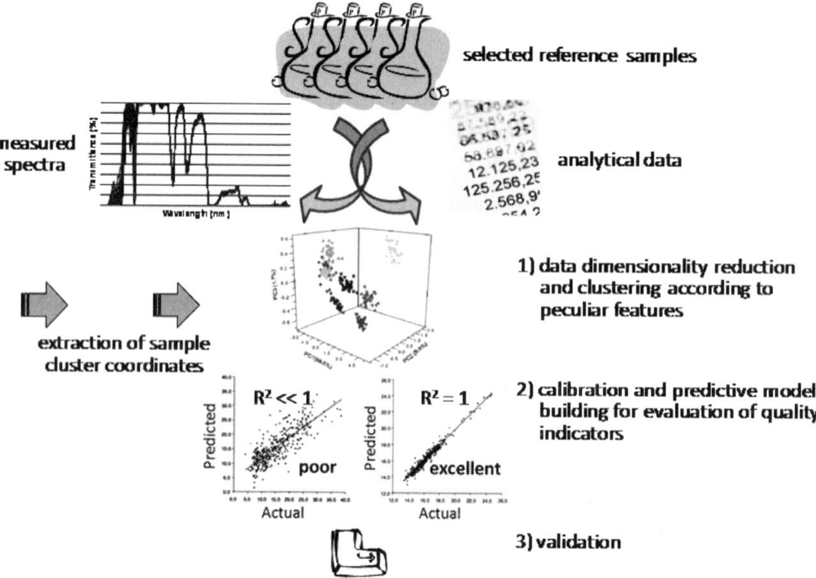


Fig. 1. Sequence of steps for multicomponent analysis using optical spectroscopy and chemometrics.

## 3.1. *Reflection/Absorption-based Sensors*

Absorption spectroscopy is used to measure liquid samples, while gels, powders and solids are analyzed by means of reflection spectroscopy. The Vis range reveals the presence of dyes and pigments[16]. The NIR range shows wide bands arising from overlapping absorptions that correspond to overtones and combinations of vibrational modes related to C-H, O-H, and N-H bonds. Consequently, it is highly informative as regards the content of water, proteins, fats, sugar, and cellulose.

For more than 30 years, Vis and NIR spectroscopy has been widely used for the online assessment of food quality during production[17,18,19].

Chemometrics enhances the effectiveness of spectroscopy, and cumbersome laboratory spectrophotometers are starting to be replaced by portable and handheld analyzers. Typically, the industrial processing of the most common foodstuffs such as meat, cereals, fruits, vegetables, oils, and dairy products, can be controlled in real time in order to verify whether the quality of the product meets with the standards required. Some examples of reflection/absorption-based sensors, complemented by chemometric processing, follows. In all cases, optical spectra were processed for data dimensionality reduction as well as for the extraction of typical features from which to create a prediction model of the parameter sought. The effectiveness of the model was then confirmed by a validation step. In all examples mentioned below, the regression coefficient, $R^2$, which correlates real and predicted values, was higher than 0.9.

*Meat.* All meats are very susceptible to spoilage, and are also expensive as compared to other food types. Hence, checking spoilage as well as the composition of the meat during meat processing is an absolute priority. Reflectance spectroscopy in the 600-1000 nm band has demonstrated its effectiveness to rapidly detect and quantify the microbial spoilage in chicken meat[20,21], while reflectance in the Vis band only has been used to control the premature browning of beef while being cooked in industrial ovens[22]. Other quality assessments of beef have involved the monitoring of quality indicators, such as fat, pH, water, and protein content. For this purpose, reflectance in the Vis/NIR bands has been successfully used for different types of meat[23,24].

*Cereals.* Compact spectrometers are particularly attractive for installation on harvesting machines for the online detection of quality indicators, such as dry matter, proteins, and starch[25]. Cereals are frequently contaminated by aflatoxins. Their detection using fluorimetric methods is described in the next section. NIR spectroscopy has also proved to be an effective technology for detecting B1-aflatoxin in maize and barley[26].

*Fruits and vegetables.* Vision-based machines based on spectroscopy are widely available commercially for the mechanical sorting of fruits and vegetables. Moreover, Vis/NIR spectroscopy or NIR alone can be used for non-destructive quality testing[27]. Typical examples are the evaluation of acidity and Brix[28], total soluble solids[29] and the sugar content of fruits[30,31,32].

*Oils.* Although several vegetal oils are used for cooking and dressings, the most expensive oil and the one most appreciated worldwide is Extra Virgin Olive Oil (EVOO), which is the only vegetal oil that is consumed as it is - i.e. freshly extracted from the fruit. Frequently, bottles of EVOO bear labels that certify the area of origin, and the oils themselves are characterized by distinctive tastes and exceptional nutritional properties. The main concern regarding EVOO is the risk of its being adulterated by the addition of cheaper oils in order to increase profits. In addition to an economic burden, EVOO adulteration is detrimental if consumers react by buying other cooking fats or dressings, convinced that EVOO cannot be trusted. The negative implications involving consumer confidence are even worse than the economic ones. Lastly, EVOO protection measures also imply the conservation of the production area, as far as landscape, tourism, and job preservation are concerned. Vis and NIR absorption spectroscopy, always assisted by chemometric data processing, provides an excellent technique for authenticating the area of origin[33], and for detecting and quantifying various types of adulterants[34,35]. For example, Figure 2-left shows the spectra of a sample of high-quality EVOO (O1), the spectra of lower quality EVOOs (F1, F2, and F5), as well as the spectra of samples with 50% adulteration. Figure 2-right shows the classification map obtained using a chemometric processing of spectroscopic data, which was able to identify adulteration and adulterant type[36]. Once the adulteration has been identified, a subsequent regression demonstrates the possibility of detecting an adulteration percentage as low as even 5%.

*Dairy products.* Colorimetry can be used for preliminary milk screening, since an abnormal color could be caused by colostrum or by blood due to udder infections. Apart from the hygienic aspects, the commercial value of milk is related to three major constituents: fat, protein, and lactose. Absorption spectroscopy carried out in the 600-

1000 nm band, and trained using a chemometric processing, is an ideal technique for measuring these constituents by means of a single

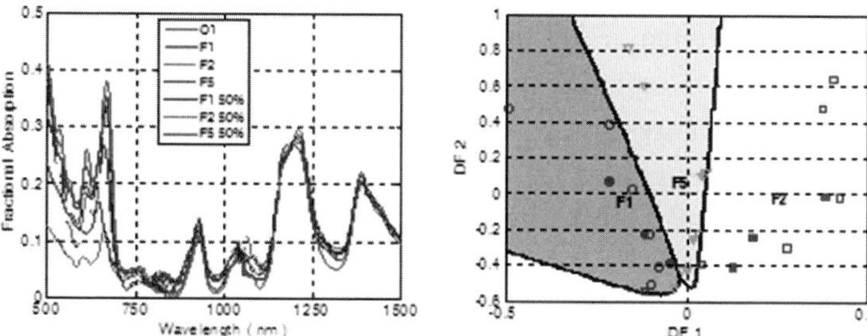

Fig. 2. Left: absorption spectra in the Vis/NIR of high quality extra virgin olive oil (O1), of other olive oils used as adulterants (F1, F2, and F5), and of their mixtures. Right: classification map obtained using a chemometric processing of spectroscopic data for distinguishing adulteration and adulterant type.

spectroscopic measurement[37,38]. In fact, it has been used in automatic milking systems for the real-time and online assessment of milk quality during milking[39]. The same spectroscopic band is also useful for the detection of adulterants in milk, typically vegetal oils, urea, and caustic soda[40].

## 3.2. *Fluorescence-based Sensors*

Intact food is intrinsically fluorescent, since many constituents such as chlorophylls, vitamins, proteins, aminoacids, pigments, etc., are naturally fluorescent[41]. The FoodFluor Database available at http://www.models.kvl.dk/foodfluor provides a snapshot of fluorescence spectra of the most common foodstuffs. A typical example is that of the monitoring of fruit ripening and maturity[42], the oxidation of dairy products[43], egg freshness[44], and the quality of beers[45] and distillates[46].

In addition to assessing quality indicators, fluorescence techniques are widely used for detecting certain important harmful factors. The monitoring of meat spoilage is a typical example[47,48]. Feces are the primary source of pathogenic organisms that cause food contamination. The automatic inspection of wide areas can be achieved with the use of hyperspectal fluorescence imaging, which advantageously combines reflectance imaging with fluorescence spectroscopy[49]. A UV or laser source is used for illumination and an intensified camera as detector. The camera is equipped with several filters for acquiring images over many wavebands. In practice, a 3D image-cube is obtained, in which the x-y dimension provides spatial information, while the third dimension represents the fluorescence spectral information. The automatic inspection of fruits, poultry skin, and pesticide residues in vegetables represents a successful food application using hyperspectral fluorescence spectroscopy[50].

Mycotoxins[51], especially aflatoxins and ochratoxins, are other toxic substances that exhibit a native fluorescence that emerges from the intrinsic fluorescence of the food. These can grow on a wide range of agricultural commodities in the fields, as well as during post-harvest operations and storage conditions, and belong to human carcinogen-group 1[52,53,54]. Conventional analytical techniques for their detection are based on extraction and subsequent HPLC or ELISA measurements[55].

Following the vision of "green analytics", fluorescence techniques have been successfully used for detecting B1-aflatoxin in alcoholic beverages such as wines and beer, and M1-aflatoxin in milk.

B1-aflatoxin in beer originates from contaminated cereals; in wines, it originates in fungi which grow on grapes in the vineyards, and is carried over into the wine when the grapes are pressed. M1-aflatoxin is the "milk" toxin: it is the main metabolite of B1-aflatoxin, and is found in the milk of lactating animals after they have ingested feed contaminated by B1-aflatoxin. These mycotoxins, which are heat-stable, remain in food indefinitely during the production process.

The European Union has a regulation regarding aflatoxins in certain food products[56]. For wine or beer, the only maximum allowable limit is the one for B1-aflatoxin in barley and malt. Fluorescence of wines excited at $\lambda$=365 nm, and detected in the 400-600 nm band makes it possible to detect B1-aflatoxin with a limit of detection of 31 and 43 ppb for white and rosé wine, respectively, and 62 ppb for beer[57].

Regulations for milk tolerate M1-aflatoxin up to 0.05 µg/Kg (50 ppt).

A novel approach using a low-cost fiber optic setup for fluorescence spectroscopy has demonstrated its effectiveness in detecting the threshold safety limit. As shown in Figure 3-top, a UV-LED emitting at

$\lambda$=360 nm was used for excitation, and a compact spectrometer for detection. A bifurcated fiber optic bundle was coupled to a flow-cell for 90°-fluorescence measurements. A simple pre-treatment of the milk sample was used to achieve a clarified sample in which the sole fluorescence effect was caused by the M1-aflatoxin concentration. Figure 3-bottom/left shows the fluorescence spectra at different M1-aflatoxin concentrations. A multivariate data processing of these data then made it possible to build a classification map in which the milk samples appeared noticeably clustered, depending on the M1-aflatoxin concentration[58].

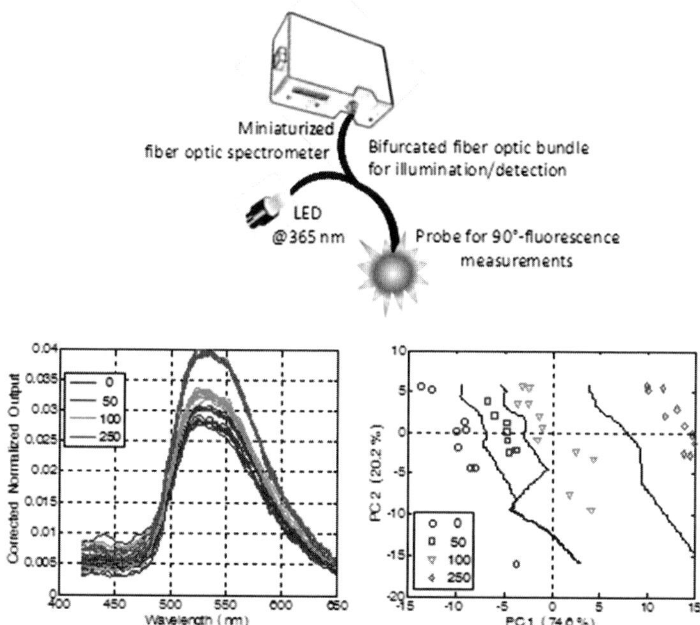


Fig. 3. Top: Diagram of fluorescence spectroscopy setup for M1-aflatoxin detection. Bottom-left: Fluorescence spectra of milk contaminated by different M1-aflatoxin concentrations. Bottom-right: classification map for identifying milk samples that have different M1-aflatoxin contaminations.

As shown in Figure 3-bottom/right, each point of the map represents a milk sample with different levels of contamination: contaminated samples at the legal limit of 50 ppt are clearly discernible from aflatoxin-free samples, and the clusters of contaminated samples are situated closer to the right side of the map. A discriminant analysis by means of the K-nearest neighbor method[59] made it possible to draw the decision boundaries that define the region of the map relative to the M1-aflatoxin concentration under consideration. The implementation of this instrument as a compact and hand-held device could be exploited by operators with minimal skills, enabling the prompt and timely discarding of contaminated stocks before their inclusion in the production chain.


### 3.3. *Raman-based Sensors*

The Raman effect occurs when monochromatic light impinges on a molecule interacting with its electron cloud. In addition to the incident light scattered with an unchanged wavelength (elastic or Rayleigh scattering), a small fraction of the order of $10^{-6}$ of the incident light is inelastically scattered, leaving the sample with longer or shorter wavelengths as compared to the incident one. Usually, the shift to longer wavelengths is considered, since it is more intense. The wavelength shift does not depend on the illumination wavelength, but depends only on the molecular vibrational levels. Hence, the scattered spectrum is a unique expression of the molecular structure, and is considered as a molecular fingerprint.

The Raman spectrum is usually expressed as the intensity versus frequency difference with respect to the illumination (wavenumbers, cm $^{-1}$). On this scale, the band positions are located at frequencies that correspond to the energy levels of different functional group vibrations. Therefore, although it is governed by different rules, the Raman spectrum appears to be complementary to the infrared absorption spectrum, and provides similar information. In some cases, Raman spectroscopy offers advantages with respect to infrared spectroscopy. In fact, it has a confocal nature, which enables sample profiling through many plastic or glass types of packaging. Moreover, water has a weak Raman signal, so that samples can be analyzed without the dehydration necessary in infrared spectroscopy[60,61].

Raman spectroscopy is used in the processing industry for detecting quality indicators, such as the glucose content of beverages[62], carotenoids in pasta or fruit[63,64], antioxidants in oils[65], the content of robusta in coffee[66], and milk powder composition[67,68]. For safety applications, Raman spectroscopy is now considered to be the quickest and most reliable analytical tool for the detection of melamine in milk powder[69,70].

Another example of safety application is the detection meat and fish spoilage[71,72], also thanks to the use of hand-held systems[73]. Since proteins and lipids have clear Raman spectra, their changes during muscle-food deterioration is an indication of changes in quality and of spoilage[74,75].

Unfortunately, some foods are intrinsically fluorescent, and a strong background fluorescence tends to overwhelm the Raman signal when visible laser lights are used. Although some background suppression algorithms can be applied[76,77], excitation using NIR wavelengths is more suitable for detecting clearer Raman spectra. When the 785 nm wavelength is sufficient for avoiding fluorescence excitation, convenient Si-based detectors can be used. However, in many cases the longer 1064 nm is needed, which implies the use of more expensive InGaAs detectors. The influence of excitation wavelengths can be seen in Figure 4. Figure 4-left shows the Raman spectra excited at 785 nm of several types of honey, while Figure 4-right shows the Raman spectra of honey and other sugars, excited at 1064 nm, where the Raman peaks clearly emerge from the fluorescence background noise.

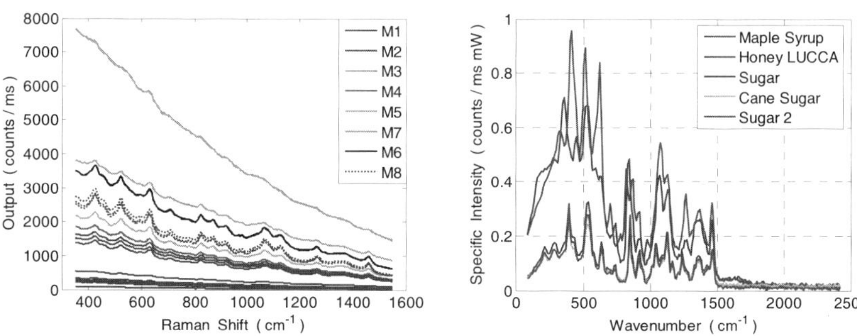


Fig. 4. Raman spectra of honeys and sugars. Left: honeys of different botanic origins, excited at 785 nm. Right: honey, maple syrup, and sugars excited at 1064 nm.

## 4. Colorimetric Spots for non-Destructive Testing of Packed Food

Modern food packaging should serve not only for wrapping food and for its physical protection. It should also have "active" features, such as indications of traceability, the capability of prolonging the shelf-life, the incorporation of anti-microbials, spoilage retarders, oxygen scavengers, and a non-destructive indication of food quality and safety. While Radio-Frequency Identification (RFID) tags are already commonly used for traceability and authentication purposes, the so-called "smart packaging" of the near future will have to include sensors[78,79].

Early sensors for food packaging were, in practice, chemically-mediated sensors: a chemical transducer was fixed to the packaging so as to interact with the food, and an external optical unit was used for the transducer interrogation. For example, the integrity of modified atmosphere packaging was monitored by using a label made of a ruthenium-based complex entrapped in a sol-gel matrix. The oxygen content inside the packaging was obtained by measuring the fluorescence quenching of the spot[80]. An array of diverse metalloporphyrin-based sensors was experimented as the precursor of a smart cap for artificial olfaction, for perceiving in non-destructive manner the onset of rancidity in bottled vegetal oils[81].

In order to provide customers with labels for a self-assessment, the most modern trend for so-called "sensorized" packaging is the development of color-based indicators, the color of which indicates the correct conservation and food quality. Time-temperature indicators are now available, based on thermochromic inks that inform the consumer when a liquid is cold enough to drink[82]. For a better indication of the cold-chain respect, dyes with selected melting points can indicate when the temperature exceeds a certain critical level, as shown in Figure 5-left[83, 84].

Other labels for smart packaging are required in order to indicate freshness. Since spoiled fish causes an increase in pH in the packaging headspace, due to the release of amines, fish freshness can be monitored simply using an indicator dye that makes pH-dependent color changes[85,86]. Another pH indicator has been used to indicate the ripeness of packaged fruit: the ripeSense® sensor changes color by reacting to the aroma released by fruit as it ripens, which is related to the atmosphere of the inside packaging. The sensor is initially red, graduates to orange, and finally yellow. By matching the color of the sensor with their eating preferences, customers can thus choose fruit at the stage of ripeness that they prefer, as shown in Figure 5-right[87].

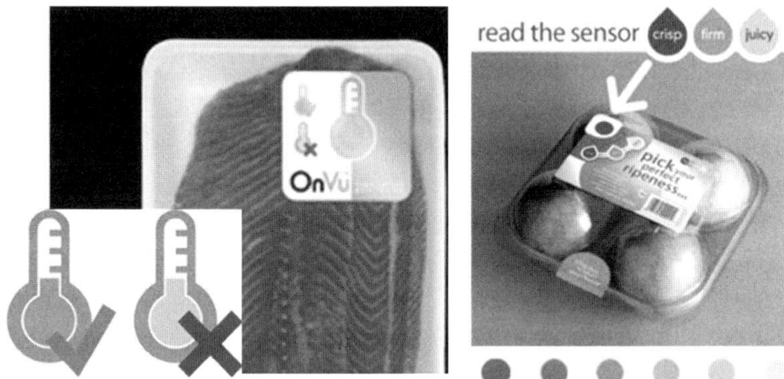


Fig. 5. Labels for smart packaging; Left: temperature check; right: fruit ripeness.

## 5.  Conclusions and Outlook

The practical applications of the spectroscopic technique to food quality and safety assessment have been reviewed, with an indication of the most successful and up-to-date examples. As far as industrial food production is concerned, conventional laboratory instruments for spectroscopy have been replaced by portable or hand-held devices. The growing availability of bright LEDs and miniaturized spectrometers will further enhance the implementation of new and more effective instruments for online measurements, for safer products and better marketing actions. From the consumer's point of view, a very modern approach is constituted by personal involvement in quality and safety assessment. "Sensorized" packagings and smart labels meet these emerging requirements.

A more visionary concept involves the use of smartphones. These can be used passively for reading bar or QR codes and for gathering additional product information from Internet-connected data warehouses, such as in the case of extended labels, traceability or product history indications. They can also be connected through the socket to small spectroscopic devices that make use of the App-functionalities of smartphones. The most futuristic use of a smartphone could be for direct spectroscopy, since the smartphone can be viewed as an embedded spectrometer, using the white LED for lighting and the CCD camera for detection[88,89]. The consumer will be able to rapid quality and safety checks simply by taking pictures of foods and using suitable Apps.

Another application of spectroscopy for food is expected to be in detecting nanomaterials. Actually, nanomaterials are increasingly being used in the production of healthier and nutraceutic foods, as well as for innovative pouches that prolong the food shelf-life. In addition to the prospective beneficial effects promised by nanomaterials, lively discussions and studies are springing up on their possible toxicological impact[90]. Regulations concerning the use and type of nanomaterials are expected soon, and, consequently, analytical techniques for their detection and quantification will be necessary[91,92]. Raman spectroscopy, which already proved effectiveness for characterizing nanomaterials[93], is now expected to be even more exploited for food applications.

## References


1. C. H. Lee and T. W. Hill. *Light-Matter Interaction* (Wiley online library, DOI : 10.1002/9783527619016. index, 2011).
2. X. Wang and O. S. Wolfbeis, *Anal. Chem.*, in press (2013).
3. O. S. Wolfbeis, *Anal. Chem.* **80**, 4269 (2008).
4. P. K. Ghosh and D. S. Jayas, *Sens. & Instrum Food Qual.* **3**, 3 (2009).
5. M. Koel and M. Kaljurand. *Green Analytical Chemistry* (Royal Society of Chemistry, Cambridge 2010).
6. S. Armenta, S. Garrigues and M. de la Guardia, *Trends Anal. Chem.* **27**, 497 (2008).
7. J. Moros, S. Garrigues and M. de la Guardia, *Trends Anal. Chem.* **29**, 578 (2010).
8. D. I. Ellis, V. L. Brewster, W. B. Dunn, J. W. Allwood, A. P. Golovanon and R. Goodacre, *Chem. Soc. Rev.* **41**, 5706 (2012).
9. L. A. Berrueta, S. M. Alonso-Salces and K. Héberger, *J. Chromatogr. A* **1158**, 196 (2007).
10. D. Toher, G. Downey and T. Brendan Murphy, *Chem. Intell. Lab. Sys.* **89**, 102 (2007).
11. R. Karoui, G. Mazerolles and E. Dufour, *Int. Dairy J.* **13**, 607 (2003).

12. J. E. Jackson. *A User's Guide to Principal Components* (John Wiley & Sons Inc., Hoboken NJ 2003).

13. T. Naes, T. Isaksson, T. Fearn and T. Davies. *A User-Friendly Guide to Multivariate Calibration and Classification* (NIR Publications, Chichester 2002).

14. M. J. Adams. *Chemometrics in Analytical Spectroscopy* (Royal Society of Chemistry, Cambridge 1995).

15. S. Wold, M. Sjostrom and L. Erikkson, *Chem. Intel. Lab. Sys.* **58**, 109 (2001).

16. R. E. Wrolstad, T. E. Acree, E. A. Decker, M. H. Penner, D. S. Reid, S. J. Schwartz, C.F. Shoemaker and P. Sporns. *Handbook of Food Analytical Chemistry, Pigments, Colorants, Flovours, Texture, and Bioactive Food Components (J.* Wiley & Sons, Hoboken 2005).

17. H. Büning-Pfaue, *Food Chem* **82**, 107 (2003).

18. H. Huang, H. Yu, H. Xu and Y. Ying, *J. Food Eng.* **87**, 303 (2008).

19. S. N. Jha. *Nondestructive Evaluation of Food Quality* (Springer-Verlag, Berlin, Germany 2010).

20. M. Lin, M. Al-Holy, M. Mousavi-Hesary, H. Al-Qadiri, A.G. Cavinato and B.A. Rasco, *Lett. Appl. Microb* **39**, 148 (2004).

21. D. Alexandrakis, G. Downey and A. G. M. Scannel, *Sens. Instrum. Food Qual.* **5**, 57 (2011).

22. M. O'Farrell, E. Lewis, C. Flanagan, W.B. Lyons and N. Jackman, *IEEE Sens J.* **5**, 1407 (2005).

23. B. Savenije, G. H. Geesink, J. G. P. van der Palen and G. Hemke, *Meat Sci.* **73**, 181 (2006).

24. N. Prieto, S. Andrés, F.J. Giráldez, A.R. Mantecón and P. Lavín, *Meat Sci.* **74**, 487 (2006).

25. J. M. Montes, H. F. Utz, W. Schipprack, B. Kusterer, J. Muminovic, C. Paul and A. E. Melchinger, *Plant Breeding* **125**, 591 (2006).

26. V. Fernández-Ibañez, A. Soldado, A. Martínez- Fernández and B. de la Roza-Delgado, *Food Chem.* **113**, 629 (2009).

27. D. Cozzolino, W. U. Cynkar, N. Shah and P. Smith, *Food Res. Int.* **44**, 1838 (2011).

28. A. Hernández-Gómez, Y. He and A. García Pereira, *J. Food Eng.* **77**, 313 (2006).

29. M. Golic and K. B. Walsh, *Anal. Chim. Acta* **555**, 286 (2006).

30. T. Temma, K. Hanamatsu and F. Shinoki, *Opt. Rev.* **9**, 40 (2002).

31. P. P Subedi and K. B. Walsh, *Photoharvest Biol. Tech.* **62**, 238 (2011).

32. H. Lin and Y.Ying, *Sens. & Instrum Food Qual.* **3**, 130 (2009).

33. O. Galtier, N. Dupuy, Y. Le Dréau, D. Olliver, C. Pinatel, J. Kister and J. Artaud, *Anal. Chim. Acta.* **595**, 136 (2007).

34. G. Downey, P. McIntyre and A. N. Davies, *J. Agric. Food Chem.* **50**, 5520 (2002).

35. J. S. Torrecilla, E. Rojo, J. C. Dominguez and F. Rodrìguez, *J. Agric. Food Chem.* **58**, 1679 (2010).

36. A. G. Mignani, L. Ciaccheri, H. Ottevaere, H. Thienpont, L. Conte, M. Marega, A. Cichelli, C. Attilio and A. Cimato, *Anal. Bioanal. Chem.* **399**, 1315 (2011).

37. H. M. Al-Qadiri, M. Lin, M. A. Al-Holy, A. G. Cavinato and B. A. Rasco, *J. Dairy Sci.* **91**, 950 (2008).

38. B. Aernouts, E. Polshin, J. Lammertyn and W. Saeys, *J. Dairy Sci.* **94**, 5315 (2011).

39. M. Kawasaki, S. Kawamura, M. Tsukahara, S. Morita, M. Komiya and M. Natsuga, *Comp. Electron. Agri.* **63**, 22 (2008).

40. S. N. Jha and T. Matsuoka, *J. Food Sci. Technol.* **41**, 313 (2004).

41. J. Christensen, L. Nørgaard, R. Bro and S. B. Engelsen, *Chem. Rev.* **106**, 1979 (2006).

42. D. Moshou, S. Wahlen, R. Strasser, A. Schenk, J. De Baerdemaeker and H. Ramon, *Biosys Eng.* **91**, 163 (2005).

43. R. Karoui, E. Dufour and J. De Baerdemaeker, *Food Chem.* **101**, 1305 (2007).

44. R. Karoui, R. Schoonheydt, E. Decuypere, B. Nicolaï and J. De Baerdemaeker, *Anal Chim. Acta* **582**, 83 (2007).

45. E. Sikorska, A. Gliszczyńska-Świgło, M. Insińska-Rak, I. Khmelinskii, D. De Keukeleire and M. Sikorski, *Anal Chim. Acta* **613**, 207 (2008).

46. J. Sádecká, J. Tóthová and P. Májek, *Food Chem.* **117**, 491 (2009).

47. A. Aït-Kaddour, T. Boubellouta and I. Chevallier, *Meat Sci.* **88**, 675 (2011).

48. S. Clerjon, F. Peyrin and J. Lepetit, *Meat Sci.* **88**, 28 (2011).

49. M. S. Kim, A.M. Lefcourt and Y.R. Chen, *Appl. Opt.* **42**, 3927 (2003).

50. R. Zhang, Y. Ying, X. Rao and J. Li, *J. Sci. Food Agric.* **92**, 2397 (2012).

51. N. W. Turner, S. Subrahmanyam and S. A. Piletsky, *Anal Chim. Acta* **632**, 168 (2009).

52. N. Magan and M. Olsen. *Mycotoxins in Food: Detection and Control* (CRC Press LLC, Boca Raton FL 2004).

53. M. Weidenborner. *Encyclopedia of Food Mycotoxins* (Springer-Verlag, Berlin, Germany 2001).

54. http://www.mycotoxins.org/.

55. G.S. Shephard, *Anal. Bioanal. Chem.* **395**, 1215 (2009).

56. http://eur-lex.europa.eu/ LexUriServ/LexUriServ.do?uri=OJ:L:2003:326:0012:0015:EN:PDF

57. C. Rasch, M. Böttcher and M. Kumke, *Anal. Bioanal. Chem.* **397**, 87 (2010).

58. A. G. Mignani, C. Cucci, L. Ciaccheri, C. Dall'Asta, G. Galaverna, A. Dossena and R. Marchelli, *SPIE Proc.* **7004**, 70045S-1 (2008).

59. T. M. Cover and P.E. Hart, *IEEE Trans. Infor. Theory* **IT-13**, 21 (1967).

60. J. R. Ferraro. *Introductory Raman Spectroscopy* (Elsevier Science USA 2003).

61. L. A. Lyon, C. D. Keating, A. P. Fox, B. E. Baker, L. He, S. R. Nicewamer, S. P. Mulvaney and M. J. Natan, *Anal. Chem.* **70**, 341R (1998).

62. I. Delfino, C. Camerlingo, M. Portaccio, B. Della Ventura, L. Mita, D. G. Mita and M. Lepore, *Food Chem.* **127**, 735 (2011).

63. O. Dóka, D. Bicanic, J.G. Buijnsters, R. Spruijt, S. Luterotti and G. Végvári, *Eur. Food Res. Technol.* **230**, 813 (2010).

64. D. Bicanic, D. Dimitrovski, S. Luterotti, C. Twisk, J. G. Buijnster and O. Dóka, *Food Chem.* **121**, 832 (2005).

65. F. Paiva-Martins, V. Rodrigues, R. Calheiros and M. P. M. Marques, *J. Sci. Food Agric.* **91**, 309 (2011).

66. T. Vermelinger, L. D'Ambrosio, B. Klopprogge and C. Yeretzian, *J. Agric. Food Chem.* **59**, 9074 (2011).

67. C. M. McGoverin, A. S. S. Clark, S. E. Holroyd and K. C. Gordon, *Anal. Chim. Acta* **673**, 26 (2010).

68. M. R. Almeida, K. De S. Oliveira, R. Stephani and L. F. C. de Oliveira, *J. Raman Spec.* **42**, 1548 (2011).

69. S. Okazaki, M. Hiramatsu, K. Gonmori, O. Suzuki and A.T. Tu, *Forensic Toxicol.* **27**, 94 (2009).

70. Y. Cheng, Y. Dong, J. Wu, X. Yang, H. Bai, H. Zheng, D. Ren, Y. Zou and M. Li, *J. Food Com. Anal.* **23**, 199 (2010).

71. K. Sowoidnich, H. Schmidt, M. Mainwald, B. Sumpf and H. D. Kronfeldt, *Food Bioprocess Technol.* **3**, 878 (2010).

72. A M. Herrero, *Food Chem.* **107**, 1642 (2008).

73. G. Jordan, R. Thomasius, H. Shröder, O. Schlüter, B. Sumpf, M. Maiwald, H. Schmidt, H.D. Kronfeldt, R. Scheuer, Schwägele and K. D. Lang, *J. Verbr.Lebensm.* **4**, 7 (2009).

74. H. W. Wong, S. M. Choi, D. L. Phillips and C. Y. Ma, *Food Chem.* **113**, 363 (2009).

75. C. M. Henry, *Chem. Eng. News Archive* **83**, 36 (2005).

76. S. J. Baek, A. Park, A. Shen and J. Hu, *J. Raman Spec.* **42**, 1987 (2011).

77. A. De Luca, M. Mazilu, A. Riches, C. S. Herrington, and K. Dholakia, *Anal. Chem* **82**, 738 (2010).

78. B. Kuswandi, Y. Wicaksono Jayus, A. Abdullah, L.Y. Heng and M. Ahmad, *Sens. Instrum. Food Qual.* **5**, 137 (2011).

79. http://kb.activepackaging.eu/ .

80. F. C. O'Mahony, T.C. O'Riordan, N. Papkovskaia, J.P. Kerry and D.B. Papkovsky, *Food Control* **17**, 286 (2006).

81. A. G. Mignani, L. Ciaccheri, A. A. Mencaglia, R. Paolesse, M. Mastroianni, D. Monti, G. Buonocore, M. A. Del Nobile, A. Mentana and M. F. Grimaldi, *Proc. IEEE-Sensors 2008 Conference* 1108 (2008).

82. http://www.ball-europe.com/704.html .

83. http://www.onvu.com .

84. http://www.3m.com/product/information/MonitorMark-Time-Temperature-Indicator.html .

85. A. Pacquit, J. Frisby, D. Diamond, K. T. Lau, A. Farrell, B. Quilty and D. Diamond, *Food Chem.* **102**, 466 (2007).
86. K. I. Oberg, R. Hodyss and J.L. Beauchamp, *Sens. Act. B* **115**, 79 (2006).
87. http://www.ripesense.com/ripesense_why.html .
88. Z. J. Smith, K. Chu, A. R. Espenson, M. Rahimzadeh, A. Gryshuk, M. Molinaro, D. M. Dwyre, S. Lane, D. Matthews and S. Wachsmann-Hogiu, *PLOS ONE* **6**, e17150 (2011).
89. http://www.kickstarter.com/projects/jywarren/public-lab-diy-spectrometry-kit.
90. http://www.efsa.europa.eu/en/faqs/faqttc.htm.
91. C. Blasco and Y. Picó, *Trends Anal. Chem.* **30**, 84 (2011).
92. K. Tiede, A. B. A. Boxall, S. P. Tear, J. Lewis, H. David and M. Hassellov, *Food Additives and Contaminants* **25**, 795 (2008).
93. T. Yamamoto, Y. Murakami, J. Motoyanagi, T. Fukushima, S. Maruyama, and M. Kato, *Anal. Chem.* **81**, 7336 (2009).

# OPTICAL BIOSENSING IN MEDICAL AND CLINICAL DIAGNOSTICS


Francesco Baldini*, Ambra Giannetti, Sara Tombelli and Cosimo Trono

*Institute of Applied Physics "Nello Carrara," National Research Council, Via Madonna del Piano 10, 50019 Sesto Fiorentino, Italy*
*E-mail: f.baldini@ifac.cnr.it*



The novel developments in the area of optical sensors for the detection of chemical and biochemical analytes in medical and clinical diagnostics are described. In particular, the new progress in the field of optical nanosensors for intracellular applications is described, with particular attention to oligonucleotide optical switches, in which the "nanodevice" can work as intracellular nanosensor and drug at the same time. Also point of care testing devices for novel clinical applications in the area of sepsis and of detection of immunosuppressants in transplanted patients are described as well as the new developments in the field of invasive optical fiber sensors for gastroesophageal apparatus.


## 1. Introduction

Optical biosensors are becoming more and more important thanks to the advantages that the optical approach can offer over other methodologies.

The new developments in nanoparticle and nanomaterial technologies are opening new perspectives in optical sensing, with the advent of optical nanosensors for the quantitative determination of bioanalytes at intracellular level. Fluorescent labels have been used under microscopy, since many years, to localize bioanalytes at cellular levels and to follow biological pathways. But in recent years, the possibility of developing optical nanosensors capable of achieving quantitative information in real time on intracellular events, such as metabolite dynamics, activated gene expressions or efficiency of administrated drugs paves to unprecedented scenario[1,2]. In the recent years, a demanding request from physicians has been given for the development of devices to be located close to the patient's bed and able to provide in a short time the measurement of bioanalytes of interest, the so called Point Of Care Testing (POCT) device[3,4]. This request derives from the necessity to perform shortly the measurement of bioanalytes, the knowledge of which can help the physicians to formulate the right diagnosis or to decide the correct therapy, avoiding to deliver the samples to the central laboratories and to wait, for hours, for the results of the analysis. In general, in POCT application, multianalyte detection of a few biomarkers is required, differently from genomics or proteomics. The last years have seen an incredible effort led by research centers and industries in the design and development of POCT devices, with fertility tests in urine, glucose in adipose tissue and measurement of blood gases and electrolytes being the first area of applications. If blood gas and electrolytes are excluded, most of the POCT devices on the market are able to measure one analyte per time, and - in the case of multiple detections - they generally make use of separate cartridges. The combination of optical biochip and microfluidics can offer the possibility of parallelization with the simultaneous measurement of different analytes at the same time on the same chip[5].

It is unquestionable that optical fibers are able to offer competitive, if not unique performances, in invasive application thanks to their intrinsic miniaturization, the absence of electric contacts and the electromagnetic immunity[6-8]. Of particular interest is the case of optical fibers sensors in gastroesophageal apparatus for the detection of chemical/biochemical parameters[9].

This chapter will describe some of the last results in these three areas of applications: optical intracellular nanosensors, POCT devices and invasive sensors in the gastroesophageal apparatus.


## 2.  Optical Intracellular Nanosensors

Nanoparticle and nanomaterial technologies in the biomedical field are significantly impacting the development of diagnostic tools. In particular, the advent of optical nanosensors for the quantitative determination of bioanalytes at intracellular level is one of the most fascinating challenges offered by the new developments in these technologies[10]. With the advent of new nanostructured materials, such as magnetic and gold nanoparticles, quantum dots and carbon nanomaterials, new perspectives were opened to drive molecules inside cells. The overall strategy for the detection of intracellular bioanalytes, with optical nanosensors inside the cell, is not a trivial problem but relies on the conjugation of nanosized structures (Figure 1) with different molecules capable: 1) to address the nanostructures to the unhealthy cells, 2) to bind marker molecules differentially expressed by these cells or interact with the chemicals of

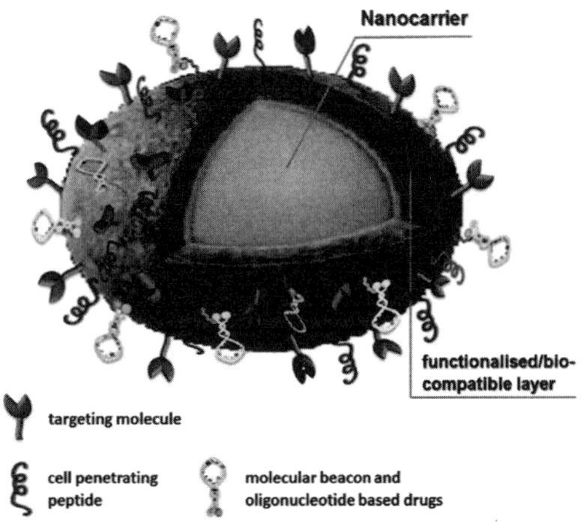


Fig. 1. Sketch of a nanosized structure for the delivery of optical nanosensors.

interest, 3) to improve the cell membrane penetration, and 4) to provide optical signals related to the amount of investigated analyte.

Several important reviews have appeared on intracellular nanosensors focused on optical fiber devices (nanotips)11 or on nanoparticle PEBBLE (Photonic Explorer for Bioanalysis with Biologically Localized Embedding) sensors12. This paragraph provides an overview on the optical intracellular nanosensors based on oligonucleotides both as biorecognition element and as signaling molecule. In particular, in the case of oligonucleotidic optical probes, the "nanodevice" can work as intracellular nanosensor and drug at the same time, by addressing specific RNA messengers and preventing the overexpression of proteins associated to pathologic conditions. Particular interest has been focused on oligonucleotide probes capable to turn on or to modify their light emission upon the molecular interaction with well-defined molecular targets13. Among this type of probes, Molecular Beacons (MBs) have been used in a variety of applications, including intracellular sensing. MBs are DNA sequences composed of one target-recognition region and two short complementary stem sequences: in the absence of the target, the sequence forms a stem–loop structure which brings a quencher and fluorophore, which are located at opposite ends of the MB, into close proximity, resulting in fluorescence quenching (Figure 2). In the presence of a target molecule, hybridization between the target and the loop sequence of the MB induces the opening of the stem resulting in the spatial separation of the fluorophore and quencher

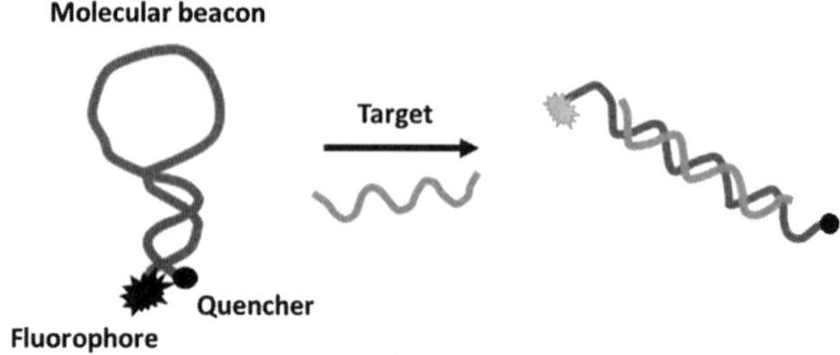


Fig. 2. Scheme of a molecular beacon. Conformation before and after the incubation with the target.

with restoration of fluorescence[14]. MBs have been used not only for the detection of oligonucleotide sequences (e.g. mRNA), but, by exploiting aptamer-based beacons with a similar hairpin design, they have been applied to the intracellular sensing of proteins or smaller molecules such as Adenosine triphosphate (ATP)[15].

The use of MBs or other similar oligonucleotidic probes in cellular environment can be critical for different reasons, such as false positive signals due to the MB degradation by endogenous nucleases, or high background signals due to the incomplete quenching of the quencher[16].

In order to avoid dramatic false-positive signals given by nuclease degradation, protein binding and thermodynamic fluctuations, MBs have been synthesized with several modifications into the structure, such as phosphorothioate and 2'-O-methyl RNA bases which provide higher stability in intracellular environment. Another possibility is offered by the use of Locked Nucleic Acids (LNAs) or Peptide Nucleic Acids (PNA), instead of DNA sequences[17,18].

Another critical factor in intracellular nanosensors design is the delivery tool, since the method used to deliver the probe into the cell can influence not only its localization but also the fate of the probe inside the cell such as its degradation or segregation into organelles. Common delivery methods that have recently been shown to be effective also in delivering MBs into living cells, include cell penetrating peptides (CPPs), or lipidic molecules such as lipofectamine, micelles or lyposomes and other commercially available lipid-based transfection agents[19,20]. Among the physical methods, electroporation/microporation and microinjection provide other possibilities for delivering oligonucleotides into cells for sensing[21]. Besides these conventional nucleic acids delivery tools, inorganic nanoparticles (NPs), such as iron oxide and gold nanoparticles, quantum dots, and other nanomaterials such as carbon nanowires and nanotubes, are gaining significant interest in the field of drug delivery and imaging[22]. The nano-size of NPs facilitates their intracellular delivery into the cytosol and provides a large surface area for various bioconjugations of biological recognition element to target specific tissues/cells and for attaching therapeutic agents[23].

In particular, gold nanoparticles (AuNPs) in combination with hairpin-shaped oligonucleotides have been reported for the detection of several mRNA targets[24,25]. For example, by using AuNPs modified with two MBs specific for two mRNA, labelled with two different fluorophores, FITC and Cy5, the simultaneous imaging of the two targets was possible[24]. The combination of AuNPs and short complementary strands has also been exploited for the detection of ATP in cells[26]: in this work, a dense layer of ATP specific aptamer was immobilized onto gold nanoparticles after its hybridization with a Cy5 labeled "reporter" oligonucleotide. These so called *aptamer nano-flares* were demonstrated to allow determination of ATP inside HeLa cells. Very recently, AuNPs have been proposed, for the first time, in combination with a DNAzyme for the intracellular detection of uranyl ion[27].

Coupling of CNTs, in particular single walled nanotubes (SWNT), with MBs has been exploited to demonstrate the ability of nanotubes to protect DNA in cellular environment[28]. The same property of DNA protection has been attributed to functionalized nanoscaled graphene oxide (GO)[29]. More recently, similar carbon-based nanomaterials have been also investigated in this area: in particular, the MB for survivin was attached to graphite nanoparticles, 4-5 nm in diameter, consisting of stacked graphene sheets[30]. In this case, the graphite nanoparticle was both used as fluorescence quencher molecule and as delivery cargo for the MB. The complex MB-graphite nanoparticle also showed lower cytotoxicity if compared to other carbon nanomaterials and a long cell imaging time.

Besides these more classical nanomaterials, other nano-objects have been coupled to MBs for intracellular sensing. In one case, glyceraldehyde-3-phosphate dehydrogenase (GAPDH) mRNA was selectively detected into a single HeLa cell by a MB attached to a silicon nanoneedle (400 nm in diameter)[31]. Cationic core–shell nanoparticles, made up of a core of polymethylmethacrylate (PMMA), surrounded by a shell bearing cationic groups, have been very recently evidenced as potential intracellular delivery vehicles for surviving mRNA-specific MB[32].

## 3. POCT devices

Coagulation, cardiac markers, glucose and infectious agents are some of the clinical applications in which POCT devices are appearing on the market or are strongly requested by physicians. One of the most evident examples of the need of POCT devices is provided by the infarction which has to be identified in a very short time. It is apparent that, in this case, saving time, also a few minutes, in making the correct diagnosis can be crucial for the patient's survival.

But the timeliness in a correct diagnosis can also be essential in many other cases. The need of discrimination between viral and bacterial inflammation in intensive care patients, or in any case the need of a fast identification of the origin of infections is a high request coming from physicians. It has been recently demonstrated that every hour delay until administration of an effective antibiotic treatment in septic shock increases mortality by 7%[33]. Although the right panel of biomarkers has not been still identified[34], procalcitonin, C-reactive protein, interleukines 6, 8 and 10 and neopterin have to be considered very promising candidates[35]. For this reason, rapid, simple and low-cost medical devices for multiple parameter screening are essential for early diagnosis and improved treatments of sepsis. One solution to this significant demand requires the advent of new and more reliable biosensors integrated within the same biochip. Differently from those devices for genomics or proteomics, where a large number of spots, of the order of thousands, are necessary, in many clinical applications the measurement of a limited number of parameters (5 - 15) is sufficient to help physicians in the diagnosis of pathologies or in the choice of the appropriate therapy. In the recent years, multianalyte optical platforms for sepsis analysis and diagnosis have been described based on fluorescence-based immunoassay. In the device developed by the authors of this chapter[36,37], the heart of the optical platform is an optical biochip constituted by a two-piece polymethylmetacrylate (PMMA) chip, with 13-microchannels (50 μm high, 600 μm width, 10 mm long) through which the analyzed sample flows (Figure 3). The sensing layer, where the immunochemical reaction takes place, is located on the upper part of each microchannel; the bottom part of the chip is in black PMMA in order to decrease the scattering of the light coming from the source. The chip is interrogated with a novel optoelectronic platform, based on fluorescence anisotropy. The fluorescence emitted by a fluorophore located at a distance from a dielectric interface smaller than or comparable with the emitted wavelength is anisotropic, mainly directed towards the denser medium and with well-defined preferential directions. Therefore, most of the fluorescence emitted by the sensing layer, located at the interface between the rarer liquid medium and the denser plastic medium, remains entrapped inside the chip. The particular comb-shape of the top of the chip implies that the fluorescence is guided along plastic waveguides, which run side by side to the microchannel. A line-shaped beam from a 635-nm laser-diode excites perpendicularly the sensing layer and the emitted fluorescence, coupled to the plastic waveguide, is collected by a

Fig 3. Photo of the PMMA chip constituted by thirteen channels for the simultaneous measurement of the sepsis biomarkers.

single plastic optical fiber and sent to an amplified photodiode, after a filtering action, which removes any signal coming from the source. A motorized translation stage allows the automatic scanning of the 13 different microchannels and a microfluidic manifold allows pumping the sample under test through the different microchannels. A channel valve allows selecting different driving, mixing and delivering configurations.

In the device described in [38], based on a fluorescence microarray, the format is a sandwich immunoassay. The assay processing and read out is done automated by the system which consists of the bioassay, the read out based on planar waveguide technique and a fluidic set-up. The biochip consists of a glass substrate, coated with a very thin high index waveguide. The microarray is spotted on the surface of this waveguide.

Through an optical diffraction grating, structured into the glass substrate at one end of the chip, the light from a semiconductor laser is coupled into this waveguide over the whole width of the chip. Due to the low thickness of the waveguide and the high difference of refractive index relative to the substrate glass, the high evanescent wave interacts with the spotted microarray. The emitted fluorescence light is detected by an uncooled CCD camera, which is located on the opposite side to the coated surface. As the sensing spots within one array will have considerably differing fluorescence intensity, typically several images with varying integration time are taken. For array analysis, the signal processing software automatically chooses the integration time that is optimal for each single spot.

The measurement of immunosuppressants and related metabolites in transplanted patients is another example in which POCT devices are strongly requested by physicians. In the last couple of years, it has become increasingly evident that improved patient outcome results from pharmacokinetic dosing strategies requiring multiple sampling with the modelling of time-concentration curves. Strategies based on sparse sampling have been developed for clinical purposes estimating the area under the concentration time curve (AUC) and have been shown to substantially improve patient outcome[39,40]. Further substantial progress can be expected from continuous measurement rather than estimation of the AUC. The aim of a recently started European project (Nanodem)[41], is the development of a novel therapeutic drug monitoring POCT device for the automatic measurement, in transplanted patients, of immunosuppressants, and in particular cyclosporine, tacrolimus, sirolimus and mycophenolic acid, characterized by a narrow therapeutic range and serious potential side effects. The patient will be connected to the device by an intravenous microdialysis catheter to allow 48-h online measurements. Based on this minimally-invasive approach, the therapeutic drugs and related metabolites will be monitored at short time intervals. The need of mixing the dialysate with the chemical reagents and the necessity of incubation times for the bioassay implementation, unavoidable procedure for bioanalyte detection, implies that a continuous measurement of such analytes is impossible, but the miniaturization down to micro- and nano-scales will lead to very short time intervals, of the order of a few minutes. Heart of the device will be a multi-parametric optical chip, which will make use of the recent developments in nanotechnology to convert the concentration changes of the analytes in detectable luminescent signals. A block diagram of the final instrument is shown in Figure 4.

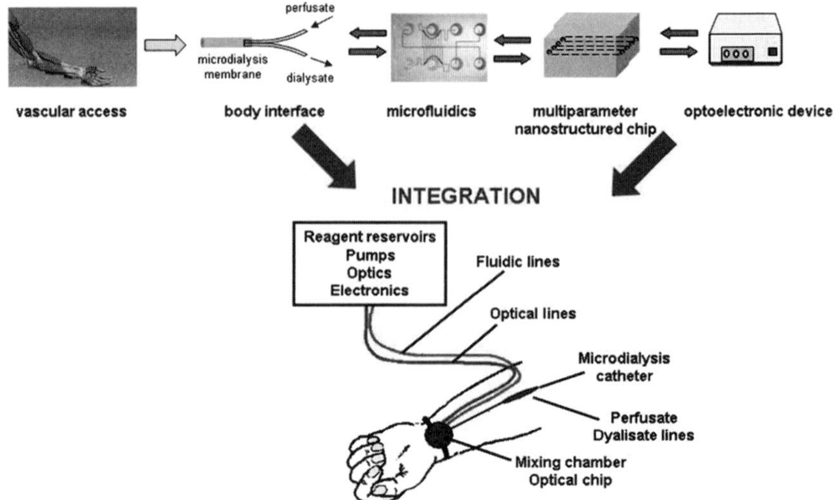



Fig. 4. Block diagram of the proposed POCT device proposed for the detection of immunosuppressants.

## 4. Optical Fiber Sensors in Gastroesophageal Apparatus

The use of optical fiber sensors for the measurement of bile containing refluxes is a practice more and more diffusing among physicians[42-44]. It is widely accepted that the presence of bile components in the gastric reflux from the stomach to the esophagus, such as biliary acids[45], can have a damaging effect on the esophageal mucosa starting from inflammation up to the formation of preneoplastic lesions.

In the past years, attempts were carried out to develop optical fiber pH sensors for in-vivo gastric and/or esophageal measurements. Physicians requirements for the gastric use of a pH sensor are: i) working range between 1 and 8 pH units, ii) accuracy $\leq 0.1$ pH units, iii) response time less than 30 seconds and continuous 24-hour monitoring. In the first examples described in literature, either fluorescence- or absorption-based sensors were proposed[46-48]. Even if two of them were successfully tested in-vivo on animals[47] and on patients[48], these sensors were all characterized by the problem of the use of two different indicators to cover the whole range of interest. This fact implies a quite complicate calibration curve with several calibration points, with a lengthening of the calibration procedure unacceptable by physicians. It should be said that gastric pH measurement is actually performed, routinely, with miniaturized glass electrodes, which need only a two-point calibration curve. A few years ago, the use of methyl red immobilized on Controlled Pore Glasses (CPG) was proposed as a suitable optical transducer for gastric applications[9]. As a matter of fact, this acid-base indicator was characterized by a wide working range, after being covalently bound to the CPGs, covering perfectly the range of interest.

Recently a novel optical fiber probe was proposed (Figure 5) with the sensor tip constituted by two plastic fibers (core diameter 200 µm) cut at their extremities at an angle larger than the total reflection angle and the CPGs with methyl red on its surface immobilized on the lateral surface of the fibers[49]. Thanks to this geometry, the light, coupled laterally from one fiber to the other one, crosses the CPG and its transmission is modulated by the pH value. The CPG immobilization is performed by heating the plastic fibers up to their melting point and pressing the CPGs

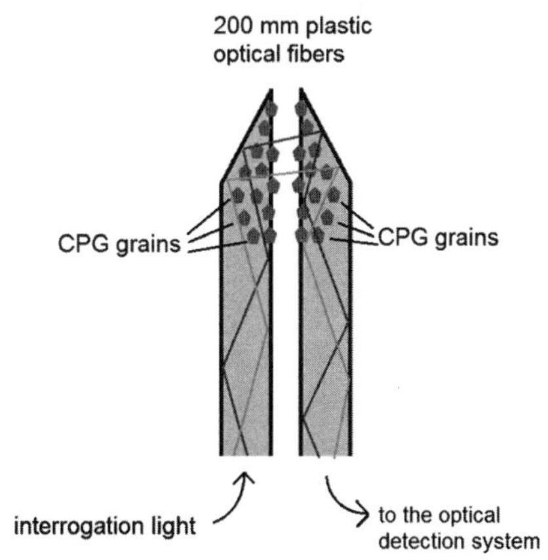


Fig. 5. Sketch of the optical fiber tip for pH detection.

on the fiber surface. The CPGs penetrates within the softened fibers and remained entrapped once the fiber cools back to the room temperature.

Laboratory tests showed good performances with the optical fiber probe, which was characterized by a working range between 1.5 and 8 pH units, an accuracy $\leq 0.1$ pH units and a response time $\leq 30$ seconds.

The probe was also characterized by a good lifetime, with a perfect reproducibility of the calibration curve repeated three times over a period of three weeks.

## 5. Conclusions

Optical chemical and biochemical sensors are finding continuously increasing applications in medical and clinical diagnostics and it is apparent that they will be a more and more important tool in the hands of physicians in the incoming future. In terms of optical intracellular nanosensors, MBs or similar probes can be a valuable and fascinating tool for intracellular sensing, having the characteristics of being extremely versatile in terms of possible structure, signaling modus and, especially, number of possible targets. In POCT applications, optics is playing a fundamental role in view of the realization of compact and transportable device to be located close to the bed of the patients, and capable to measure several parameters at the same time, with the applications to sepsis detection and diagnosis and to immunosuppressant detection in intensive care patients as last examples of their importance in medicine. At the same time optical fiber sensors are still offering unique possibilities as invasive sensors, thanks to their easy handling, high miniaturization and total electromagnetic compatibility.

## Acknowledgments

The authors wish to thank the European Community for the funding to the 5-year project CARE-MAN (HealthCARE by Biosensor Measurements and Networking, contract nr IP 017333) and to the 5-year project NANODEM (Nanophotonic device for multiple therapeutic drug monitoring, contract nr 318372). The authors also thanks the regional and national institutions for the support achieved by the regional project NANOCELL (PAR FAS REGIONE TOSCANA Linea 1.1.a.3) and by the national flagship project NANOMAX.

# References


1. M. J. Ruedas-Rama, J. D. Walters, A. Orte, and E. A. H. Hall, *Anal. Chim. Acta,* **751** 1-23 (2012).
2. J. Kneipp, H. Kneipp, B. Wittig, and K. Kneipp, *Nanomedicine: Nanotechnology, Biology, and Medicine,* **6** 214 (2010).
3. P. B. Luppa, C. Müller, A. Schlichtiger and H. Schlebusch, *Trends Anal. Chem.,* **30** 887 (2011).
4. V. Gubala, L. F. Harris, A. J. Ricco, M. X. Tan, and D. E. Williams, *Anal. Chem.,* **84** 487 (2012).
5. T. Vo-Dinh, In: Vo-Dinh T (Ed) *Biomedical photonics handbook,* CRC, Boca Raton, Fla. (2003).
6. G. Mignani and F. Baldini, *Rep. Prog. Phys.,* **59** 1 (1996).
7. K. T. V. Grattan, and T. Sun, *Sens. Actuat. A: Phys.,* **82** 40 (2000).
8. F. Baldini, A. Giannetti, A. G. Mignani and C. Trono, *Curr. Anal. Chem.,* **4**, 378 (2008).
9. F. Baldini, *Anal. Bioanal. Chem..* **375** 732 (2003).
10. T. Vo-Dinh, P. Kasili and M. Wabuyele, *Nanomedicine,* **2**, 22 (2006).
11. T. Vo-Dinh and Y. Zhang, *Wiley Interdiscip. Rev. Nanomed. Nanobiotechnol.,* **3**, 79 (2011).
12. Y. E. Lee, R. Smith and R. Kopelman, *Annu. Rev. Anal. Chem.,* **2**, 57 (2009).
13. K. Huang and A. A. Martí, *Anal. Bioanal. Chem.,* **402**, 3091 (2012).
14. K. Wang, *Angew. Chem. Int. Ed.,* **47**, 2 (2008).
15. X. Tan, T. Chen, X. Xiong, Y. Mao, G. Zhu, E. Yasun, C. Li, Z. Zhu and W. Tan, *Anal. Chem.,* **84**, 8622 (2012).
16. S. Wu, L. Peng, M. You, D. Han, T. Chen, K. R. Williams, C. J. Yang, and W. Tan, *Int. J. Mol. Imaging,* **2012**, 501579 (2012).
17. E. Catrina, S. A. Marras, D. P. and Bratu, *ACS Chem. Biol.,* **7**, 1586 (2012).
18. Y. Wu, C. J. Yang, L. L. Moroz and W. Tan, *Anal. Chem.,* **80**, 3025 (2008).
19. N. Nitin, P. J. Santangelo, G. Kim, S. Nie and G. Bao, *Nucleic Acids Res.,* **32** e58 (2004).
20. W. J. Kang, Y. L. Cho, J. R. Chae, J. D. Lee, K. J. Choi and S. Kim *Biomaterials,* **32**, 1915 (2011).
21. J. Perlette and W. Tan, *Anal. Chem.,* **73**, 5544 (2001).
22. Z. P. Xu, Q. H. Zeng, G. Q. Lu and A. B. Yu, *Chemical Engineering Science,* **61**, 1027 (2006).
23. G. Liu, M. Swierczewska, S. Lee and X. Chen, *Nano Today,* **5**, 524 (2010).
24. G. Qiao, Y. Gao, Z. Yu, L. Zhuo, and B. Tang, *Chem. Eur. J.,* **17**, 11210 (2011).

25. E. Prigodich, P. S. Randeria, W. E. Briley, N. J. Kim, W. L. Daniel, D. A. Giljohann and C. A. Mirkin, *Anal. Chem.,* **84**, 2062 (2012).

26. Zheng, D. S. Seferos, D. A. Giljohann, P. C. Patel and C. A. Mirkin, *Nano Letters,* **9**, 3258 (2009).

27. P. Wu, K. Hwang, T. Lan and Y. Lu, *J. Am. Chem. Soc.,* **135**, 5254 (2013).

28. Y. Wu, J. A. Phillips, H. Liu, R. Yang and W. Tan, *ACS Nano,* **2**, 2023 (2008).

29. H. Lu, C. L. Zhu, J. Li, J. J. Liu, X. Chen and H. H. Yang, *Chem. Comm.,* **46**, 3116 (2010).

30. Y. Piao, F. Liu and T. S. Seo, *ACS Appl. Mater. Interfaces,* **4**, 6785 (2012).

31. T. Kihara, N. Yoshida, T. Kitagawa, C. Nakamura, N. Nakamura and J. Miyake *Biosens. Bioelectron.,* **26**, 1449 (2010).

32. A. Giannetti, S. Tombelli, C. Trono, M. Ballestri, G. Giambastiani, A. Guerrini, G. Sotgiu, G. Tuci, G. Varchi and F. Baldini, *Proc. SPIE 8596, Reporters, Markers, Dyes, Nanoparticles, and Molecular Probes for Biomedical Applications V,* 85960U (2013).

33. A. Kumar, D. Roberts, K. Wood, B. Light, J. Parrillo, S. Sharma, R. Suppes, D. Feinstein, S. Zanotti, L. Taiberg, D. Gurka, A. Kumar and M. Cheang, *Crit. Care Med.,* **34**, 1589 (2006).

34. A. Pierrakos and J. L. Vincent, *Critical Care,* 14 (2010).

35. U. Sauer, P. Domnanich, and C. Preininger, *Anal. Biochem.,* **419**, 46 (2011).

36. F. Baldini, L. Bolzoni, A. Giannetti, M. Kess, P. M. Krämer, E. Kremmer, G. Porro, F. Senesi and C. Trono, *Anal. Bioanal. Chem.,* **393**, 1183 (2009).

37. F. Baldini, A. Carloni, A. Giannetti, G. Porro and C. Trono, *Sens. Actuat. B,* **139**, 64 (2009).

38. M. Kemmler, B. Koger, G. Sulz, U. Sauer, E. Schleicher, C. Preininger and A. Brandenburg, *Sens. Actuat. B,* **139**, 44 (2009)

39. R. G. Morris, G. R. Russ, M. J. Cervelli, R. Juneja, S. P. McDonald and T. H. Mathew, *Ther. Drug Monit.,* **24**, 479 (2002).

40. M. Hadjibabaie, I. Vazirian, M. Iravani, S. A. Moosavi, K. Alimoghaddam, A. Ghavamzadeh and S. Rezaee, *Ther. Drug Monit.,* **33**, 673 (2011).

41. http://nanodem.ifac.cnr.it/

42. P. Bechi, F. Pucciani, F. Baldini, F. Cosi, R. Falciai, R. Mazzanti, A. Castagnoli, A. Passeri and S. Boscherini, *Dig. Dis. Sci.,* **38**, 1297 (1993).

43. M. F. Vaezi, R. G. Lacamera and J. E. Richter, *Am. J. Physiol.,* G1050 (1994).

44. Pohl and R. Tutuian, *Best Pract. Res. Clin. Gastroent.,* **23**, 299 (2009).

45. B. S. Kaur, R. Ouatu-Lascar, M. B. Omary and G. Triadafilopoulos, *Am. J. Physiol. Gastrointestin. Liver Physiol.,* **278**, G1000 (2000).

46. H. E. Posch, M. J. P. Leiner and O. S. Wolfbeis, *Fresenius J. Anal. Chem.* **334** 162 (1989).

47. J. Netto, J. I. Peterson, M. McShane and V. Hampshire, *Sens. Actuators B,* **29** 157 (1995).
48. Baldini, P. Bechi, S. Bracci, F. Cosi and F. Pucciani, *Sens. Actuators B* **29** 164 (1995).
49. Baldini and C. Trono, Italian Patent FI2010A000237 (2010).

# PHOTONICS FOR FORENSIC APPLICATIONS


Antonella Tajani

*National Research Council of Italy, Via dei Taurini, 19, 00185, Rome, Italy*
*E-mail: antonella.tajani@cnr.it*



This chapter illustrates a few of the numerous techniques, based on electromagnetic waves, employed in forensic science investigations by Reparto Carabinieri Investigazioni Scientifiche in Rome. Although it is not possible to provide an exhaustive description of the large number of examples where light sources can be used for forensic applications, the reader will hopefully gain an appreciation of the complex nature of this fascinating field, the current state of knowledge and the areas where research remains to be done.


## 1. Introduction

Over the last few years, interest in forensic science has increased, partly as a result of popular television crime movies.

The successful investigation and prosecution of crimes require, in most cases, the collection, preservation and forensic analysis of evidence.

Therefore, nowadays, most criminal cases strongly rely on forensic science to bring evidence in a court of law, contributing to the successful prosecution and conviction of criminals as well as to the exoneration of innocent people. Forensics is often crucial in judicial or criminal investigation processes, retracing or building the sequence of events as they might have occurred during the actual crime. Due to the variety of complex problems, "forensic science" encompasses a broad range of forensic disciplines, each with its own set of technologies and practices.

According to the categorization used by the American National Institute of Justice, these are[1]: general toxicology, firearms/toolmarks, questioned documents, trace evidence, controlled substances, biological/serology screening (including DNA analysis), fire debris/arson analysis, impression evidence, blood pattern analysis, crime scene investigation, medico-legal death investigation and digital evidence.

Some of them are laboratory based (e.g., nuclear and mitochondrial DNA analysis, toxicology and drug analysis); others rest on expert interpretation of observed patterns (e.g., fingerprints, writing samples, toolmarks, bite marks and specimens, such as hair).

Many are the analytical techniques, based on electromagnetic radiation, in use at forensic laboratories or at the scene crime; just to name a few: optical microscopy, (fluorescence or Raman) spectroscopy, luminescence, fluorescence.

| ULTRA VIOLET | | VISIBLE SPECTRUM measured in nanometers (nm) | | | | | | INFRARED |
|---|---|---|---|---|---|---|---|---|
| 190-290 | 290-400 | 400-430 | 430-490 | 490-575 | 575-590 | 590-620 | 620-700 | >700 |
| SHORTWAVE UV | LONGWAVE UV | VIOLET | BLUE | GREEN | YELLOW | ORANGE | RED | IR |

INCREASING WAVELENGTH →

← INCREASING ENERGY

| ITEM | SEARCH | GOGGLE | CAMERA FILTER |
|---|---|---|---|
| SHOEPRINTS | WHITE (OBLIQUE ANGLE) | CLEAR OR YELLOW | NONE |
| TREATED MUD SHOEPRINTS POROUS SURFACE | 535, TREAT WITH DFO | RED | RED BP/LP BP600 |
| TREATED MUD SHOEPRINTS NON-POROUS SURFACE | TREAT W/Safranin O OR 455/CSS – Basic Yellow | ORANGE | ORANGE BP550 |
| HAIR (UNTREATED – BLACK) | WHITE (OBLIQUE ANGLE) | CLEAR | NONE |
| HAIR (TREATED OR RED/BLONDE) | 415/CSS | YELLOW ORANGE | YELLOW ORANGE |
| BONE | 455/CSS/515 | ORANGE | ORANGE |
| TEETH | 455/CSS/515 | ORANGE | ORANGE |
| FINGERNAILS | 455/CSS/515 | ORANGE | ORANGE |
| BODY FLUIDS (START) | CSS | ORANGE | 1-2 ORANGE |
| (Dark surfaces and for saliva) | UV | CLEAR/YELLOW | YELLOW |
| (Dark surfaces show 'crusty' spot) | WHITE (OBLIQUE ANGLE) | CLEAR | NONE |
| BITE MARK/BRUISE (FRESH) | 415/445 | YELLOW | 1-2 YELLOW |
| TO | 455/CSS/515 | ORANGE | 1-2 ORANGE |
| (OLDER) | 535/555/575 | RED | RED BP600 |
| GSR: GUN SHOT RES. | 455/CSS | ORANGE | 2 ORANGE/BP550 |
| | CSS | ORANGE | 2 ORANGE/BP550 |
| BLOOD (UNTREATED) | 415 | CLEAR OR YELLOW | NONE/BP415 |
| BLOOD (TREATED) W/FLUORESCEIN | 455 | ORANGE | ORANGE/BP550 |
| DFO PRINTS | 455 | ORANGE | 2 ORANGE/BP550 |
| (ON FLUORESCING BACKGROUNDS) | 535/555 | RED | 1-2 RED BP600 |
| (ON NON-FLUORESCING BACKGROUNDS) | SP575 | RED | 1-2 RED BP600 |
| NINHYDRIN | 555/575/600/630 OR WHITE | CLEAR | NONE |
| NINHYDRIN/ZnCL | 515/CSS | CLEAR | NONE/BP515 |
| BASIC YELLOW | 445 | YELLOW | 2 YELLOW |
| | 455/CSS | ORANGE | 2 ORANGE |
| RHODAMINE-6G | 515 | ORANGE | 2 ORANGE/BP550 |
| ARDROX | UV | CLEAR | UV Blocking |
| | 415 | YELLOW | 1-2 YELLOW/BP500 |

Fig. 1. Portion of the electromagnetic spectrum that matters to forensics[2].

The methods based on those techniques have, among others, the main advantage of identifying even latent traces or body fluids in a non-destructive manner, which is the crucial point when samples, such as

DNA, must be preserved during time. Figure 1 shows the electromagnetic spectrum of interest for forensic investigations together with some of the typical applications[2].

Many are the types of evidence considered in forensic investigations, but for the sake of conciseness, only some of the above mentioned aspects will be illustrated in this chapter. The examples described in the following pages represent few criminal cases faced by the Italian Reparto Carabinieri of Rome, RIS.

## 2. Detection of Biological Evidence

Analysis of biological material recovered at crime scenes or on a person, clothing, or weapon is among the most important types of evidence in forensics. Some evidence - for example, pet hairs, insects, seeds or other botanical remnants - comes from the crime scene or from an environment through which a victim or suspect has recently traversed. Others come from victim's or suspect's specimens.

The most common body fluids are blood, semen and saliva, but others, such as vaginal fluid, urine, hair, tissue, bones and sweat can also play important roles including the contribution of valuable DNA evidence, essential in the identification of victims and suspects and in disaster victim identification or missing persons investigations. Many body fluid stains are either invisible to the naked eye[3] or similar in appearance to other fluids or substances. Even when the identity of a stain may seem obvious, absolute confirmation is necessary in order for the evidence to be used in court to either prove or disprove a fact in a case. This is especially important with the possible occurrence of mixtures. A great number of different methods[4], either presumptive or confirmatory tests, are generally used to identify this biological evidence; presumptive are screening tests, confirmatory tests conclusively identify the species of the particular evidence[5].

As in many cases small quantities of biological material are available, non-destructive, reproducible analysis is essential to preserve these valuable samples.

Driven by the importance for forensic applications, body fluid identification methods have been extensively developed in recent years[6-9].

Significant advances in laser technology and the development of novel light detectors have dramatically improved spectroscopic methods for molecular characterization over the last decade.

Biospectroscopy for forensic purposes opens new and exciting opportunities for the development of on-field, non-destructive, confirmatory methods for body fluid identification at a crime scene. In addition, biospectroscopy methods, based on Raman and fluorescence spectroscopy, are universally applicable to all body fluids and their dry traces, unlike the majority of current techniques which are valid for individual fluids only. Moreover, they represent a rapid, confirmatory, non-destructive identification test of a body fluid at a crime scene[3].

Compared to fluorescence, Raman spectroscopy has a much higher selectivity and peculiarity to chemical and biochemical species. Its vibrational signature[10] correlates with the known composition of the fluid[11]. In spite of lower sensitivity, it is extremely indicated when mixtures of multiple body fluids need to be resolved[12].

Portable Raman spectrometers have largely been used in forensics in the last years to detect narcotics, explosives, powders, chemical weapons, industrial chemicals[13,14] and for homeland security applications at larger extent[15].

Though Raman spectroscopy has been used for fingerprint analysis to detect, for example, trapped drugs or explosives[16,17], this paragraph will not deal with it, focusing instead on the use of light source[18], from ultra-violet, visible up to the infrared components, such as crimescope, in a few forensic applications.

For example, in the case of blood, alternate light sources are generally used at 415 nm, the wavelength under which bloodstains absorb light and are thus more visible to the naked eye.

As we will see in the next pages, most commonly, though, a catalytic chemical test that turns color or luminesces in presence of blood is used as screening check. Scene investigators may also use Luminol, fluorescein, or crystal violet to identify areas at the scene where attempts were made to clean a bloody crime scene.

Luminol, based on chemoluminescence, is a very sensitive technique also for the analysis of residues, old stains in walls, carpets, upholstery, wooden floors or painted surfaces.

## 2.1. *Biological Fluids and Latent Prints Analysis*

The use of forensic light sources allows investigators to observe forensic findings such as latent fingerprints, body fluids, hair and fibers, bruises, bite marks, wound patterns, shoe and foot imprints, drug traces, bone fragment detection at a crime scene. They have more sensitivity than traditional methods thus increasing the amount of evidence uncovered and the quality of the evidence photographed and collected.

Generally, a forensic light source is made up of a powerful xenon or metal-halide lamps, emitting electromagnetic waves ranging from the UV up to IR (figure 2).

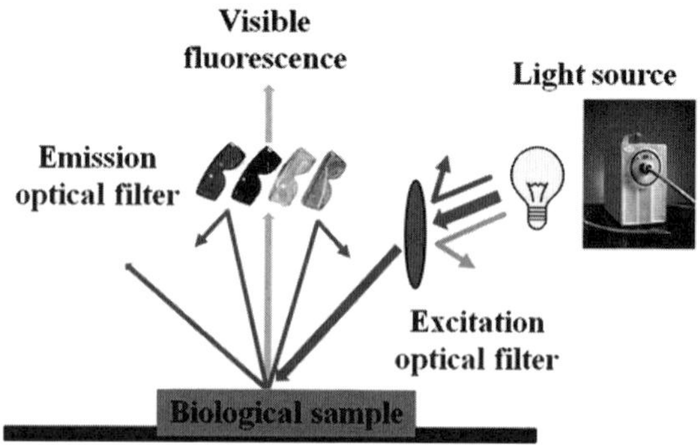


Fig. 2. Schematic of a biological sample analysis by a light source and different filters.

By using filters, selected wavelengths can be used to enhance the visualization of evidence by light interaction techniques including fluorescence (evidence glows), absorption (evidence darkens), and oblique lighting (small particle evidence revealed).

For example, as shown in figure 3 (left top and bottom images), no biological evidence is visible when examining the scenes by naked eye.

When illuminating the areas of interest by using blue light and yellow filters, in both cases, clear traces of a glowing substance, later classified as semen, are detected on the wall and on the surface of a WC (figure 3, right images) inspected at the crime scenes.

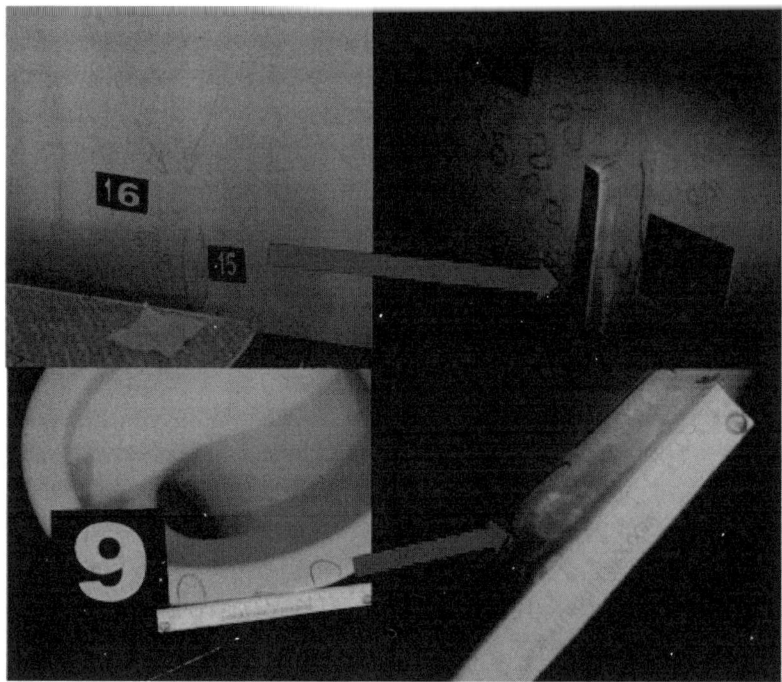

Fig. 3. Left images top and bottom: two different crime scenes as visible by naked eye. When illuminated with the appropriate wavelength, latent biological traces (seminal fluid) are visible (details zoomed and indicated with a red arrow on the top and bottom right).

A typical forensic light source is shown in figure 4. The device, consisting of a Xenon 500W lamp, is adequately small and versatile to be taken to the crime scene as well as to fit comfortably in the laboratory.

The Crimescopes, CS-16-500 of the SPEX Forensics[19] covers the UV - VIS range and it can work up to IR. It has 15 filters mounted on a wheel and white light. The degree to which various substances become visible when using different filters depends on the state of the substance (trace) and the surface on which this substance exists. The conditions of the crime scene often will determine which wavelength will be the most effective.

Since body fluids like semen, saliva, urine, sweat and vaginal secretions have a low absorption coefficient, they are naturally fluorescent when using a high intensity UV light.

Therefore, the use of light source in this range offers a unique method for the identification of these particular body fluids. Dried biological samples will then glow under the light source illumination, as illustrated in figure 5 where saliva, vaginal fluid and semen stains (figure 6), hardly visible to the naked eye, are clearly observable when using emitting blue light in the range 430 - 455 nm and dressing yellow filter goggles.

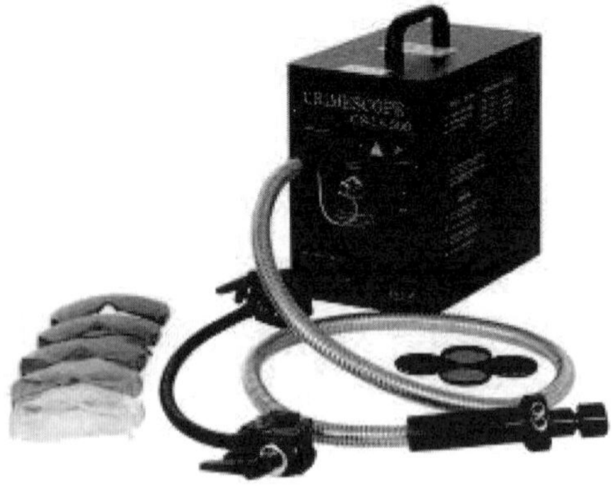

Fig. 4. Crimescope CS- 16-500 in use at RIS in Rome.

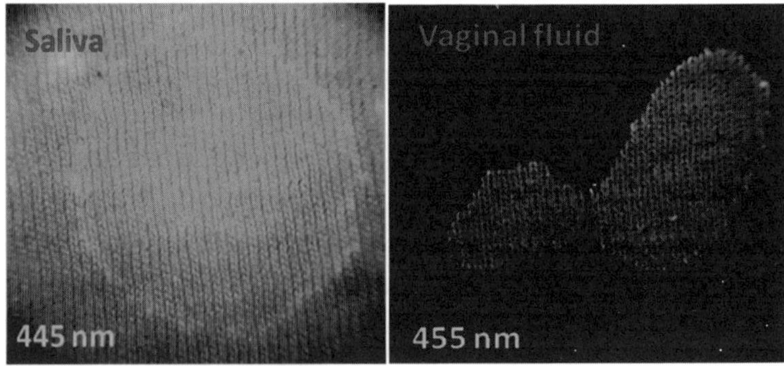


Fig. 5. Specimens of saliva (emitting light set at 445 nm)  and vaginal secretion (emitting light set at at 455 nm).

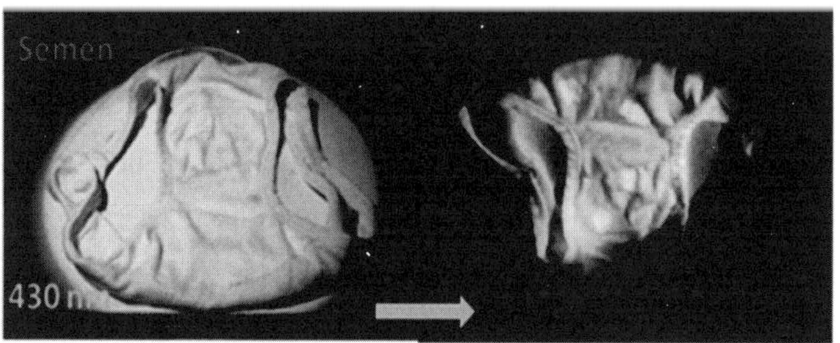


Fig. 6. Fluorescence of semen traces, invisible by naked eye (left), glowing as greenish (right).

The detection of latent fingerprints is among the primary applications of a forensic light source, such as the CS-16-500. A fingerprint is an individual characteristic. It is yet to be found that prints taken from different individuals possess identical ridge characteristics. Furthermore, fingerprints remain unchanged during an individual's lifetime and have general characteristic ridge patterns that permit them to be systematically classified. Because of this unique characteristic, a print is a conclusive evidence and a valuable tool among advanced technologies even today.

Latent prints cannot be seen by naked eyes as left on an object by the body's natural greases and oils. Powders, chemicals, and even lasers are used to make them visible on the crime scene evidence. Thanks to the much higher sensitivity (10-100 times more) compared to the conventional method of black powder dusting and lifting, the use of electromagnetic radiation has been successfully utilized for revealing latent prints on various types of textured surfaces, backgrounds which mask ridge details, fragile or contaminated surfaces. Different wavelengths are required for processing different surfaces making a forensic light source with tunable or multiple wavelengths an ideal tool for any crime scene investigator. In case the background surface glows under light source illumination, it is necessary to tune the light so that only the print of interest glows.

Figure 7 shows latent fingerprint identified by mean of a secondary method which allows photographing the fluorescent print by using, in this case, the crimescope. The left specimen was obtained thanks to the use of DFO (1,8-diazafluoren-9-one), a chemical compound which makes fingerprints to glow when hit by laser or blue-green light. This organic agent reacts with amino acids present in the fingerprint perspiration to form highly fluorescent derivatives. The right example has been developed by employing the crimescope in association with a fluorescent dyer, normally used on prints previously treated with cyanoacrylate.

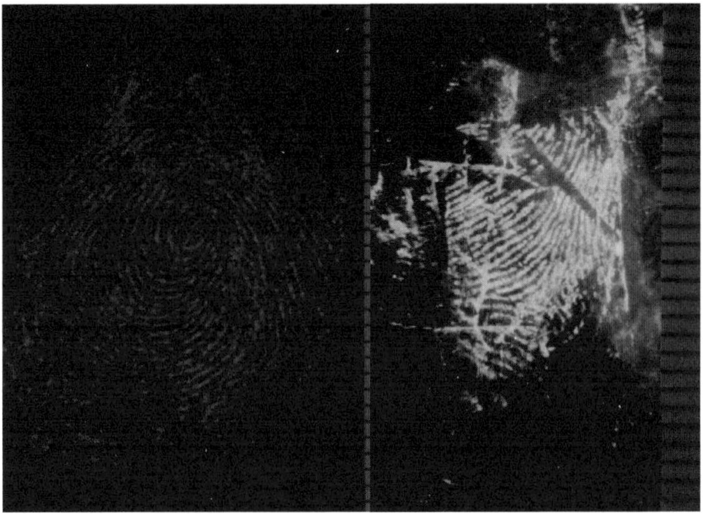

Fig. 7. Latent fingerprints developed, respectively, left image: with DFO (diazafluoren) glowing at $\lambda_{em}$= 580 nm (illuminated at $\lambda_{ecc}$= 560 nm) and, right image: with basic yellow, emitting at $\lambda_{em}$= 495 nm (excitation $\lambda_{ecc}$= 445 nm).

## 2.2. *The Forensic Luminol Test for Blood*

Blood is one of the most common physical evidences in investigations of violent crimes. Forensic analysis of blood, found at a crime scene, provides valuable information that can be decisive in solving a crime.

Generally, the use of forensic serological techniques can answer most of the questions, such as: "Is it really blood? Is the blood human? Which is the source of the blood (identity and sex)? How did the blood get there?.....". Blood stains can be hard to see on different substrates and difficult to separate from other materials. Since half of the 19th century, many different blood tests have been described. Often these tests are based on the ability of hemoglobin to catalyze the oxidation of a chromogenic compound which produces a color change[20]. All of them produce false positive and negative results to some degree, therefore they are only presumptive.

Since 1928, when its forensic potential was first reported[21], Luminol has been utilized for the detection of latent bloodstains during investigations involving violent crime. This test is based on chemiluminescence[22-26]. Luminol powder ($C_8H_7O_3N_3$) is mixed with hydrogen peroxide ($H_2O_2$) and a hydroxide (e.g., KOH). The Luminol solution is sprayed where suspected blood stains might be present. Only a tiny amount of iron is required to catalyze the reaction as the sensitivity of this test is very high and minute traces of blood can be easily identified even when attempts have been made to wash away the incriminating evidence.

The iron from the hemoglobin in the blood serves as a catalyst for the chemiluminescence reaction that causes Luminol to glow (blue light).

The half-life of the emission is rather variable, depending mainly on both the quantity and the quality of the catalyst, given constant concentrations of Luminol, the oxidant and the base[27-29]. In most cases the half-life of chemiluminescence from blood has been observed to be about 20 - 40 s, although detectable emission may be viewed for up to 3 min[30,31]. In typical conditions of Luminol forensic testing, the intensity of the light is primarily proportional to the concentration of the metal ion present, given both the oxidizer and the reductant (Luminol) at a constant concentration[26,32]. The key events in the discovery, study and use of Luminol as a forensic reagent are presented in figure 8. Over the last 20 years, Luminol has become one of the most commonly exploited chemiluminescent reagents for applications to molecular biology and analytical chemistry. It has been used as the basis for a large number of sensitive and selective detection methods, including High Performance Liquid Chromatography (HPLC), immunoassay, DNA probes, DNA typing and as substrate in western blot detection[26,31-39].

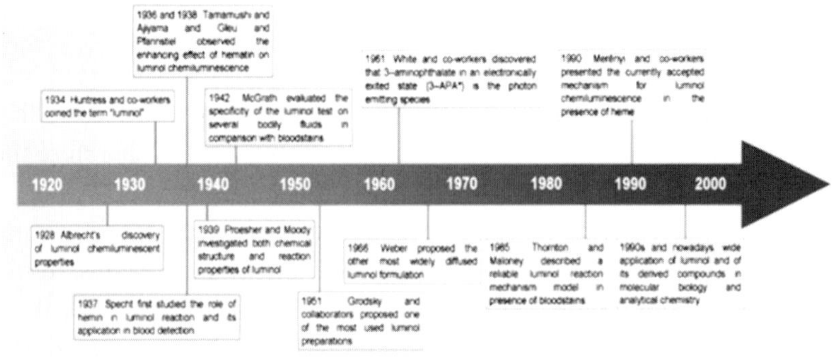


Fig. 8. Key dates in the use of Luminol in forensics.

In the following imagines, a few examples of blood analyses carried out by the RIS of Rome are displayed. All crime scenes show how the use of Luminol is a key tool to identify biological traces otherwise not visible when looking at them by naked eye. Recent applications have also been successfully carried out[40,41] in historical and archaeological studies, disclosing an interesting new field for Luminol-based assays.

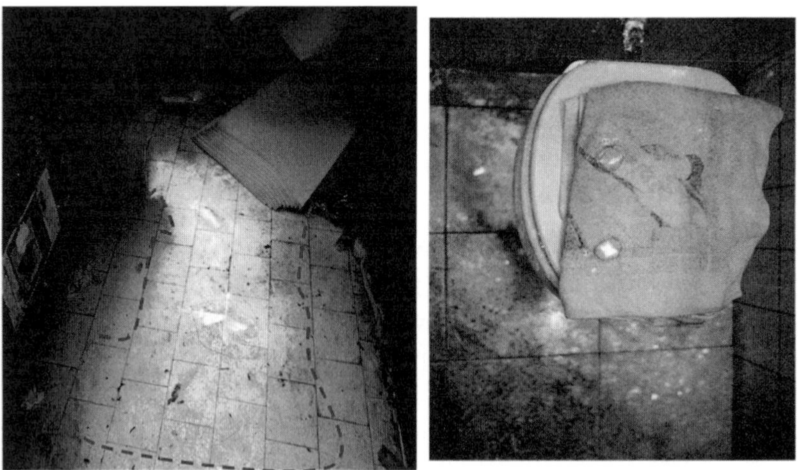

Fig. 9. Luminol test performed on tiles floor (left and right) and on a carpet (right); latent blood washed is still visible by the Luminol test as a bright blue light emission.

Fig. 10. Washed bloodstains on a car upholstery; the presence of metal objects interfering with the Luminol test is visible as a marked light emission resembling the object shapes.

For further details and a comprehensive review of forensic applications of Luminol as a presumptive test for latent blood detection, the reader can refer to 42 and 43.

## 3. Gunshot Residue Analysis

### 3.1. *Gunshot Residue Formation*

When a gun is discharged, a mixture of vapors and particulate material is expelled from the muzzle, ejecting port, barrel/cylinder gap or open breech. These products of firearm discharge, originating from the combustion process, are collectively called Gunshot Residues (GSR).

Research and recent development in technology have led to improvements and refinements in GSR examinations over the past several years[44].

The detection and identification of GSR particles on suspect's body, such as skin, on an entrance wound of a victim, on objects or on other target materials at the crime scene, such as clothing or surfaces (vehicle interiors and exteriors), have always represented an important information in investigation besides being a determinant evidence in the courts of justice. Usually, analysis of GSR particles is carried out in order to find out whether or not a suspect is involved in shotgun activities or to determine a bullet entrance hole or to estimate a firing distance[45,46].

Indeed, the discharge of a firearm, particularly a revolver, can deposit residues even to persons at close proximity, so interpretations as to who fired the weapon should be made with caution[47].

GSR from firearms discharge consists of many different components: burnt and unburnt particles from the propulsive charge, the primer, the bullet, the cartridge case and the firearm itself[48-51]. When the primer mix burns, it escapes from the gun in the form of a vapour or plume of airborne particulate and it then solidifies into fine particles (different in shape and size - from submicron to over 100 μm), settling into the hands, clothing and surfaces in the immediate proximity of the discharged weapon. Depending on the weapon, on its condition, on the cartridges, on the calibre, as well as on the environmental conditions (indoor or outdoor), particle dispersion and its flight may be different.

According to the ASTM standard[52], it is considered as "characteristic of GSR" spheroid-like shape particles (or shape characteristics with having been molten) between 0.5 and 5 μm in diameter, composed by lead (Pb), antimony (Sb) and barium (Ba). Those composed only by antimony and barium are considered "consistent with GSR". In fact, most primers consist of lead styphnate (Pb) as an initiating explosive, barium nitrate (Ba) as an oxidizer, and antimony sulfide (Sb) as a fuel; therefore, a combination of these elements in a single particle proves very significant.

The morphology of primer-derived particles (see figure 11) is consistent with the theory of rapid cooling from extreme temperatures (1500 to 3600°C) and high pressures (1400 to 40000 psi)[53,54]. The process is schematically shown in figure 12.

Fig. 11. GSR particle originated by sublimation of Pb, Ba and Sb and rapid condensation.

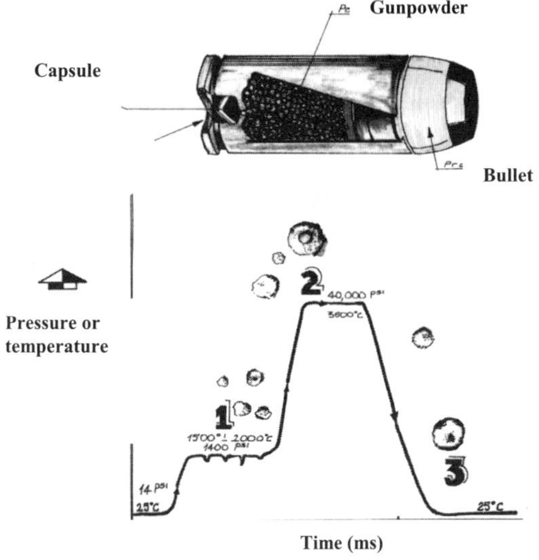


Fig. 12. Gunshot residues from a firearm through the 3-phase explosion of the primer capsule.

Indeed, bullet ejection makes temperature and pressure to lower till environmental values. As melting points of the three elements are 328 °C for lead, 631 °C for antimony and 725 °C for barium, respectively, antimony and barium are the first to solidify. They will then constitute the core of the GSR grain with possible external lead nodules jacket. The grains are made of Ba, Sb and Pb, either in single particles, or associated among each other together with other elements.

Identification of GSR can be very critical as contamination via secondary transfer and environmental and occupational particles can occur. They might interfere the identification of GSR and can originate from stud guns, cap guns, pyrotechnics, brake linings, lead smelting, lead-acid battery and lighter flints.

Particles containing lead and bromine are found in the emissions from the combustion of leaded petrol. Lead is found in plumbing materials, in battery plates, in type metal, in solder, in glass and in paint. Antimony is found in several alloys, often with lead, and its oxide is used as a fire retarding in cotton and polyester blend fibers. Barium is found in paint, in automobile grease and barium sulphate from paper is probably the dominant source of environmental barium on hands[55]. A study of barium and antimony distribution on the hands of 269 non-shooters found that activities, such as working as an auto mechanic, electrician or construction worker, had a tendency to result in higher levels of Ba and Sb on the hands[56]. Therefore, forensic examiners must be very cautious when analyzing samples.

The International Scientific Community agrees in identifying two main GSR particles classes, depending on their chemical composition, morphology and size, characterized by a peculiar homogeneous distribution originated from the very fast condensation process:
  - "Characteristic (or unique)": made of  Pb-Sb-Ba, occasionally in combination with some other elements such as Ca, Si, with traces of S. Cu and Zn are absent or in traces;
  - "Consistent with (but not-unique):" made of aggregates of Ba-Sb, Pb-Sb, Pb–Ba, Pb, Ba, if S is absent, or present only as a trace, Sb (rare).
  - In both "characteristic" and "consistent" with GSR particles, one or several of the following and only the following elements could also be present: Si, Ca, Al, Cu, Fe, S, P (rare), Zn (only if Cu is also present), Ni (rare and only with Cu and Zn), K and Cl.

## 3.2. *Gunshot Residue Analysis*

Since the 1980s, GSR analysis techniques have been based on the use of the Scanning Electron Microscopy/Energy-Dispersive Spectrometry (SEM/EDS), a non-destructive technique, which provides information about the unique elemental composition and on the morphological characteristics of the GSR material being analyzed[54, 57-59]. This represents the most powerful tool for forensic scientists to determine the proximity to a discharging firearm and/or the contact with a surface exposed to GSR. It has become the preferred method of analysis over bulk techniques, such as atomic absorption, because it provides increased specificity, as well as the ability to conduct analysis without chemicals.

Particle analysis, in fact, can identify individual gunshot residue particles through both morphological and elemental characteristics.

Moreover, minimal sample preparation is required.

The spheroidal morphology, typical of these particles, discriminates them from the general debris lifted from the hand and other surfaces under the secondary electron imaging mode. The higher atomic numbers of their characteristic elements allow them to be found as brighter particles among the darker background under the backscattered electron imaging mode.

A feasible technique employed to identify GSR particles is X-ray mapping[59,60]. Elemental profile is the most definitive characteristic of such particles, thus micrometer-sized spheroidal particles of various origins other than GSR are usually easily excluded as being composed of elements other than lead, barium, and antimony and do not constitute a problem[61]. Thanks to the capability of representing the bi-dimensional spatial distribution of the energy emission of the chemical element of the sample, x-ray mapping technique, coupled with SEM/EDS, can offer a new fundamental evaluation parameter in analysis of gunshot residues, making particle analysis quicker and easier.

X-ray mapping technique in SEM/EDS analysis consists in the use of pseudo-colors to represent the bi-dimensional spatial distribution of the energy emission of the chemical elements present in the sample.

Generally, prior spectrographic analysis of the chemical components is carried out and a different color is then successively associated to each of them. Examples of x-ray mapping and association of colors to elemental composition are given in the following pictures.

Figure 13 (a) shows the analysis of a GSR particle by x-ray mapping.

The frequency of the Pb - M line is associated to the red color, while the Sb - and Ba - L line are ascribed to the yellow and green colors, respectively.

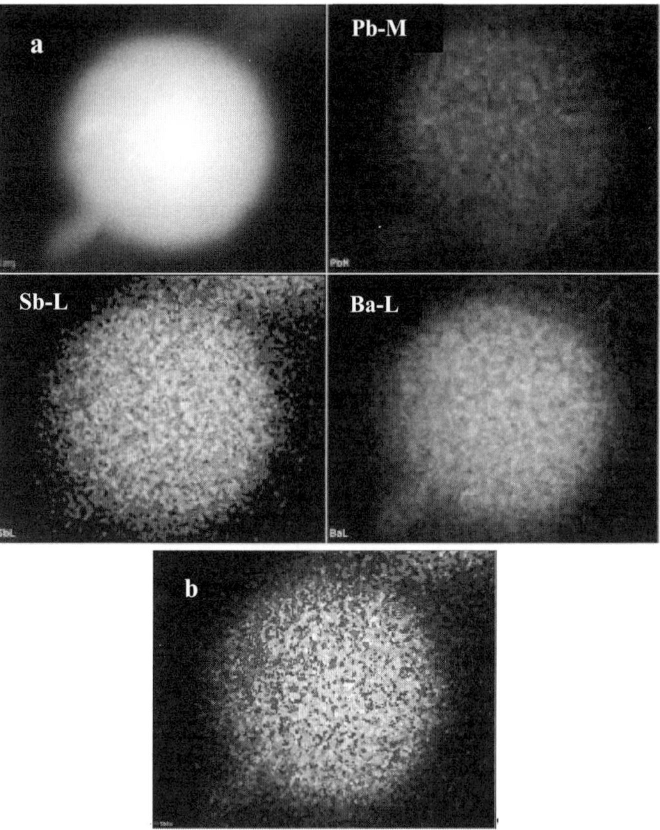


Fig. 13. a) X-ray mapping of a spheroidal GSR particle. The red map represents the Pb-M line; the Sb - and Ba - L lines correspond to the yellow and green colors, respectively; b) Superposition of the three frequencies of the three elements homogeneously distributed in the specimen.

Figure 13 (b) represents the superposition of the frequencies of the three elements which appear homogeneously distributed. An example of SEM/EDS analysis of GSR coupled with x-ray mapping is shown in figure 14. The analyzed particle (a) shows the typical emission lines in the fluorescence spectrum (b). Yellow and red colors correspond to the Pb -L and -M lines, respectively (c and d). Lead nodules can be easily observed all around the particle. Images e and f represent the same particle where blue and green correspond to frequencies of the other two components, Ba-L and Sb-L lines respectively.

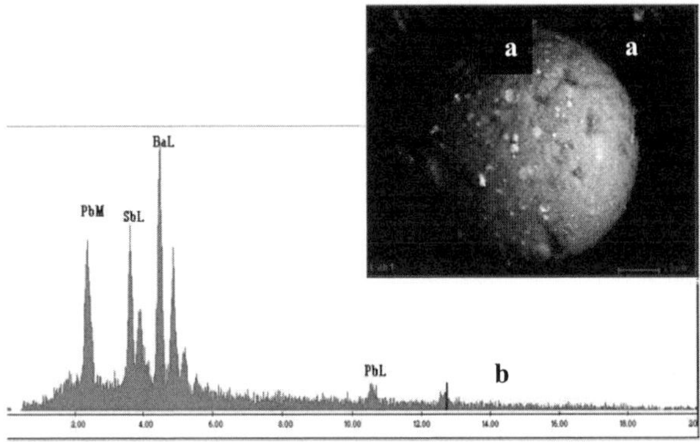

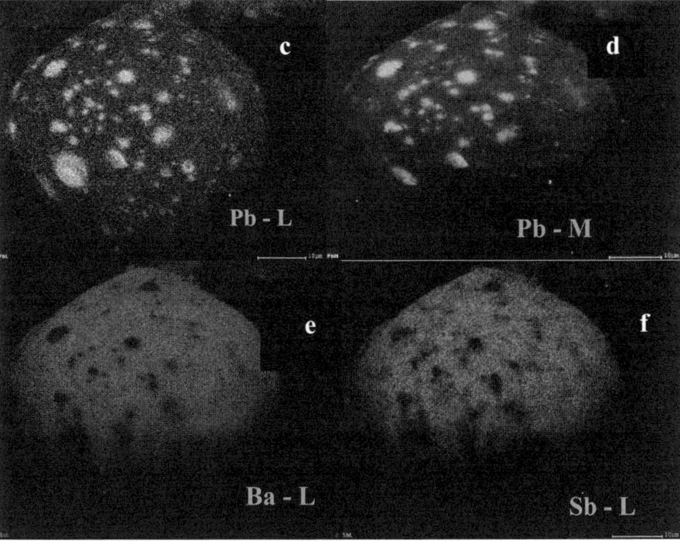

Fig. 14. SEM/EDS analysis coupled with X-ray mapping of a spheroidal GSR particle (a), containing  Pb, Sb, and Ba; (b) X-ray spectrum; X-ray maps of single emissions of Pb-L (c), Pb-M (d), Ba-L (e) and Sb-L (f) lines.

As  mentioned earlier, identification of GSR particles can be critical. Previous research has shown that brake linings and their wear, certain detonated fireworks and exploded air bags, can have particles with GSR-like elemental composition or morphology[62,63].

A comprehensive study by RIS[59,60] was performed to discriminate gunshot residue particles from aggregates of environmental and occupational origin  by means of emission line spectra analyses coupled with x-ray mapping. The results of this study have shown that an analytical approach, based on a very careful interpretation of the spectra (such as the fore mentioned bi-dimensional spatial distribution of the energy emission of the chemical elements of the sample), can be considered a very useful tool to properly distinguish the particles under examination, in particular those from brake pads and fireworks.

In the following picture (figure 15), X-ray mapping technique has been used to analyze the distributions of the various chemical elements within the sample, such as car paints.

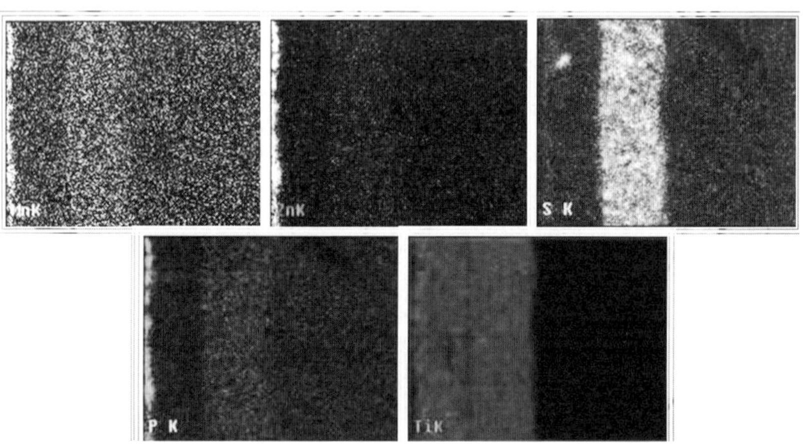


Fig. 15. X-ray maps of car paints.

## 4.   3D Reconstruction by Laser Scanner

Laser scanning is a precise method of 3D data acquisition. It represents a rapidly growing technology that captures the digital shape of physical objects. It can be used to process, measure and present 3D scans of complex environments and geometries.

The great potential of laser scanning, mainly due to its ability to enhance the design process and reduce data collection errors, makes it an attractive alternative to traditional data collection techniques, offering many advantages over traditional surveying methods. It is a very high precision, high resolution instrument which makes possible to reproduce virtually and accurately (with preservation of both dimensions and proportions) existing conditions in any environment. It represents an ideal device for 3D building documentation, civil surveying, interior design, construction and renovation, reverse engineering, historic preservation or forensic crime scene reconstruction.

Indeed, while technology is relatively new, 3D laser scanning is a revolutionary breakthrough for the law enforcement and federal crime agencies, and major police departments all over the world as it gives the possibility to go back, even years later, to explore detailed images and accurate measurements of the original scene.

Therefore, it represents a very powerful tool, not available until a few years ago, for the prosecution and defense teams, able to capture and show crime scene information into the courtroom. Indeed, the crime scene is frozen forever and it can be analyzed any time later.

The acquisition procedure consists of two steps:
1) the environment of interest is scanned by using the laser scanner;
2) different pictures of the examined area are captured by using digital cameras and dedicated procedures (lenses and number of shots) which provide a very accurate reconstruction even from a chromatic point of view.

### 4.1. *Operating Procedures*

The scanner used by the RIS of Rome for their work is the HDS6200[64] (Leica), shown in figure 16. The scanner consists of a tripod mounted phase-based laser which, through an internal rotating mirror, is fired in known directions. The laser beam travels out, hits the objects of interest and is reflected back to the scanner, generating a 360°, three dimensional "point-cloud" of survey information. The beam is therefore swept over the surrounding area, virtually blanketing the selected scene with laser measurements.

When the scanner detects the return signal, it calculates a distance measurement based on the phase shift or flight time of this reflected signal. Combining these distance measurements with internal angle measurements of the scanner's rotating mirrors, relative X, Y, Z co-ordinates can be established for each point on a surface that the beam hits.

Fig. 16. Leica HDS6200 ultra-high speed laser scanner.

A laser scanner collects hundreds of thousands of closely spaced surface geometry measurements (points) with high accuracy in just a few minutes. These surface points, combined with a sophisticated and dedicated software, are displayed on a computer, resulting in a "point cloud" looking like a dense, 3D bit-map image of the surveyed area which can be viewed and navigated much like a 3D model.

The software plays a critical role in handling and viewing high-definition point clouds and helps in the speedy extraction of data.

The device also measures the intensity of the reflected beam, giving the possibility to calculate a parameter related to the reflectance of the hit material. By combining the point cloud with the colored map of the different reflectances, a colorimetric reproduction of the scanned object (or area) can be obtained, as shown in figure 17, where a selected part of a wardrobe has been virtually reproduced by using the 3D laser scanner.

The information collected and stored as point cloud allows to re-build the analyzed environments and to make precise measurements between any point with high accuracy, thanks to the possibility to overlay the scan data over the photographic data. Once virtually reconstructed the environment of interest, it is possible to be part of it, move inside it and view the scene from infinite points of view.

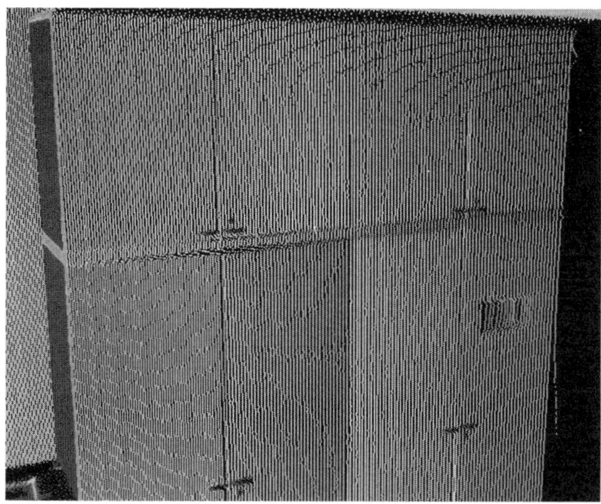

Fig. 17. Cromatic reconstruction wardrobe by a 3D laser scanner.

Therefore, it is possible to create multiple views of the same situation in order to show what somebody may or may not have been able to view of an incident based on where they were standing.

One of the possible applications of this in the forensic area is the reconstruction of a witness viewpoint. Thanks to the ability of the instrument to capture the scene from different angles, one has the possibility to settle, virtually, the person in the position where he/she was at the time of the crime event, and "play again" what happened from that point. This gives the possibility to verify, without doubt, the declaration of the witness.

Inside a house one can show a room or scene from any view, proving the jury what a homeowner would see looking out of the window, as well as what a person outside would see looking in from the street, preserving the geometry.

Moreover, if new evidence comes up some time later, one can make new measurements from the original data.

Another example is to determine where a gunshot originated from through the laser scanning of ballistic trajectory rods. Figure 18 shows a car involved in a shooting event with the blue line representing the trajectory reconstructed by using rods inserted in the bullet perforation.

Fig. 18. Reconstruction of a ballistic trajectory (blue line) inside a car involved in a shooting accident.

Therefore, the ballistic reconstruction, through the trajectory scans for bullet path and point of impact, gives the possibility to study and understand precisely the dynamic of any incident, especially when the sequence of events is not clear and a simulation can reasonably help in reproducing what happened in reality.

And, by exploiting the capability of the device to virtually relocate objects/persons not anymore available when the crime scene was analyzed by the investigators, one can reproduce with extreme accuracy the entire accident for further evidence of the facts.

For example, in case the car was moved after the shooting, just by scanning the crime scene and the vehicle, separately, one can virtually place the car exactly where it was, so that investigators can make additional measurements and examinations, including verification of witness viewpoints without having to physically move anything into the scene or even return to the place.

A similar ballistic reconstruction is possible in case of blood drops trajectories and for the determination of the centers of origin of the bloodstains. For crime scene investigation in cases of homicide, the pattern of bloodstains as well as the ballistic approximation of the trajectories of the blood drops at the site are of critical importance. The morphology of the bloodstain pattern serves to determine the approximate blood source locations, the minimum number of blows and the positioning of the victim.

Advantages of the Blood Pattern Analysis (BPA) method are: short preparation time on site, non-contact measurement of bloodstains and high accuracy of the bloodstain analysis. This technique offers accurate results on the number and position of the areas of origin of bloodstains, allowing relevant forensic conclusions regarding the course of events reconstructed by the ballistic bloodstain pattern analysis.

## 5. Questioned Documents Examination

The forensic document examination refers to the analysis of documents potentially counterfeit or altered which may become disputed in a court of law. Nowadays, forensic document examiners are faced with a growing number of challenges, including: forgeries, alterations, image and page substitution and hidden security features in printing technology.

Moreover, documents such as passports, visas, financial/ fiduciary documents, driving licenses, identity cards, banknotes, insurance documents, wills and cheques might be counterfeited or fraudulently modified.

On the other hand, it is sometimes necessary to decipher information that has been obscured, obliterated or erased.

The most suitable technique for this type of analysis is based on a Video Spectral Comparator (VSC), a digital imaging device, representing a valuable and compact laboratory tool for any document examination.

This is a device, available at the majority of advanced forensic laboratories, which allows to analyze inks, visualize hidden security features and reveal alterations and irregularities on the document of interest.

A VSC is a multi-spectral comparison instrument, capable of viewing and recording the response of documents/inks when exposed to light of varying wavelengths including: visible, infrared (lighting and luminescence), ultraviolet, transmitted, coaxial and oblique lighting.

This variety of light spectrum allows for the detailed examination and comparison of inks, security features (such as holograms), altered or obliterated information, and invisible personal information (such as personal data encoded into the passport photo, linking the picture to the owner and to the document).

One of the possible examples, among the forensic features typical of this device, consists in the possibility of reading and deciphering the two typeface lines printed at the bottom of the biographical data page – the so called Machine Readable Zone (MRZ) - available in all passports. These data, consistent with the International Civil Aviation Organization (ICAO) standards, have letters, numbers and hatch marks.

Being able to read these lines, the machine is able to make a fast check of the consistency of the data.

Another kind of investigations where VSC is generally used is the verification of the currency's authenticity. Security features and other intentional markers can luminesce in the presence of UV radiant energy.

The presence of these features on non American currency, for example, can assist in identifying fake banknotes.

In Italy, since Euro was introduced, the reference authority for this analysis is the European Central Bank. However, generally, forged bills are not very difficult to be identified by other means, unless very sophisticated counterfeit.

The most common devices used for these analysis are the VSC (Foster & Freeman)[65] and the Projectina Docucenter NIRVIS[66], representing comprehensive digital imaging systems providing the questioned document examiner with an extensive range of facilities for detecting irregularities on altered and counterfeit documents. They are equipped with a high-resolution colour camera and zoom lens, a range of viewing filters and multiple illumination sources (tungsten, halogen, and fluorescent lamps). An integral microspectrometer allows to perform measurement of reflectance, transmission, and fluorescent features.

The different excitation filter positions range in bandwidth from 80 - 140 nm within the 380 - 1000 nm portion of the spectrum. They also have two different types of barrier filters, long-pass and band-pass, which can be used alone or in combination with one another.

The performing optics, high resolution digital camera, fully controlled working sequences and functions permit investigations in short UV and in the IR luminescence range.

Zoom optics allow for magnification up to 80 - 100 times. In particular, a digital imaging filter detects and enhances latent security images on documents. Thanks to an integral 6 nm resolution microspectrometer, measurements of reflectance, transmission, and fluorescence can be performed.

Forerunners of the last generation of video spectral comparators, these devices are based on the hyperspectral imaging method (combination of spectroscopy and digital imaging) which gives the possibility to optimize the spectral differences between two different inks on a document.

Thanks to a sophisticated software, it is possible to visualize, with a simple command, images of the whole document collected at different wavelengths (from 350 to 1000 nm). As hyperspectral imaging is a non-destructive technique, the sample is preserved for subsequent chemical analysis.

The complex integration of so many different components represents the ideal tool to store loads of information and visualize it in a very easy and straightforward way, which is a key point when showing data in front of a court.

Some of these devices are also equipped with worldwide documents databases, comprehensive forensic libraries, gathering authentic documents imagines, which allow to identify and compare specific features present in the analyzed material.

When using the IR radiant energy source and filters, the examiner is able to see through inks to reveal objects that are obscured to the naked eye. This feature can be used alone or in combination with other lighting options, such as transmitted light, to visualize concealed watermarks.

The picture in figure 19 has been reproduced by employing this particular feature. The document shows part of text written by two different black inks of the gel pen type. Although the two inks appear similar in color to the naked eye, they can have different color components revealed by color spectrum analysis. By using the video spectral comparator, the  specimen was illuminated with the visible light through IR fluorescence (green excitation - 510 nm and 750 nm cut off filter). Under these excitation conditions, the altered text, initially hidden, came out showing a "1" before the "4" in the numerical part and "teen" after "four" where the figures are indicated: $ 4.000 dollars have become  $ 14.000.

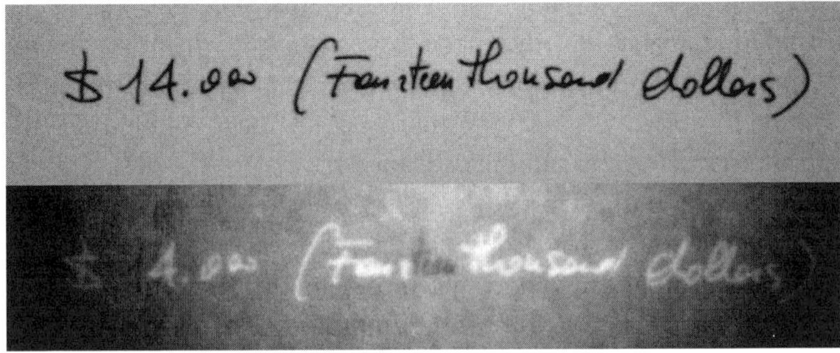

Fig. 19. Writing by two different inks observed through IR luminescence by a video spectral comparator. The numbers and the figures have been fraudulently added to increase the value of the inspected document (bottom image shows the number and the letters added in the original document to falsify the check).

Beside the advantage of obtaining the result with extreme rapidity, the visual impact in forensics of this tool, based on a multi-spectral imaging system, is impressive as, in case of disputed documents, it gives the possibility to show evidence in a court in front of a non technical jury.

## Acknowledgments

The author gratefully acknowledges Reparto Carabinieri Investigazioni Scientifiche in Rome, in particular Tenente Colonnello Luigi Ripani, Tenente Colonnello Davide Zavattaro, Maggiore Paolo Fratini, Capitano Filippo Barni and Tenente Rita Termini who collaborated at various levels to this work through fruitful discussions, valuable material, bibliography and constructive comments.

## References


1. Status and Needs of Forensic Science Service Providers: A Report to Congress, National Institute of Justice, 2006, www.ojp.usdoj.gov/nij/pubs-sum/213420.htm.
2. Forensic Light Source Applications: Wavelengths and Uses, http://www.horiba.com/fileadmin/uploads/Scientific/Documents/Forensics/fls.pdf
3. D. L. Shenkenberg, *Photon Spect.*, **43** (2009).
4. L. W. Chen and K. B. Ee, *Malaysian J. Forensic Sci.*, **1**, 1 (2010).
5. K. Virkler and I. K. Lednev, *Forensic Sci. Int.*, **188** (2009).
6. N. S. Soukos, K. Crowley, M. P. Bamberg, R. Gillies, A. G. Doukas, R. Evans and N. A. Kollias, *Forensic. Sci. Int.*, **114** (2000).
7. L. S. Powers and C. R. Lloyd, *Method and apparatus for detecting and imaging the presence of biological materials,* US Patent 7186990, March 6, 2007.
8. C. Estes, A. Duncan, B. Wade, C. Lloyd, W. Ellis Jr. and L. Powers, *Biosens. Bioelectron.*, **18** (2003).
9. K. De Wael, L. Lepot, F. Gason and B. Gilbert, *Forensic Sci. Int.*, **180** (2008).
10. J. Grasselli, Chemical Applications of Raman Spectroscopy, John Wiley & Sons, New York (1981).
11. K. Virkler and I. K. Lednev, *Foresic Sci. Int.*, **181** e1- e5 (2008).
12. V. Sikirzhytski, K. Virkler and I. K. Lednev, *Sensors*, **10** (2010).
13. E. Kiriakous, *Forensic. Sci. Int.*, **202** (2010).
14. M. Jacoby, *Chem. Eng. News*, **86** (2008).
15. G. Mogilevsky, L. Borland, M. Brickhouse and A. W. Fountain III, *Int. J. of Spectroscopy*, **2012** (2012).
16. H. G. M. Edwards and J. S. Day, *J. of Raman Spectroscopy*, **35**, 7, (2004).
17. H. G. M. Edwards and J. S. Day, *Vibrational Spectroscopy*, **41**, 2, (2006).
18. C. Lennard and M. Stoilovic, *The practice of crime scene Investigation*, United States of America, J. Horswell editor, CRC Press, pp 97-123 (2004).
19. http://www.crimescope.com/march%2015/crimescope.htm .
20. S. H. James, P. E. Kish and T.P. Sutton, *Principle of bloodstain pattern analysis: theory and practice*, CRC press Taylor & Francis Group, ISBN 978-1-8493-2014-9 (2005).
21. H.O. Albrecht, Zeitschrift für Physikalische Chemie, **136**, (1928).
22. E. Wiedemann, *Ann. Phys. Chem.,* **34** (1888).
23. A. M. García-Campaña, W. R. G. Baeyens, L. Cuadros-Rodríguez, F. Alés Barrero, J. M. Bosque-Sendra and L. Gámiz-Gracia, *Curr. Org. Chem.*, **6** (2002).
24. J. W. Hastings and C. H. Johnson, *Methods Enzymol.*, **360** (2003).

25. A. Roda, M. Guarigli, E. Michelini, M. Mirasoli and P. Pasini, *Anal. Chem.*, **75** (2003).

26. N. W. Barnett and P. S. Francis, *Chemiluminescence: Liquid-Phase, Encyclopedia of Analytical Science*, 2$^{nd}$ ed., Elsevier Academic Press, London (2005).

27. E. H. White and M. M. Bursey, J. *Am. Chem. Soc.*, **86** (1964).

28. D. F. Roswell and E. H. White, *Methods Enzymol.*, **57** (1978).

29. M. J. Cormier and P. M. Prichard, *J. Biol. Chem.* , **243** (1968)

30. R. King and G. M. Miskelly, *Talanta*, **67** (2005).

31. E. J. M. Kent, D.A. Elliot and G. M. Miskelly, *J. Forensic Sci.*, 48 (2003)

32. N.W. Barnett and P. S. Francis, Chemiluminescence: Overview, Encyclopedia *of Analytical Science*, 2$^{nd}$ ed., Elsevier Academic Press, London, (2005).

33. L. J. Kricka, P .E. Stanley, G. H. G. Thorpe and T. P. Whitehead, *Proceedings* of the 3$^{rd}$ International Symposium on Bioluminescence and *Chemiluminescence,* Academic Press, New York, (1984).

34. T. Nieman, in: J.W. Birks (Ed.), *Chemiluminescence and Photochemical Reaction Detection in Chromatography*, VCH, New York, pp. 99–123, (1989).

35. A. M. García-Campaña, W. R. G. Baeyens and Y. Zhao, *Anal. Chem.* **69** (1997).

36. J. Yuan and A. M. Shiller, *Anal. Chem.,* **71** (1999).

37. A. Roda, P. Pasini, M. Guardigli, M. Baraldini, M. Musiani and M. Mirasoli, Fresenius J. Anal. Chem., **366** (2000).

38. A. M. García-Campaña,W R. G. Baeyens, *Chemiluminescence in Analytical Chemistry*, CRC Press, Boca Raton, (2001).

39. M. Yamaguchi, H. Yoshida and  H. Nohta, *J. Chromatogr., A* **950** (2002).

40. A. L. Vish and T. E. Yeshion, *N. Am. Archaeol.*, **25** (2004).

41. A. Tug, Y. D. Alakoc and I.H. Hanci, *Forensic Sci. Int.*,**153** (2005).

42. F. Barni, S. W. Lewis, A. Berti, G. M. Miskelly and G. Lago, *Talanta* , **72**, (2007).

43. S. W. Lewis and F. Barni,  Luminol in A. Jamieson and A. Moenssens *(eds.),* Encyclopedia of Forensic Science, John Wiley & Sons Ltd, Chichester, UK:1645 - 1656 (2009).

44. W. Matty, Assoc. Firearms Tool Mark Examiners J., **19** (1987).

45. W. Lichtenberg, *Forensic Sci. Rev.*, **2** (1990).

46. K. Sellier, *Forensic Sci. Progress*,  **4**, Springer-Berlin (1991).

47. J. I. Thornton, D. Crim, P.J. Cashman, *J. Forensic Sci.*, **31** (1986).

48. K. M. Pun and A. Gallusser, *Forensic Sci Int.*,  **175** (2007).

49. B. A. Bydal, Assoc. Firearms Tool Mark Examiners J., **22** (1990).

50. J. S. Wallace, Assoc. Firearms Tool Mark Examiners J., **22** (1990).

51. A. R. Calloway, P. F. Jones, G. L. Loper, R. S. Nesbitt and G. M. Wolten, Final Report on Particle Analysis for Gunshot Residue Detection, The Aerospace Corporation, El Segundo, CA (1977).

52. Standard Guide for Gunshot Residue Analysis by Scanning Electron Microscopy/ Energy-Dispersive Spectrometry, ASTM Standards, Designation E 1588-08 (2008).

53. G. M. Wolten, R. S. Nesbitt, *J. Forensic Sci.*, **25** (3)(1980).

54. S. Basu, *J. Forensic Sci.*, **27** (1) (1982).

55. J. Andrasko, A. C. Maehly, *Forensic Sci.,* **22** (2) (1977).

56. D.G. Havekost, C.A. Peters and R.D. Koons, *J. Forensic Sci.,* **35** (1990).

57. G. M. Wolten, R. S. Nesbitt, A. R. Calloway, G. L. Loper and P. F. Jones, Equipment System Improvement Program-Final Report on Particle Analysis for Gunshot Residue Detection, Report ATR-77(7915)-3, The Aerospace Corporation, El Segundo, CA (1977).

58. J. S. Wallace, J. McQuillan, *J. Forensic Sci.,***24** (5) (1984).

59. B. Cardinetti, C. Ciampini, C. D'Onofrio, G. Orlando, L. Gravina, F. Ferrari, D. Di Tullio and L. Torresi, *Forensic Sci. Int.*, **143** (2004).

60. L. Garofano, P. Fratini, M. Minardi, G.P. Bizzaro, F. Tortora and L. Manna, *Scanning*, **25** (2) (2005).

61. L. Garofano, M. Capra, F. Ferrari, G.P. Bizzaro, D. Di Tullio, M. Dell'Olio and A. Ghitti, *Forensic Sci. Int.,* 103 (1) (1999).

62. C. A. Torre, *J. Forensic Sci.* **47** (2002).

63. R. E. Berk, *J. Forensic Sci.*, **1** (2009).

64. http://hds.leica-geosystems.com/en/Leica-HDS6200_64228.htm.

65. http://www.fosterfreeman.com/.

66. http://www.forensictechnology.com/nirvis/.

# FUTURE TRENDS

Mauro Varasi

*Finmeccanica*
*Piazza Monte Grappa 4, 00195 Roma, Italy*
*E-mail: mauro.varasi@finmeccanica.com*

The growing demand in terms of safety and security generates challenging requirements in the various areas of systems, ranging from sensors to processors, up to the actuators. Photonics is able to provide appropriate responses and in the future to offer radically innovative solutions in each of these areas, particularly for sensing. This key role has been recognized by the European Community by including photonics among the six Key Enabling Technologies (KET) having the greatest potential to strengthen Europe's capacity for innovation and strategic industrial development. The trends of development in the field of photonics for safety and security are supported on the one hand by the research on basic technologies and devices, often oriented parallel to those for other applications, including communications, and on the other, at the system level, by specific applications. Among the main technological trends, those with greater impact to safety and security are geared towards miniaturization, hybrid integration and quantum photonics, as well as the development of embedded systems and networked solutions, key factors in the safety and security applications. The trends in application driven solutions will be examined in some major areas such as autonomous systems, smart and critical infrastructure, platform protection, anti terrorism, secure communications, environmental monitoring and transportation.

## 1. Introduction

The development of photonic technologies that have been in recent decades, has resulted in a pervasive application in the main operational functions of the systems: sensors, processors, and actuators. Along with this generic trend, photonics is called to play a key role in the future to sustain a safe and secure society through environmentally and economically sustainable solutions. This key role has-been recognized by the European Community which included photonics among the six KETs (Key Enabling Technologies)[1], having great potential to strengthen Europe's capacity for innovation and strategic industrial development.

This chapter will deal with an analysis of macro technological trends with high potential impact to safety and security applications, identified primarily as inertial evolution of researches and development going on to date, and a survey of the main solutions expected to respond to particularly critical operational needs.

Among the technological macro trends that characterize the photonic technologies in general, some can be identified as particularly relevant to safety and security applications:

 (i) The integration through miniaturization technologies;
(ii) The growing "intimacy" between photons and electrons, through on-chip more and more integrated approaches, and between photonics and mechanics, through MOEMS and NOEMS solutions;
(iii) The exploitation of "quantum" concepts and solutions;
(iv) The growing impact of "embedded system" concepts, as well as the evolution towards "system embedded" photonics;
 (v) The networking through digital networks, wired and wireless, which enables the cooperation among subsystems (swarm) and the distributed and higher-level processing (data fusion).

Beyond these technological macro trends, the evolution of the photonic solutions aimed at safety and security are driven by some relevant application scenarios.

In the following of the chapter, these main trends will be discussed in some details, along with some relevant applications such as in the fields of the autonomous systems, the smart and critical infrastructure, the platform protection, anti terrorism, secure communications, environmental monitoring and transportation. The trends in photonic solutions and in the related enabling technologies will be discussed, taking into account the operational functions that provide, case by cases, the most significant development opportunities.

## 2.  Technological Trends

The technological trends characterizing photonics applied to safety and security do not differ substantially from those of photonics in general. On one hand, trends driven by the whole of the most innovative technologies, ranging from photonic crystals to plasmonics, including various quantum phenomena, which, in various combinations, give rise to what is generally called Nanophotonics. As part of nano photonics, a few strands of innovation can be identified at the device level, including those oriented to miniaturization, hybrid integration and quantum photonics.

On the other hand, some trends largely application-driven, including the development of concepts of embedded systems, aimed at smart structures, and the integration of distributed elements, sensors and processors, in cooperating networks.

### 2.1. *Miniaturization*

The way leading to the pervasive introduction of photonic technologies in safety and security systems, as already demonstrated by the development of components for optical fiber communications, necessarily passes through the development of increasingly miniaturized and multifunzional[2] solutions, both to reduce dimensions, weights and consumption, and to implement advanced features otherwise difficult or impossible to be implemented in a reliable and cost effective way.

Necessary elements for the development of these solutions are the technology platforms for multi-functional planar circuit integration.

Among the various possible platforms, those silicon- and polymer-materials based emerge for the potential to integrate at relatively low cost a wide range of passive and active functions. The silicon photonic technology is the most mature among the planar photonic platforms, widely used in the production of devices for optical fiber communications[3]. The integration of very low propagation losses silica optical circuits with electronics on the same silicon substrate, allows a rich toolbox enabling for a wide range of device solutions. The limits of silicon as substrate for the active device realization, such as laser sources and detectors, are overcome by hybrid integration of heterogeneous elements, as in the case of laser sources through the surface bonding of III-V semiconductor chips. But, the feasibility of silicon has been demonstrated only for lasers built by four wave mixing and stimulated Raman scattering[4] effects, even though both require optical pumping.

Polymers play an important role as photonic platform too, capable of integrating, at low-cost, devices with optical, electrical and mechanical characteristics, modulated through the molecular functionalization of the polymers. For example, periodical nanostructures[5] can be easily integrated in the polymer planar optical circuit through self-assembling and photolithographic processes, being these nanostructures the basic element for the realization of photonic crystal structures.

The photonic crystal structures, regardless if realized in silicon or polymeric photonic platforms, allow the realization of a wide range of devices, such as very compact phase/amplitude modulators and switches[6], to so-called "slow-wave structures". These make possible to control the group velocity of the light signal. The latter is extremely interesting for the realization of innovative devices for sensing, microwave signal processing, optical communications and computing[7].


## 2.2. *Integration*

The aim of developing new generation devices, enabling, for example, "all optical" circuits or new sensors with very high sensitivity, in the field of safety and security, opened up a strong trend pointing to a deeper integration of photonics and electronics, mainly at chip level, but also at board and system levels, and between photonics and micro electro mechanical systems.

A first significant step towards the functional integration between photonics and electronics concerns the integration among components, as in the silicon photonics technology, where the optical circuit is easily combined with the C-MOS electronics[8]. Or, as in the case of "all fiber-based optoelectronics", by incorporating[9] crystalline semiconductor devices directly into optical fiber waveguides.

Sources and detectors are fabricated within the optical fibers via a novel high pressure microfluidic chemical deposition technique, inside the microscale to nanoscale holes of silica Microstructured Optical Fibers (MOFs). The all fiber circuit would be of strategic relevance in fiber optic communication systems as well in the fiber optic sensor systems.

Going deeper in the direction of physical integration between photons and electrons, numerous research activities have been launched and others are expected in the future, such as those on the excitons and the high nonlinearity materials, to realize, for example, the "all optical" switches. In a simplified manner, an exciton can be defined as the electron-hole pair generated in a semiconductor by the absorption of a photon that, in turn, generates a photon at the time of recombination.

Experimental proof of principle for all-optical excitonic transistors, where a second light source controls the re-emission of the primary photon by using exciton as an intermediate medium, has been demontrated[10]. Another frontier for integration between photons and electrons is plasmonics[11]. "Surface plasmons" are collective oscillations of free electrons in the metal at the metal/dielectric interface which can be excited by optical waves. The integration of plasmonic waveguide in optical circuits allows a new generation of advanced devices with high potential in communication and sensor applications.

The arena of the integration between photonics and MEMS spans over a wide range, from consumer projectors to nanosensing. MEMS diffractive gratings and micromirror arrays allow spatial light modulation for laser projection displays, potentially integrated in smart phones.

But the integration is going much deeper, as in the case of optically controlled mechanical forces between nanostructures[12] in which optical radiation is exploited to induce forces behaving like the Casimir force, creating a new generation of "photonic nanomachines" exploitable in sensing and quantum computing.


### 2.3. Quantum Photonics

The implementation of concepts of quantum physics in the photonic field has had an impact both in the engineering of solid state devices, and in the development of architectural concepts for signal processing, of big impact in security applications.

In recent years, the technology of solid state devices for the generation and the detection of photonic signal has increasingly exploited quantum concepts in a wide range of configurations, ranging from quantum wells to quantum dots, with a trend that has not yet reached its limits. Emblematic examples of this trend are the focal plan arrays for IR imaging, where the Quantum Well IR Photodetectors (QWIP) and Quantum Dot IR Photodetectors (QDIP) have already demonstrated results competitive with the more conventional solutions based on CMT. In particular, the QDIP that, due to the 3-dimensional confinement, may offer substantial advantages including higher operating temperatures and surface-normal direct absorption, and QCL (Quantum Cascade Lasers), which now allow to have sources tunable in the IR without the need to be cryo-cooled, with a trend oriented towards increasing the power and the extension of the tunability range.

Progress on a nearly daily basis is currently underway in developing new schemes and setups for quantum information manipulation with substantial advantages over its classical counterparts. Logic gates and wires are becoming smaller and soon they will be made out of only a handful of atoms. If this process is to continue in the future, the new quantum technology can replace or supplement what we currently have.

Quantum Information Technology can entirely support new schemes of information processing based on the principles of quantum mechanics.

Its eventual impact may be as great as or greater than that of its classical predecessor.

A promising line of research in modern, applied cryptography consists in using novel and secure technologies to support existing, well-established protocols. In particular, the adoption of quantum channels endows an encrypted communication session with theoretically unbreakable security and robustness against eavesdropping, thanks to the basic principles ruling quantum mechanics. As already described in chapter 13, the primary task of cryptography is to enable two parties (commonly called Alice and Bob) to mask confidential messages such, that the transmitted data are illegible to any unauthorized third party (called Eve). Usually this is done using shared secret keys. However, in principle it is always possible to intercept classical key distribution unnoticed. The recent development of quantum key distribution (standard BB84) can cover this major loophole of classical cryptography.

It allows Alice and Bob to establish two completely secure keys by transmitting single quanta (qubits) along a quantum channel. The underlying principle of quantum key distribution is that nature prohibits gaining information on the state of a quantum system without disturbing it. Therefore, in appropriately designed schemes, no tapping of the qubits is possible without showing up to Alice and Bob[13].

These concepts feed growing interest for sensing solutions, as in the case of "quantum radar"[14], where entangled states are exploited.

Quantum entanglement is a form of quantum superposition of two or more objects which have to be described with reference to each other, even though each object may be spatially separated. It results in the apparently instantaneous transmission of perturbations between entangled objects. In the concept of "quantum radar", one of the two entangled photons is sent towards the target and the successful interaction is observed through the correlated photon maintained in memory.

Last but not least, the application in the optical computing field, where a new quantum optics toolbox, opened up by "coherent photon conversion"[15], promises to lead to a nonlinear optical quantum computer.

All-optical quantum computing became feasible with the discovery that scalable quantum computing is possible using only nanoscale elements, such as photon sources, linear optical elements, single-photon detectors and quantum teleportation.

"If powerful quantum computers could be built, most asymmetric cryptographic protocols in use today would no longer be secure, which would present a serious challenge for open networks and cryptographers should be prepared for this situation" (C. Bennet).

Beyond the Classical Information-Theoretic Key Establishment (CITKE) schemes, the fast-growing knowledge accumulated on quantum computation can be used to design new public-key schemes and study their resilience to quantum computing attacks. One can indeed construct classical public-key schemes based on the lattice shortest-vector problem. Such a public-key scheme is extremely inefficient in terms of performance. However, since this problem is in quantum NP, it is not threatened by any potential speed-up on a quantum computer.

The post-quantum computing cryptography is an extremely rich and stimulating research field, on which close collaboration among computer scientists and physicists, both interested in quantum information, will continue to be extremely fertile, as it has already proven to be over the past years[16, 17].

These and many other applications of entangled states concept require the development of specific technologies in terms of sources and detectors, as well as planar integrated circuits.


## 2.4. *Embedded Systems*

The overcoming of the heterogeneity of the individual components at device and subsystem levels, towards more and more integrated structures, including computing and processing capabilities, represents the general trend towards "embedded systems" enabling for the next generation of smart sensors and systems.

Photonic sensors are representative of this trend, through the increasing integration with local computing platforms performing tasks previously concentrated in central processing resources, remote from the sensor. Example of this trend is the evolution of visible and IR image sensors towards the intimate integration with processors which provide control functions necessary for the operation of the sensor, such as calibrations and image pre-processing, such as the interpolation or the uniformity correction.

But, if on one side the photonic subsystems develop concepts of "embedded system", on the other, the same subsystems would be more and more "system embedded", as for the use of fiber optic for the distribution of digital and analog signals in large systems, like aerial and naval platforms. Large electronic systems, as for example radar as well as EW (Electronic Warfare) systems, are increasingly pervaded of photonics beyond the fiber optic networks, executing tasks and specific functions like waveform generation or signal sampling.

## 2.5. *Networking*

The trend towards integration of heterogeneous and distributed elements for the implementation of advanced sensing functions towards cognitive approaches, it is increasingly important also in view of the cloud environment.

In this perspective, photonic elements of the systems must include capabilities for the networking through digital networks, wired and wireless, which enables the cooperation among distributed subsystems (swarm) and the higher-level real time processing (data fusion).


## 3.   Application Drivers

Beyond the trends of the enabling technologies, the future of photonics in safety and security is driven by some critical applications that, though with very challenging requirements, offer large room for the development of photonic solutions.


## 3.1. *Unmanned / Unattended systems*

The role of unmanned and unattended systems in security and safety applications is rapidly and pervasively growing.

Examples are the extensive use of UAV (Unmanned Aerial Vehicles) for ground and maritime surveillance missions, UGV (Unmanned Ground Vehicles) in high risk activities, as in C-IED, UUV (Unmanned Underwater Vehicles) in activities impossible to man, as for the deep water operations during the oil spill in the Gulf of Mexico in 2011, and UGS (Unattended Ground Sensors) systems, used, for instance, to protect soldiers with a remote persistent unattended surveillance capability for improved situational awareness and actionable intelligence.

The exploitation of photonics in unattended/unmanned systems is a strong enabling factor to support the trend towards the progressive reduction of the human presence in the loop, or in other words the increase of autonomy of the systems. In this respect sensors, and in particular photonic sensors, as input of every cognitive process necessary for the autonomy of the systems, are called to play a growing role, along with the fiber optic digital networks for on board high data flow rate connectivity and photonic solutions for on board "real time" signal processing capability.

## 3.2. *Infrastructure*

In the new scenario human society is facing, Smart Infrastructures are among those with highest potential both in terms of investment and complexity and for opportunities for advanced photonic technology developments.

The quest for a better quality of life, minimizing the level of vulnerability to various risks, finds a natural framework in the broad environment of Smart Infrastructures.

Many aspects of the concept of Smart Infrastructure, from smart buildings to smart cities, from smart mobility to smart grid, create challenging requirements for the issue of safety and security. Even in this case, photonics can play decisive roles in meeting these requirements and in making possible solutions otherwise unattainable.

The broad spectrum of photonic sensors, ranging from distributed fiber optic sensors for structural or perimeter monitoring to visible or infrared vision systems, constitute an essential asset for the development of "smart" system concepts. These represent an essential part of cognitive processes, enabling systems to an advanced interaction with humans. Beyond those, photonic technologies also open up completely new scenarios.

An example of this widespread utilization is represented by photonic interfaces for mobile telephone terminals.

Today, billions of users have mobile phones with photonic interfaces enabling high impact applications both for the user and for system management as a whole. Integrated cameras and projectors, high definition displays and other photonic components can, through the network to which the mobile terminal is connected, interact and serve as input and output ports for the most disparate smart systems, by gathering data to improve users' health care or by managing users' interfaces with smart infrastructures.

### 3.3. *Active Protection*

The need for protection of both platforms, human and sensitive sites, implies the use of active protection devices in which photonic technologies, and in particular lasers, can provide significant opportunities and advantages.

The spectrum of possible applications is wide and goes from vision jamming (Laser Dazzlers) to threat destruction (Direct Energy Weapon - DEW), resulting in challenging requirements in terms of tunability, power and cost of the laser sources.

At the lowest energy level, (hundreds mw) there are non lethal laser systems designed to control crowds or potential attackers through visual jamming. These are called **laser dazzlers**, i.e., a person's vision is temporarily impaired so that driving a vehicle or grasping a weapon is not possible. Green lasers are typically used for this purpose, with trends towards compact devices, integrated into multifunctional hand held systems (including range finders, designators, …).

When increasing the power output into the watt to tenth of watts range, one finds the **infrared countermeasure**.

More than 90% of all military aircraft losses over the past 25 years were caused by IR seeker (working in the 2-5 µm band) guided missiles.

Lots of attacks by terrorists against civil aircrafts have been also reported in the same period, using IR guided MANPADS (Man Portable Air Defence Systems) and the range of the threats including IR guided ammunitions is increasing. Therefore, it is clear that the protection against such threats is of paramount importance.

A very effective answer to this threat are the so-called DIRCM18 (Directed Infra Red Counter Measures) systems, composed of UV and IR sensors that identify the event of launch of the threat (Missile Approach Warner - MAW). This is a passive IR system for threat search and tracking during its approaching flight, having a system for directing the laser on the IR seeker sensor with scopes ranging from IR sensor "dazzling" (saturation of the detector) to its destruction.

Much of the DIRCM technology relevant to aerial platforms can be transferred to the protection of ground vehicles and ships. The evolution of DIRCM systems, in addition to the need to maximize the probability of threat identification, aims toward configurations enabling the awareness and protection. This, regardless of the direction of arrival of the threat, presently limited to the lower half space looking to the ground and toward lower cost systems.

In order to have spherical coverage protection, distributed system configurations are proposed. In these systems, each element includes independent sensors, the delivery of a common central laser source. The elements distribution on the platform surface allows the full space coverage. For the laser sources, important targets for future trends are: faster and wider tunability, in the range of seekers sensitivity in order to circumvent protection filters and maximize the killing probability and higher power (tens of watts), in the minimum size and maximum efficiency. In these directions, QCL (Quantum Cascade Laser) technology offers significant advantages in terms of tunability and compactness[19] compared to more conventional sources based on solid state OPO sources.

**Laser DEW (L-DEW)** systems are at higher power levels with the aim of physically destroying targets. Despite the announced reduction (-33 %) proposed for the fiscal year 2013 in US for direct energy and laser weapons and the end of the ALTB (Airborne Laser Test Bed) experiment (last flight  from Edwards air base on February 14, 2012), a strategic trends is still in place. This includes the exploitation of high energy lasers in the few kw range for C-IED (Counter Improvised Explosive Device), in the 10 kw range for C-RAM (Counter Rocket  Artillery and Mortar) and up to multi 100 kw for anti-missiles, including ICBM (Inter Continental  Ballistic Missile) and anti ground light vehicles and small ships missions[20, 21].

After having shot down in February 2010 a threat representative ballistic missile, the ALTB (a modified Boeing 747-400, housing a megawatt class Chemical Oxygen Iodine Laser – COIL at 1315 nm, supplied by Nortrhop Grumman) definitely demonstrated that laser direct energy is a viable technology to destroy targets, but also that efforts are necessary to improve efficiency to more than 35%, in order to reduce laser volume and weight and to overcome the hazardous chemicals handling.

Other chemical lasers like Hydrogen or Deuterium Fluoride (2,6-3,3 µm and 3,5-4.2 µm) demonstrated multi hundreds kw emissions, but with similar problems as for COIL.

Therefore, the tendency is towards Solid State Laser (SSL) solutions and, more ambitiously, Free Electron Lasers (FEL), to produce 100 kw class sources. For this purpose, US Army is funding the Joint High Power Solid State Laser (JHPSSL). Nd-YAG SSL demonstrated good beam quality in the multi tens kw range[22], but thermal management of the cavity is still critical. Fiber laser seems to offer viable solutions. IPG Photonics have reported operation of a 10 kw single mode ($M^2<1.3$) $Yb^{3+}$ doped silicate fiber around 1 µm wavelength. Exploiting the easy scalability offered by the fiber sources, a 50 kw multimode beam has been demonstrated by the combination of such fiber lasers, for DEW applications by Raytheon (Waltham, MA). Further work on the fiber is needed to improve the maximum output power of the single cavity, overcoming the non linear limiting effects. Photonic Crystal fibers and new doping materials (Thulium - $Tm^{3+}$) seems to offer room for improvements. The FEL, among other advantages, is enabling to "all electrical" platforms. An Office of Naval Research (ONR) sponsored FEL project is expected to demonstrate around 2018 a full scale prototype FEL working in the 100 kw range.


### 3.4. *Chemical, Biological Radiation, Nuclear and Explosives (CBRNE) Sensors*

The growing demand in terms of security that arises from civil society encourages more and more innovative developments in the multiple reference scenarios. The detection and inspection of unauthorized goods at public locations, such as airports and international borders, is one of the important issue for public security.

Innovative photonic applications are requested to make such inspections more efficient, in terms of sensitivity and minimum false alarms. The IED (Improvised Explosive Devices) threats are now the main source of causalities in asymmetric theaters, such as in Afghanistan, and are intended to be in the future in theaters that can also be urban. The chase between the development of new IED threats and solutions for their identification and neutralization is developing in a spiral in which the photonic sensors are called to determine breaks in favor of the protection of the on-ground operators.

However, in all most relevant contexts, particularly in non-cooperative environments, the possibility of making the search and the identification of the threat from "stand-off", i.e., without contact or the need of positioning of hardware in the vicinity of the threat, is of paramount importance for the security of on-ground operators. In parallel, the solutions must evolve in order to reduce false alarms, making more effective the countermeasures to be implemented. To this end, in addition to the greater reliability of the individual technological solutions, multi-sensory integration and data fusion will offer opportunities for significant improvements. The sensing in stand-off is carried out either in passive manner, through hyper / multi-spectral analysis of the scene or processing of the IR image, or in active manner, using laser radiation as a query tool of systems generically included in the family of the LIDAR (Light Detection and Ranging). The passive solutions are the subject of constant and interesting developments that already offer interesting solutions, but further developments will improve both the sensitivity and reliability.

IR and multi / hyperspectral images provide useful information to both change detection analysis and in terms of spectroscopic investigation.

Among the "active" solutions, the most relevant ones will be those able to allow spectroscopic analysis in reflection condition and with sufficient sensitivity to detect traces of the compound on the outer surfaces of the casing, intentionally made not transparent by the attacker. In this direction the trend of the studies indicates various solutions ranging from simple absorption spectroscopy, possibly supported by tunable QCL sources to the DIALs (Differential Laser Absorption Spectroscopy: exploiting the difference in the return signals of paired pulses at two wavelengths closely spaced, one of which at the absorption peak), from LIF (Laser Induced Fluorescence) to the LIBS[23] (Laser Induced Breakdown Spectroscopy: spectroscopy emission from microplasmas generated by absorption of laser radiation) and finally, very promising, the Raman spectroscopy[24], also in configurations integrated with hyperspectral imaging.


## 3.5. *Biometrics*

To ensure security, biometric data controls at national borders (airports, seaports, crossings), critical infrastructures and big events, are becoming indispensable. In the near future most of these controls (face & iris recognition, fingerprints, pathogens, etc.) will be performed by smart and highly sensitive optical sensors that, hopefully,  will be able to detect all the necessary data remotely positioned, i.e. in "stand off" condition, in passive illumination condition, in real time and without the cooperation of the subject to be inspected.

These sensors being embedded into more complex systems developed for image recognition and interpretation, motion detection, gesture analysis and detection of energetic or hazardous materials.

The facial recognition[25], among the biometrics, sets the most significant challenges for applications both of safety and security, both in terms of technologies that of algorithms for processing the data. The ability to identify and recognize people from a distance could be decisive in a wide range of cases relevant to safety. From access control to the persistent area analysis for the identification of suspects sought.

The ability to analyze in real-time facial features and their changes will allow effective behaviometrics aimed at the identification of abnormal behavior in the crowd and then to prevent riots and, in extreme cases, suicide bombings. Advanced solutions to the stand off facial recognition also opens up interesting perspectives for safety applications, ranging from the recognition of critical situations of fatigue in those who are driving vehicles, in order to generate alarms that call attention to the driver, until the implementation of complex interactions between humans and machines, enabling the latter to better understand the human response and making more interactive the virtual environments.

An emerging trend in this respect is the three-dimensional (3D) face recognition, in which 3D imaging sensors are used to extract a wide range of information from single frame, single aperture, stand off passive systems.

Solutions such as Spatial Phase Imaging (SPI)[26], already allow to capture monochrome 3D image without the need of structured light or scanned from distances conditioned only by the operational needs and optics used.

## 3.6. *Communications*

Security in transmission of information is powerfully affecting modern society at all levels. Considerable investments are on the horizon for projects that relate to the Cyber Sphere[27], and in particular the physical and information security of the communications. But also in tactical environments, such as on the battlefield or in the operations of natural disaster recovery, the security of the communication remains a crucial point for the execution of the mission.

In the field of optical fiber transmission, the quantum cryptography has made extraordinary progresses in recent years, and lately is testing technology solutions that can be implemented at network level[28].

In parallel, the so called chaos-based communications have been proposed as broadband information carriers with the potential of providing a high level of robustness and privacy in data transmission[29]. Beyond the fiber wired communications, the Free Space Optical Communications (FSOC) offer interesting solutions for security applications. Experiences in space, show potential for applications in airborne, ground and under water environments[30,31]. But despite the potential shown, the FSOC require developments to bring technology to the consumer world, overcoming psychological barriers, such as those against the use of lasers in urban environments.

## 3.7. *Environmental Monitoring*

Photonic sensors and sensor networks have an important impact in environmental monitoring, and will have an increasing relevance in safety and security challenges, significantly contributing to more efficient use of resources and thus a reduction of greenhouse gas emissions and other sources of pollution, as well as to a persistent control to prevent natural disaster and to support environmental emergency management.

In this scenario the multiple implementations of the LIDAR technology play a central role in conjunction with multi / hyper spectral sensors and advanced imaging, both in the IR as well as in the visible ranges, operated by ground, airborne, ground and space platforms. Examples such as those experienced in the monitoring of the cloud of ash emitted by the eruption of the volcano Eyjafjallajokull (2010, Island), in order to ensure the safety of aircraft in flight, or in the operations support after the hearthquake in Haiti (2010, Haiti) or at L'Aquila (2010, Italy), are only representative of the huge potential of the LIDAR technology in the natural disaster management.

Even the multi and hyperspectral technologies, from satellite or airborne, will play an important role in the missions of surveillance and mapping of the territory, from the identification of contaminants, on land and sea, to the detection of buried mines, through diversity induced on the surface vegetation.

### 3.8. *Transportations*

Including in transport, both in terms vehicles and of infrastructure, from automotive to rail transport, from sea to air, photonics, and in particular the photonic sensors, may be decisive for the safety and security.

In road transportation the implementation of intelligent DVE (Driver Vision Enhancement) solutions using cameras, IR and Visible, and image processing to combine images to eliminate blind spot and allow the driver a 360° awareness, is enabling for safer driving and the autonomous driving of future "smart cars", along with advanced photonic OWS (Obstacle Warning Systems) implemented in passive as well active configurations.  At infrastructure level, the current systems for car plate recognition, used for traffic control, should move towards more intelligent systems able to monitor the behavior of the vehicle and the integration into networks that can support the concepts of smart mobility at the local and national level.

Similarly the impact of photonic sensors is already and will be increasingly important in the rail, where fiber optic sensors located along the lines allow monitoring of traffic in terms of speed, number and weight of the individual elements of the convoy. Just as in marine and aviation, where, together with the wide range of photonic sensors, from the implementation of concepts of fly-by-wire are expected significant benefits in terms of safety of flight.

## 4. Conclusions and Outlook

Photonic technologies are likely to enjoy a bright future in safety and security applications.

The security challenge has been taken up at European level and it has been a specific priority in the EU's Seventh Framework Programme, along with the the key role of photonics recognized by including it among the six Key Enabling Technologies (KET), having the greatest potential to strengthen Europe's capacity for innovation and strategic industrial development. Photonic sensors and their networks will have an important impact in meeting challenges in multiple fields such as unmanned/autonomous systems, counter terrorism, environmental monitoring, communications and smart infrastructure, significantly contribute to the safety and the security of the systems.

## References


1. *"Final Report of the High Level Group: Key Enabling Technologies"*, http://ec.europa.eu/enterprise/sectors/ict/files/kets/hlg_report_final_en.pdf (2011)
2. H. Rong, R. Jones, A. Liu, O. Cohen, D. Hak, A. Fang and M. Paniccia, *Nature* **433**, p. 725 (2005).
3. R. Soref, *IEEE J. of Selected Topics in Quantum Electronics*, Vol. 12, No. 6, p. 1678 (2006).
4. A. Boisen et al., in *Detection and Sensing of Mines, Explosive Objects, and Obscured Targets XV*, SPIE Conference Proceedings Vol. **7664**, p. 1 (2010).
5. C. Paquet and E. Kumacheva, *Materials Today*, Vol.**11**, No. 4, p. 48 (2008).
6. M. Santagiustina et al., *IEEE Photonics Society Newsletter*, Vol. **26**, No. 1, p. 5 (2012).
7. D. M. Beggs, T. P. White, L. O'Faolain, and T. F. Krauss, *Optics Letters*, Vol. **33**, No. 2, p. 147 (2008).
8. R. Stommer, *Photonik International*, p.108 (2007).

9.  J. V. Badding et al., in *Integrated Optics: Devices, Materials and Technologies XI*, SPIE Conference Proceeding Vol. **6475**, p.1 (2007).

10. Y. Y. Kuznetsova, M. Remeika, A. A. High, A. T. Hammack, L. V. Butov, M. Hanson and A. C. Gossard, *Optics Letters* **35** (10), p.1587 (2010).

11. S. A. Maier, *IEEE J. of Selected Topics in Quantum Electronics*, Vol. 12, No. 6, p. 1214 (2006).

12. J. Roels, I. De Vlaminck, L. Lagae, B. Maes, D.Van Thourhout and R. Baets, *Nature nanotechnology*, **4**, p. 510 (2009).

13. N. Gisin, G. Ribordy, W. Tittel and H. Zbinden, *Reviews of Modern Physics*, Vol. **74**, No. 1, p. 145 (2002).

14. E. H. Allen and M. Karageorgis, *"Radar systems and method using entangles quantum particles"*, U. S. Patent No 7, 375, 802 B2 (2008).

15. N. K. Langford, S. Ramelow, R. Prevedel, W. J. Munro, G. J. Milburn and A. Zeilinger, *Nature* **478**, p. 360 (2011)

16. A. Politi, M. J. Cryan, J. G. Rarity, S. Yu, J. L. O'Brien, *Science*, Vol. **320**, p. 646 (2008).

17. A. Politi, J. C. F. Matthews and J. L. O'Brien, *Science*, Vol. **325**, p. 1221 (2009).

18. D. H. Titterton, in *Mid Infrared Semiconductor Optoelectronic*, ed. A. Krier (Springer, Heidelbergf ), p. 635 (2005).

19. J. Wagner et al., in *Technologies for Optical Countermeasures V,* SPIE Conference Proceeding Vol. **7115**, p.1 (2008).

20. D. H. Titterton, in *Fiber lasers V: Technology, Systems and Applications,* SPIE Conference Proceeding Vol. **6873**, p.12 (2008).

21. Y. Kalisky and O. Kalisky, *Opt. Eng.* **49,** p. 091003 (2010).

22. K. N. Lafortune, R. L. Hurd, E. M. Joansson, C. B. Dane, S. N. Fochs and J. M. Brase, in *High-Power Laser Beam Control*, SPIE Conference Proceeding Vol.**5333**, p.53 (2004).

23. D. A. Cremers and R. C. Chinni, *Applied Spectroscopy Reviews*, Vol. **44**, No. 6, p. 457 (2009).

24. G. Mogilevsky, L. Borland, M. Brickhouse and A. W. Fountain III, *International J. of Spectroscopy*, Vol. **2012**, Article ID 808079, doi:10.1155/2012/808079 (2012).

25. H. Ostermak, M. Nordberg and T.E. Carlsson, *Applied Optics*, Vol. 50, No. 28, p. 5592 (2011).

26. S. B. Thotat, S. K. Nayak and J. P. Dandale, *International Journal of Computer Science and Information Security,* Vol. **8**, No. 1, p. 325 (2010).

27. B. A. Barbour and D. S. Ackerson, *"3D Visualization System"*, U.S. Patent Application 2011/0134220.

28. Y. Yan, Y. Qian, H. Sharif and D. Tipper, "Communications Survey & Tutorials, IEEE", Vol. **P**, issue 99, p.1 (2012).

29. F. A. Bovino and M. Giardina, *Proceedings of the Third International Symposium on Applied Sciences in Biomedical and Communication Technologies* (ISABEL 2010), Rome, Italy, November 2010.

30. A. Uchida, *"Optical Communication with Chaotic Lasers: Applications of Nonlinear Dynamics and Synchronization"*, John Wiley & Sons (2011).
31. H. Brundage, *"Designing a Wireless Underwater Optical Communication System"*, Massachusetts Institute of Technology, Dept. of Mechanical Engineering (2010).

# Index